T0245267

CAMBRIDGE LIBRARY COLLECTION

Books of enduring scholarly value

Mathematical Sciences

From its pre-historic roots in simple counting to the algorithms powering modern desktop computers, from the genius of Archimedes to the genius of Einstein, advances in mathematical understanding and numerical techniques have been directly responsible for creating the modern world as we know it. This series will provide a library of the most influential publications and writers on mathematics in its broadest sense. As such, it will show not only the deep roots from which modern science and technology have grown, but also the astonishing breadth of application of mathematical techniques in the humanities and social sciences, and in everyday life.

Principles of Geometry

Henry Frederick Baker (1866–1956) was a renowned British mathematician specialising in algebraic geometry. He was elected a Fellow of the Royal Society in 1898 and appointed the Lowndean Professor of Astronomy and Geometry in the University of Cambridge in 1914. First published between 1922 and 1925, the six-volume *Principles of Geometry* was a synthesis of Baker's lecture series on geometry and was the first British work on geometry to use axiomatic methods without the use of co-ordinates. The first four volumes describe the projective geometry of space of between two and five dimensions, with the last two volumes reflecting Baker's later research interests in the birational theory of surfaces. The work as a whole provides a detailed insight into the geometry which was developing at the time of publication. This, the fourth volume, describes the principal configurations of space of four and five dimensions.

Cambridge University Press has long been a pioneer in the reissuing of out-of-print titles from its own backlist, producing digital reprints of books that are still sought after by scholars and students but could not be reprinted economically using traditional technology. The Cambridge Library Collection extends this activity to a wider range of books which are still of importance to researchers and professionals, either for the source material they contain, or as landmarks in the history of their academic discipline.

Drawing from the world-renowned collections in the Cambridge University Library, and guided by the advice of experts in each subject area, Cambridge University Press is using state-of-the-art scanning machines in its own Printing House to capture the content of each book selected for inclusion. The files are processed to give a consistently clear, crisp image, and the books finished to the high quality standard for which the Press is recognised around the world. The latest print-on-demand technology ensures that the books will remain available indefinitely, and that orders for single or multiple copies can quickly be supplied.

The Cambridge Library Collection will bring back to life books of enduring scholarly value (including out-of-copyright works originally issued by other publishers) across a wide range of disciplines in the humanities and social sciences and in science and technology.

Principles
of Geometry

VOLUME 4:
HIGHER GEOMETRY

H.F. BAKER

CAMBRIDGE
UNIVERSITY PRESS

CAMBRIDGE UNIVERSITY PRESS

Cambridge, New York, Melbourne, Madrid, Cape Town, Singapore,
São Paolo, Delhi, Dubai, Tokyo, Mexico City

Published in the United States of America by Cambridge University Press, New York

www.cambridge.org
Information on this title: www.cambridge.org/9781108017800

© in this compilation Cambridge University Press 2010

This edition first published 1925
This digitally printed version 2010

ISBN 978-1-108-01780-0 Paperback

This book reproduces the text of the original edition. The content and language reflect
the beliefs, practices and terminology of their time, and have not been updated.

Cambridge University Press wishes to make clear that the book, unless originally published
by Cambridge, is not being republished by, in association or collaboration with, or
with the endorsement or approval of, the original publisher or its successors in title.

PRINCIPLES OF GEOMETRY

CAMBRIDGE UNIVERSITY PRESS
LONDON : FETTER LANE, E.C. 4

LONDON : H. K. LEWIS AND CO., Ltd.,
136, Gower Street, W.C. 1
NEW YORK : THE MACMILLAN CO.
BOMBAY
CALCUTTA } MACMILLAN AND CO., Ltd.
MADRAS
TORONTO : THE MACMILLAN CO. OF
CANADA, Ltd.
TOKYO : MARUZEN-KABUSHIKI-KAISHA

ALL RIGHTS RESERVED

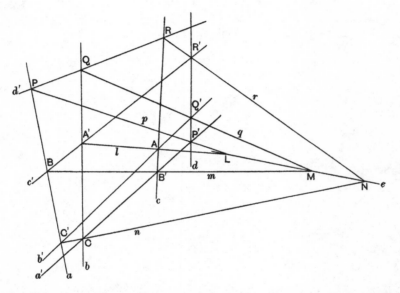

THE FIGURE OF FIFTEEN LINES AND FIFTEEN POINTS, IN SPACE OF FOUR DIMENSIONS
(See Ch. V.)

PRINCIPLES OF GEOMETRY

BY

H. F. BAKER, Sc.D., LL.D., F.R.S.,

LOWNDEAN PROFESSOR OF ASTRONOMY AND GEOMETRY, AND FELLOW OF
ST JOHN'S COLLEGE, IN THE UNIVERSITY OF CAMBRIDGE.

VOLUME IV

HIGHER GEOMETRY

BEING ILLUSTRATIONS OF THE UTILITY OF THE
CONSIDERATION OF HIGHER SPACE, ESPECIALLY
OF FOUR AND FIVE DIMENSIONS

CAMBRIDGE
AT THE UNIVERSITY PRESS

1925

PRINTED IN GREAT BRITAIN

PREFACE

THE present volume, the first written and the most revised, of the book, for which indeed, mostly, the earlier volumes were undertaken, still bears many marks of the difficulty of compressing the matter into brief compass. But the writer hopes that it may seem to the reader as remarkable as it does to him, that it should be possible to comprehend under one point of view, and that so simple, the introduction to nearly all the surfaces ordinarily studied in the geometry of three dimensions, as well as the usual line geometry. Chapters v, vi, vii seek to make clear that this is so. To these the earlier chapters are auxiliary. But Chapters ii and iv have been introduced as much for their own interest as for their illustrative value; the results obtained in these two chapters are not required in the subsequent pages. It is hoped that the Table of Contents, and the Index, may make it easy to use the volume. It will of course be understood that the volume is throughout intended to be introductory and illustrative; hardly anywhere is it complete.

To the Staff of the University Press grateful acknowledgments are due for their continued courtesy and care.

H. F. B.

1 *June* 1925.

TABLE OF CONTENTS

CHAPTER I. INTRODUCTORY. RELATIONS OF THE GEOMETRY OF TWO, THREE, FOUR AND FIVE DIMENSIONS

CHAPTER III. THE PLANE QUARTIC CURVE WITH TWO DOUBLE POINTS

CHAPTER IV. A PARTICULAR FIGURE IN SPACE OF FOUR DIMENSIONS

CHAPTER V. A FIGURE OF FIFTEEN LINES AND POINTS, IN SPACE OF FOUR DIMENSIONS; AND ASSOCIATED LOCI

CHAPTER VI. A QUARTIC SURFACE IN SPACE OF FOUR DIMENSIONS. THE CYCLIDE.

Contents

CHAPTER VII. RELATIONS IN SPACE OF FIVE DIMENSIONS. KUMMER'S SURFACE

CHAPTER I

INTRODUCTORY. RELATIONS OF THE GEOMETRY OF TWO, THREE, FOUR AND FIVE DIMENSIONS

THE present chapter consists of various examples of the interest and importance of the comparison of the geometry of spaces of different dimensions. The first section (pp. 1—32) is concerned with relations between theorems in two and in three dimensions. The second section (pp. 32—40) deals with the representation in four dimensions of some results belonging to ordinary space of three dimensions. The last section (pp. 40—64) deals with the employment of space of five dimensions for the consideration of properties arising both in three and in two dimensions. Some few references occur to space of any number of dimensions.

SECTION I. THEOREMS OF TWO AND THREE DIMENSIONS

The conics touching the fives from six arbitrary lines of a plane. Let three lines, p, q, r, be given in a plane, as well as a fourth line containing two points, I, J; let any conic be drawn touching the four lines; let σ be the conic, through the points I, J, which contains the three intersections of the lines p, q, r; then this conic σ passes through the point, S, in which intersect the tangents from I, J to the former conic. Or, in other words (Vol. II, p. 81), the circle through the intersections of three tangents of a parabola contains the focus of this parabola. Thus if four lines be given, beside the line which contains the points I, J, the conic touching the five lines being then definite, the four conics, all through I, J, each containing the intersections of three of the four given lines, meet in a point, namely, the point, S, in which the tangents from I, J to the former conic intersect (Vol. II, p. 82); namely, these are four circles meeting in the focus of the parabola. If now, finally, five lines be given, beside the line containing the points I, J, there will be five parabolas, each a conic touching the last line and four of the others, and five foci, S_1, S_2, ..., S_5. It is the case that the circle containing any three of these foci passes through the other two, that is, that the seven points S_1, ..., S_5, I, J lie on a conic. Of this theorem a proof was given by Clifford ("A synthetic proof of Miquel's theorem," *Math. Papers*, 1882, p. 38), with the help of certain particular cubic curves.

A circle through the focus, S, of a parabola is a conic containing
the intersections of three tangents of the parabola, namely IJ, IS,
JS, and may be said to be triangularly circumscribed to the parabola. Conversely, any conic triangularly circumscribed to a conic
and meeting a tangent of this in points I, J, contains the intersection, S, of the tangents to this from I and J; it may, therefore, be
regarded as a circle through the focus of a parabola, when I and J
are taken as the Absolute points. Thus the theorem above referred
to may be stated by saying that, if a, b, c, d, l' and l be six arbitrary
lines given in a plane, and I, J be two arbitrary given points of the
line l, and five conics be defined each as touching l and four of the
lines a, b, c, d, l', then there exists a conic passing through I and J
which is triangularly circumscribed to these five conics. The symmetry may suggest that this latter is also triangularly circumscribed
to the conic which is defined by touching the five lines a, b, c, d, l';
this is in fact the case. We thus have six conics, each touching five
of the six given lines; and each of the six lines touches five of the
conics. There cannot be two conics through the points, I, J, of l,
both triangularly circumscribed to the six conics; such a conic, if
existent, is defined by the points I, J and three of the conics
touching the line IJ, namely as containing the intersections,
S_1, S_2, S_3, of the tangents from I, J to these three conics, respectively.

To prove this symmetrical result we may proceed as follows:
Denote the plane of the six given lines, a, b, c, d, l, l' by ϖ. Draw
through each of the lines a, b, c, d an arbitrary plane, denoting the
intersections of these in threes by A, B, C, D, of which D is the
intersection of the planes through a, b, c, and so on. It is assumed
that A, B, C, D are not in a plane. Through the six points consisting of A, B, C, D and the two arbitrary points I, J, of the line
l, there can be put a definite cubic curve, which we denote by γ.
This will meet the plane ϖ in another point beside I and J; say, in
E. Then, through the five points A, B, C, D, E there can be drawn
another cubic curve, γ', to have the line l' for chord, meeting this,
suppose, in the points I', J' (Vol. III, p. 139). We prove that the
conic, ω, containing the five points I, J, E, I', J', is triangularly
circumscribed to the six conics touching the fives of the six given
lines a, b, c, d, l, l'. It is thus independent of the planes drawn
through a, b, c, d.

The conic ω is the intersection with the plane ϖ of the quadric,
Ω, defined by the nine points A, B, C, D, E, I, J, I', J'; the cubic
curves, γ, γ', each meeting Ω in seven points, lie on this quadric.
Cubic curves lying on a quadric are of two families, since such a
curve meets all generators of the surface, of one system of generators,
in one point, and all generators of the other system in two points;

two cubics of the same family have four common points, but two cubics of different families have five common points (Vol. III, p. 139). Thus the curves γ, γ' are of different families on Ω.

We prove now, first, that the conic ω is triangularly circumscribed to the conic touching the lines a, b, c, d, l; namely by obtaining triads of points of ω whose joins touch this conic (a, b, c, d, l). Let P be any point of ω; through this, and the other four points, A, B, C, D, of Ω, there can be drawn, lying on Ω, a cubic curve of the same family as γ (Vol. III, p. 129). This curve will meet ω in two further points, say Q and R. Either of these, with A, B, C, D, determines the cubic curve, and so determines the other two of the three points, P, Q, R, of ω. Thus, as P varies on ω, the triads P, Q, R form an involution of sets of three points thereon, and the lines QR, RP, PQ all touch a conic (Vol. II, p. 135); this conic, which we may denote by λ, is evidently triangularly inscribed in ω. We prove that λ is the conic touching a, b, c, d, l, by considering different positions of P. When P is at E, the line QR is the line l; thus λ touches l. In general the cubic curve through A, B, C, D, P, Q, R is projected from P by a quadric cone, which may be defined as that containing PA, PB, PC, PD and a particular one of the two generators of Ω at that point P. But when P is at one of the two intersections of the line d with ω, this cone degenerates, becoming the aggregate of the plane ABC, which contains d, and the plane joining PD to the particular generator at P spoken of; one of the two lines PQ, PR, in which the cone meets the plane ϖ, is thus the line d. Therefore the conic λ touches d. By a similar argument it touches a, b, c.

To prove that the conic ω is triangularly circumscribed to the conic touching a, b, c, d and l', we describe a cubic curve through a varying point, P, of ω and through A, B, C, D, lying on Ω, but of the same family as the curve γ'.

That the conic touching a, b, c, and both of l and l', is triangularly inscribed to ω, we likewise prove by obtaining an involution of sets of three points lying on ω. For this, let the points in which the lines DA, DB, DC meet the plane ϖ be denoted, respectively, by A_1, B_1 and C_1, and consider the conics drawn through the four points A_1, B_1, C_1 and E. The sets of three points other than E in which these conics meet ω are then sets of such an involution (Vol. II, p. 138), and the three joins of the points of such a set are tangents of a conic, which we denote by δ. One conic through A_1, B_1, C_1 and E consists of the two lines B_1C_1 and A_1E; thus δ touches B_1C_1, which is the line a. Similarly δ touches b and c. Again, the conics through A_1, B_1, C_1 and E may be defined by quadric cones, of vertex D, containing DA, DB, DC and DE; one such cone, however, is the cone which contains the cubic curve γ, and this meets the conic ω in the points I, J. Thus the conic δ touches the line l. Another such cone is that containing the curve γ', which meets ω in I' and J'. Thus δ also touches l'. It is thus shewn that the conic touching a, b, c and l, l' is triangularly inscribed in ω. An analogous argument proves ω to be triangularly circumscribed to the *three* conics touching l, l' and, respectively, b, c, d; c, a, d; a, b, d.

The theorem stated is thus completely established. It has been remarked that the conic obtained, triangularly circumscribed to the six primary conics, is uniquely determined by its intersections, I, J, with one of the six given lines, l. It is clear, however, from the reasoning given that the conic is also determinate when E is given; its intersections with l being on the tangents from E to the conic touching a, b, c, d, l, and its intersections with l' being on the tangents from E to the conic touching a, b, c, d, l', while the conic passes through E. Thus, whatever be the planes, drawn through a, b, c, d, by which A, B, C, D are determined, the cubic curve drawn through E, A, B, C, D, to have l for chord, meets l in the same two points, I, J; and similarly for l'.

Ex. 1. If E, I, J be three points of a cubic curve in space, of which A, B, C, D are four other points, the two axial pencils of planes joining EI and IJ to A, B, C, D are related to one another; and the former is related to the range on EI determined by the four planes containing A, B, C, D, the latter being similarly related to the range which these four planes determine on IJ. Wherefore, there is a conic, in the plane EIJ, touching the four lines in which this plane is met by the four planes containing A, B, C, D, and also touching EI, IJ. By parity of reasoning this same conic touches also the line EJ.

Ex. 2. If, in the construction above, the lines a, b, c, d, l, lying in the plane $t = 0$, be taken to be

$$x = 0, \quad y = 0, \quad z = 0, \quad ax + by + cz = 0, \quad lx + my + nz = 0,$$

the planes DBC, DCA, DAB, ABC being

$$x = 0, \quad y = 0, \quad z = 0, \quad ax + by + cz - t = 0,$$

and the point E be $(\xi, \eta, \zeta, 0)$, prove that the points I, J lie on the conic in the plane $t = 0$ given by

$$x^{-1}\xi\,(mb^{-1} - nc^{-1}) + y^{-1}\eta\,(nc^{-1} - la^{-1}) + z^{-1}\zeta\,(la^{-1} - mb^{-1}) = 0 \,;$$

this lies on the cone, of vertex D, containing the cubic curve through A, B, C, D, E which has the line l for chord.

Ex. 3. Hence infer that if the six given lines of the original plane be of equations

$$x = 0, \quad y = 0, \quad z = 0, \quad a_r x + b_r y + c_r z = 0, \quad (r = 1, 2, 3),$$

then the general conic which is triangularly circumscribed to the six conics touching the fives of these lines has for equation

$$\Delta\,(a_1x + b_1y + c_1z)\,(a_2a_3xYZ + b_2b_3yZX + c_2c_3zXY)$$
$$+ (A_1X + B_1Y + C_1Z)\,(A_2A_3Xyz + B_2B_3Yzx + C_2C_3Zxy) = 0,$$

where Δ is the determinant (a_1, b_2, c_3), and A_1, B_2, ... are the minors of a_1, b_2, ... therein, while (X, Y, Z) is an arbitrary point, such that $(a_1^{-1}A_1X, b_1^{-1}B_1Y, c_1^{-1}C_1Z)$ is the point E, or (ξ, η, ζ), of the foregoing theory. This equation, it may easily be seen, is unaltered by cyclical change of the suffixes, X, Y, Z being unaltered; it is derived from a form obtained algebraically by Mr F. P. White (*Camb. Phil. Proc.* xxii, 1924, p. 11). If we take

$$X' = A_1X + B_1Y + C_1Z,\quad Y' = A_2X + B_2Y + C_2Z,\quad Z' = A_3X + ...,$$

and also

$$x' = a_1x + b_1y + c_1z,\quad y' = a_2x + ...,\quad z' = a_3x + ...,$$

the equation is also capable of the form

$$x\,(B_1C_1x'Y'Z' + B_2C_2y'Z'X' + B_3C_3z'X'Y')$$
$$+ \Delta X\,(b_1c_1X'y'z' + b_2c_2Y'z'x' + b_3c_3Z'x'y') = 0,$$

and of two other forms, obtained from this by cyclical interchange of a, b, c, without change of X', Y', Z'.

Further, if we take the three points P_1, P_2, P_3, where P_r is of coordinates $(a_r^{-1}A_rX, b_r^{-1}B_rY, c_r^{-1}C_rZ)$, and denote the line $a_rx + b_ry + c_rz = 0$ by l_r, and the points $(1, 0, 0), (0, 1, 0), (0, 0, 1)$ by A_1, B_1, C_1, and then define a conic, S_1, as that through $(A_1, B_1, C_1, P_2, P_3)$, a conic, S_2, as that through $(A_1, B_1, C_1, P_3, P_1)$ and a conic, S_3, similarly, it will be found that the three pairs of intersections (S_1, l_1), (S_2, l_2), (S_3, l_3) lie on a conic through P_1, P_2, P_3, this conic, through these nine points, being the conic triangularly circumscribed to the six primary conics.

The reader may also be reminded of Taylor's theorem (Vol. ii, p. 61), that the poles of an arbitrary line in regard to the six primary conics lie on another conic.

Ex. 4. Any six lines of a plane can in fact be regarded as the projections of six chords of a cubic curve, from a point of this curve, these chords having the property that every five of them have a common transversal. For, taking the six conics touching the fives of the given lines, let the conic which is triangularly circumscribed to these have its equation put in the form $xz - y^2 = 0$, as is possible in an infinite number of ways. Then with an arbitrary point for the intersection of planes $x = 0$, $y = 0$, $z = 0$, this conic is on the cone projecting from this point the cubic curve whose points are given by $(\theta^3, \theta^2, \theta, 1)$. The tangent line of any conic given in the original plane, $t = 0$, may then be defined by a plane,

$$ux + vy + wz = 0,$$

through the point $(0, 0, 0, 1)$, and the tangential equation of the conic may be taken in the form

$$c'u^2 + a'v^2 - cw^2 - bvw - (a + a')\,wu - b'uv = 0,$$

in which u, v, w are tangential coordinates. The chord of the cubic curve joining the points, other than $(0, 0, 0, 1)$, in which the plane meets the curve, has for coordinates the six

$$wu - v^2, \quad uv, \quad -u^2, \quad wu, \quad vw, \quad w^2\,;$$

thus the equation of the conic expresses that the tangent line of this is the projection of a chord of the cubic curve which belongs to the linear complex (a, b, c, a', b', c'). In particular, if

$$aa' + bb' + cc' = 0,$$

the conic is triangularly inscribed to the conic $xz - y^2 = 0$, as we easily verify (Cremona, *Giorn. d. Mat.*, x, 1872, p. 47). Whence, six conics triangularly inscribed to $xz - y^2 = 0$ can be regarded as having the equations $(r = 1, ..., 6)$

$$n_r'u^2 + l_r'v^2 - n_rw^2 - m_rvw - (l_r + l_r')\,wu - m_r'uv = 0,$$

each being obtained by projecting the chords of the cubic curve which meet a line of coordinates $(l_r, m_r, n_r, l_r', m_r', n_r')$. When the six conics touch the fives of six lines, so that every five of them have a common tangent, these lines are obtained by projection from six chords of the cubic, forming with the lines $(l_r, m_r, ...)$ a double six of lines. (Cf. Wakeford, *Proc. Lond. Math. Soc.*, xv, 1916, p. 340.)

It may be remarked, too, that the conic, $(l_r, m_r, ...)$, touching the projections of the chords of the cubic which meet the line $(l_r, m_r, ...)$, has, beside five of the originally given six lines, as a sixth tangent, the line $l_r'x + m_r'y + n_r'z = 0$. If we take the intersection of this with the remaining one of the six original lines, we obtain one of six points which lie on a conic. (Cf. Vol. III, p. 136, Ex. 8, and p. 201.)

Representation of a plane upon a quadric. Consider a quadric surface, Ω ; and, upon this, a point, U, which is to be taken as centre of projection. Consider also a plane, ϖ. Let the generators at U, of the quadric Ω, meet the plane ϖ in the points I and J. Any plane meets the generators; and, thus, the conic, in which Ω is met by any plane which does not pass through U, is projected from U into a conic of the plane ϖ which passes through I and J. Regarding I and J as Absolute points, we may then speak of this conic as a circle. Conversely, any conic, in the plane ϖ, which passes through I and J, lies on a quadric cone of vertex U; this cone contains the generators UI, UJ; its remaining intersection with Ω is thus a curve of the second order, namely a plane section. Thus plane sections of the quadric Ω, not containing U, and circles of the plane ϖ, are transformable into one another by projection from U. Sections of Ω by planes through U project into lines of ϖ, and conversely.

Let α be any conic section of Ω not passing through U; and let A be the pole of the plane of α, in regard to Ω. Let α' be the circle of the plane ϖ which arises by projection of α from U. We prove that the line UA meets the plane ϖ in the pole, in regard to α', of the line IJ; that is, that the projection of A is the *centre* of the circle α'. Let the line in which the tangent plane at U, of the quadric Ω, is met by the plane of α, be denoted by u; and the point in which the line UA meets the plane of α be denoted by A_0. The line UA is the polar line of u, in regard to Ω; thus the polar plane of A_0 contains u, and A_0 is the pole of u in regard to the conic α; so that any line in the plane of α, through A_0, meets u in the harmonic conjugate of A_0, in regard to the two points in which this line meets the conic α. This relation persists after projection from U; and this proves the statement made. Conversely, however, in the correspondence between the points of the plane ϖ, and the points of the quadric Ω, the centre of the circle corresponds to the point, other than U, in which the line UA meets Ω.

We have explained (Vol. II, p. 166) how to measure the angle between two lines of a plane, say of equations $P+\lambda Q=0, P+\mu Q=0$, with respect to two other lines $P=0$, $Q=0$, by means of the ratio λ/μ. In particular, the angle between two lines lying in the tangent plane of the quadric Ω, at a point O, and meeting at this point, may be measured with respect to the generators of the quadric at this point. As each of these generators meets one of the generators at U, if O project from U into O' on the plane ϖ, the generators at O project into the lines $O'I$, $O'J$. Thus the angle in question is that between the lines in the plane ϖ, into which the two original tangents of Ω at O project, when measured with respect to the two Absolute points I, J of the plane ϖ. If O be one of the intersections of two plane sections of the quadric Ω, the other being Q, and the lines taken at O be the two tangents of these sections at O, the flat pencil in the tangent plane at O, consisting of these lines and the generators at O, is the section by the tangent plane at O of the axial pencil consisting of the two planes, which meet in OQ, and the two tangent planes of Ω drawn from the line OQ; for each of these tangent planes contains a generator at O and a generator at Q. Thus the angle in question is that of the planes of the two sections measured in regard to the two tangent planes of Ω drawn from OQ. If, in particular, the planes of the two sections are conjugate to one another, each containing the pole of the other, then they are harmonic in regard to the two tangent planes of Ω drawn through their line of intersection. Thus we reach the results that two circles in the plane ϖ intersect at the same angle at both their common points, this being the angle between the corresponding planes of the quadric Ω when suitably measured, and that the circles cut at right angles when these planes are conjugate to one another (cf. Vol. II, p. 193). By considering sections of Ω through U, all conjugate to a chosen plane section, we obtain the result that all lines through the centre of a circle cut this at right angles; by considering plane sections all passing through a point not on Ω, this being the pole of a certain section, we obtain the aggregate of all circles in the plane ϖ which cut a certain circle at right angles; by considering sections of Ω by planes through a line, we obtain a system of coaxial circles, whose limiting points are the projections of the two points in which Ω is met by the polar line of the given line, while sections by planes through this polar line give circles cutting at right angles those of the original system ; thence, two points of Ω, lying on a line through the pole of a chosen section of Ω, project into two points which are conjugate in regard to the circle into which the chosen section projects; and so on, all the familiar properties of circles being easily interpretable. It may be remarked that the angle between two plane sections of Ω, which we have identified with the angle between two circles, is also the same as the interval between the points which are the poles of these plane sections, measured with respect to the intersections with Ω of the join of these poles.

The generalised Miquel theorem. We have given (Vol. II, p. 70) the theorem that if D, E, F be any points respectively on the joins, BC, CA, AB, of three points of a plane, then the three circles AEF, BFD, CDE meet in a point. We consider this result in particular from our present point of view. There is a corresponding theorem for any number of dimensions, capable of similar proof.

(Roberts, *Proc. Lond. Math. Soc.*, XII (1881), p. 117; *ibid.*, XXV (1893), p. 306; Grace, *Trans. Camb. Phil. Soc.*, XVI (1897), p. 168; Haskell, *Arch. d. Math. u. Phys.*, V (1903), p. 278.)

Suppose we have, in space of three dimensions, any three planes passing through a point, O, say $x = 0$, $y = 0$, $z = 0$, and take three arbitrary points: A on the line of intersection of $y = 0, z = 0$; B on the line of intersection of $z = 0, x = 0$, and C on the line $x = 0, y = 0$; and then take, on the plane OBC, the point D ; on the plane OCA, the point E, and on the plane OAB, the point F ; the planes AEF, BFD, CDE will meet in a point, say Q. If O is taken on an arbitrary given quadric, and A, B, C are the intersections with this quadric of the lines $y = 0$, $z = 0$, etc., while D, E, F are on the sections of this quadric by the planes $x = 0$, $y = 0$, $z = 0$, respectively, then the point Q also lies on this quadric. For, all quadrics through the seven points O, A, B, C, D, E, F have in common a further point; of such quadrics a degenerate one consists of the two planes $OBCD$, AEF; another of the planes $OCAE$, BFD, and a third of the planes $OABF$, CDE. As these degenerate quadrics all contain the point Q, so does the original. When, now, we project the points of the given quadric from O, on to an arbitrary plane, the plane $x = 0$ gives a line containing the projections of the points B, C, D, and so on, and the section of the quadric by the plane AEF gives a circle; the three such circles meet in the point which is the projection of Q. This proves the theorem of Miquel.

It is equally the case that, in space of three dimensions, if A, B, C, D be any four points, and points, P, Q, R, P', Q', R' be taken arbitrarily, respectively on the lines DA, DB, DC, BC, CA, AB, then the four spheres, each containing one of the four points A, B, C, D and also the three points on the joins of this to the other points, meet in a point. These are the spheres $APQ'R'$, $BQR'P'$, $CRP'Q'$, $DPQR$. A proof is as follows. In space of four dimensions, consider four threefolds, $x = 0$, $y = 0$, $z = 0$, $t = 0$, meeting in a point O; any three of these threefolds will meet in a line through O, and upon each of the four lines so obtained a point may be taken; let the point on the line $y = 0$, $z = 0$, $t = 0$ be A, the others being B, C, D, of which, for example, D is on $x = 0$, $y = 0$, $z = 0$. Upon the plane $y = 0$, $z = 0$, which contains the lines OA, OD, let the point P be taken, arbitrarily; and, similarly, the points Q and R be taken on the planes OBD, OCD, respectively, as also P', Q', R' on the respective planes OBC, OCA, OAB. Then consider the threefolds $APQ'R'$, $BQR'P'$, $CRP'Q'$, $DPQR$, which we denote, respectively, by $\xi = 0$, $\eta = 0$, $\zeta = 0$, $\tau = 0$; and let T be their point of intersection. It can then be shewn that T lies on any quadric threefold constructed to contain the points O, A, B, C, D, P, Q, R, P', Q', R'. In fact the equation of a quadric threefold, in fourfold space, contains fifteen terms; thus a quadric through the eleven specified points will be of the form $\lambda_1 U_1 + \ldots + \lambda_4 U_4 = 0$, where $\lambda_1, \ldots, \lambda_4$ are arbitrary, and $U_1 = 0, \ldots, U_4 = 0$ are four such quadrics, which we suppose to be linearly independent. It follows that any quadric through the eleven specified points will also pass through the remaining $2^4 - 11$, or five, common points of these four quadrics. This assumes that these four quadrics have no common curve, or surface, and intersect in sixteen points. We can, however, at once specify four degenerate quadrics through the eleven points, namely $x\xi = 0$, $y\eta = 0$, $z\zeta = 0$, $t\tau = 0$; for instance $x = 0$ contains O, B, C, D, P', Q, R, and $\xi = 0$ contains A, P, Q', R'. Thus, every quadric through the eleven points contains the point, T, common to $\xi = 0$, $\eta = 0$, $\zeta = 0$, $\tau = 0$. This is the result desired. The four other common points of these four quadrics, which equally lie on any quadric through the eleven points, are the points such as $\eta = 0$, $\zeta = 0$, $\tau = 0$, $x = 0$. It can be shewn now that, by projection from O upon any threefold space, the theorem enunciated in regard to the four spheres is obtained. But, for convenience, the discussion of the derivation of a sphere by projection from space of four dimensions is deferred to a later section of this chapter (p. 36). A similar proof holds in higher space; this depends on a theorem that, in space of n dimensions, all quadric $(n-1)$-folds through $\frac{1}{2}n(n+1) + 1$ general points are expressible by n such quadrics, and pass through $2^n - 1 - \frac{1}{2}n(n+1)$ other points.

From the Miquel theorem in a plane we were able to infer (Vol. ii, p. 71) that, if four arbitrary lines be given, the circles each containing the intersections of three of these lines, four in all, would meet in a point; namely, by supposing the points D, E, F, in the enunciation given above, to be in line. This theorem, ascribed to Wallace (Scoticus, *Leybourn's Math. Repos.*, i, 1806, p. 170), will be considered below (p. 18). If, in the corresponding theorem above considered, for four spheres in space of three dimensions, we suppose the six points P, Q, R, P', Q', R' to be in a plane, the four spheres there taken will still meet in a point, but this point will be on the plane containing the six points. This is obvious by considering the Wallace theorem for the four lines in this plane which are obtained by the intersections of this plane with the four planes BCD, CAD, ABD, ABC; the four spheres of the theorem meet this plane in circles. We cannot, therefore, go on to infer that the sphere through A, B, C, D passes through the point of intersection of the first four spheres, as in the plane case; the five spheres meet in fours in points lying one on each of the five planes involved. It will be seen below (p. 59) that the generalisation of the Wallace theorem, which thus does not hold in space of three dimensions, holds nevertheless in space of four dimensions, and in space of any even number of dimensions (see Grace, *as above*, p. 163).

Ex. 1. Through five points, A, B, C, D, E, in three dimensions, can be drawn five linearly independent quadrics. The conics in which these meet an arbitrary plane, ϖ, are also linearly independent, and determine a definite conic, σ, inpolar to all of these. Thus every quadric through A, B, C, D, E meets the plane ϖ in a conic which is outpolar to σ. Such a quadric is formed by the plane ABC taken with any plane through the line DE; hence the point in which ϖ is met by DE is the pole, in regard to σ, of the line in which ϖ is met by the plane ABC. Thus the system of ten points and lines, in which the joining lines and planes of A, B, C,

D, E meet the plane ϖ, is self-polar in regard to the conic σ. A similar proof holds for the theorem obtained similarly by taking $(n+2)$ points in space of n dimensions (cf. Vol. II, p. 218), the number of linearly independent quadric $(n-1)$-folds which pass through $(n+2)$ general points, namely $\frac{1}{2}n(n+1)-1$, being one less than the number of terms in the equation of a quadric $(n-2)$-fold in a space of $(n-1)$ dimensions.

Ex. 2. A cubic curve through five points, *A, B, C, D, E*, meets an arbitrary plane, ϖ, in a triad which is self-polar in regard to the conic, σ, determined in ϖ by these five points, as in Ex. 1. For the three linearly independent quadrics, through the curve, meet ϖ in conics, outpolar to σ, all passing through the points of this triad. The corresponding theorem for a rational curve of order n, through $(n+2)$ points of space of n dimensions, can be similarly proved.

Ex. 3. The dual of the theorem referred to in Ex. 1 is that, if we have five planes, α, β, γ, δ, ϵ, and consider the ∞^4 quadrics touching these planes, then the enveloping cones to these quadrics, drawn from any point, *E*, are all inpolar to a particular quadric cone, say Σ, of vertex *E*, determined by the five given planes and the point *E*. Let the intersections of the four planes α, β, γ, δ be denoted by *A, B, C, D*, the point *D* being the intersection (α, β, γ), and so on. Then the line *ED* is the polar line of the plane joining *E* to the line (δ, ϵ), in regard to the cone Σ; and the line joining *E* to the point $(\beta, \gamma, \epsilon)$ is the polar line of the plane joining *E* to the line (α, δ); that is, the line joining *E* to the point where *AD* meets ϵ is the polar line of the plane *EBC*, in regard to Σ. Thus, considering the conic where the plane ϵ is met by the cone Σ, this conic, σ, is that before described, in regard to which the joining lines and planes of the five points *A, B, C, D, E* determine a polar system in the plane ϵ in which σ lies.

Regarding σ as an Absolute conic, the point *E* is the intersection of perpendiculars drawn from the points *A, B, C, D*, respectively to the planes α, β, γ, δ; and each pair of opposite joins of *A, B, C, D*, such as *BC* and *AD*, are at right angles to one another (cf. Vol. III, p. 78, Ex. 7). In regard to σ, the relation of the five points *A, B, C, D, E* is symmetrical.

Ex. 4. Five points of a given cubic curve, *A, B, C, D, E*, determine as above a conic, σ, in a given arbitrary plane. Consider what is the aggregate of all such conics when the five points of the cubic curve have all possible positions thereon. If the curve meet the plane in the points *P, Q, R*, any such conic, σ, must be such that *P, Q, R* are a self-polar triad in regard to σ, as we have seen. The five quadrics through *A, B, C, D, E* may be regarded as consisting of three quadrics containing the curve, together with two degenerate quadrics each consisting of the plane *ABC* taken with a plane through the line DE. Thus the point F, in which the line DE meets the plane PQR, must be the pole, in regard to σ, of the line in which the plane ABC meets the plane PQR. Conversely, then, taking any conic, σ, in regard to which P, Q, R are a self-polar triad, we can find five points A, B, C, D, E of the cubic curve, for which this is the appropriate conic. Namely, by taking A, B, C arbitrarily upon the curve, then finding the pole, F, of the plane ABC in regard to σ, and then drawing the chord, DE, of the curve which passes through F. The aggregate of the conics, σ, for all possible sets of five points of the curve, is thus the same as the aggregate of the conics of the plane PQR in regard to which P, Q, R are a self-polar triad.

Ex. 5. The proof of Miquel's theorem in a plane which was given in Vol. II, p. 70, was in effect by the polar system determined on the Absolute line by the conics through four points of the plane. This proof can be used also for the Miquel theorem for the four spheres through $APQ'R'$, $BQR'P'$, $CRP'Q'$, $DPQR$, in the notation used above. For it can be shewn that, if O be any further independent point, in the space of three dimensions, and we consider the four conics, in any plane ϖ, such as the conic σ in Ex. 1 above, arising, respectively, from the four sets of five points $OAPQ'R'$, $OBQR'P'$, $OCRP'Q'$, $ODPQR$, then these conics are linearly connected. Denote these conics by σ_1, σ_2, σ_3, σ_4. Thus, if three quadrics drawn, respectively, through $APQ'R'$, $BQR'P'$, $CRP'Q'$ have in common a conic, S, on the plane ϖ, then S, being outpolar to each of σ_1, σ_2, σ_3, is outpolar to σ_4; from this the proof can be completed.

Ex. 6. The theorem referred to in Ex. 5 is equivalent to the following : Let x, y, z, t be four lines in a plane, and A, B, C, D be four points of the plane, such that the line AD contains the point (y, z), the line BC contains the point (x, t), and so on, the six joins of the four points containing the six intersections, properly chosen, of the four lines. Further, let P, Q, R, P', Q', R' be arbitrary points respectively on the lines AD, BD, CD, BC, CA, AB. Then the triads P, Q, R and x, y, z, being in perspective from D, are polars of one another in regard to a conic, say σ_4; similarly, the triads P, Q', R' and y, z, t are polars of one another in regard to a conic, σ_1; and there are two other conics, σ_2, σ_3, determined similarly. These four conics are linearly connected.

The theory of inversion in a plane. Consider a particular correspondence of two points of space in three dimensions, which is of importance. Let L be a fixed point, and λ a fixed plane. To any point, P, we can then make correspond the point, P', of the line LP, which is harmonically conjugate to P, with respect to L and the point of LP which lies on λ; then, conversely, the point corresponding to P' is the point P itself. By this correspondence, to the points of a line, PQ, correspond the points of another line, $P'Q'$, and these two lines meet on the plane λ; also, to the points of a plane, PQR, correspond the points of another plane, $P'Q'R'$, these two planes meeting in a line of the plane λ. Hence, also, as a particular result, to the points of a quadric cone, of vertex O, correspond the points of another quadric cone, of vertex, say, O', these two cones having in common the points of a conic in the plane λ. The two cones have then in common, also, the points of another conic; the points of this second conic must then correspond to one another in pairs, since the cones have no points in common other than those of these two conics, and the points of the conic in the plane λ correspond each to itself. The plane of the second conic common to the two cones must, therefore, pass through L; and, hence, the tangent planes of the two cones at the points where the second conic meets the plane λ, which are points of contact of the cones, must pass through L, as well as through the vertices of both cones.

Now suppose that the plane λ is the polar plane of L in regard to a certain quadric, Ω. Then the point, P', which corresponds to a point, P, lying on the quadric, is the second intersection of LP with Ω; and, to a plane section of Ω, corresponds another plane section, lying on the quadric cone joining L to the first section, the two sections meeting in two points of the polar section, λ. Unless, indeed, the first section is in a plane passing through L, in which case the second section coincides with it. Let, now, U be a definite point of the quadric Ω, and ϖ a definite plane, and let us project the points of Ω from U upon ϖ, as before. Then the polar section, λ, of L, projects into a definite circle of ϖ, say $[\lambda]$, with centre at the projection of L, as we have seen; two points, P, P', of Ω, on a line through L, project into two points of ϖ, say $[P]$ and $[P']$; as every plane through the line LPP' gives a section of Ω which is conjugate to the polar section λ, the points $[P]$, $[P']$ lie on an infinite number of circles cutting the circle $[\lambda]$ at right angles; thus the points $[P]$, $[P']$ lie on a line through the centre of $[\lambda]$, and are inverse points in regard to this circle, being harmonic conjugates in regard to the points where the line $[PP']$ meets $[\lambda]$. This *inversion* is the same as that defined in Vol. ii, p. 67. A line, or a circle, of the plane ϖ, is clearly changed by this process into a circle, or, in particular, into a line, the circles inverse to one another in regard to $[\lambda]$ being those corresponding to two plane sections of Ω which lie on a quadric cone of vertex L. If, on Ω, P and P' be two points of a line through L, either generator at P meets one of the generators at P', the point of meeting being the point of contact, of a tangent plane of Ω drawn through the line PP', and, therefore, on the polar section

λ; thus these generators correspond to one another in regard to
L; from this it is easily seen that two circles in the plane ϖ
intersect at the same angle as their inverse circles in regard to $[\lambda]$,
the angle being measured as explained above. Again, any plane
section, μ, of Ω, its pole M, and the enveloping cone of Ω along μ,
whose vertex is M, are changed by the correspondence in regard to
L, respectively, into a section μ', its pole M', and the enveloping
cone of Ω along μ'; thus, in the plane ϖ, any circle and two points
which are inverses to one another in regard to this circle, are
changed, by inversion in regard to $[\lambda]$, into a circle and two points
inverse to one another in regard thereto. It may also be remarked
that if, instead of L, we take another point, L_1, lying on the line
UL, and its corresponding polar section, λ_1, as base for the correspondence, the difference in the plane ϖ is that inversion takes
place in regard to another circle, $[\lambda_1]$, having the same centre
as $[\lambda]$.

Ex. 1. Taking coordinates in which the plane ϖ is $t = 0$, the
point of projection, U, is $x = 0$, $y = 0$, $z = 0$, with $z = 0$ for the
tangent plane of Ω at this point, the generators being $x \pm iy = 0$,
$z = 0$, the equation of Ω is of the form $F = 0$, where

$$F = x^2 + y^2 + 2gzx + 2fyz + cz^2 - 2zt.$$

If, then, L be (x_0, y_0, z_0, t_0), the point, P, of Ω, be (ξ, η, ζ, τ), and
P' be $(\xi', \eta', \zeta', \tau')$, such that $\xi' = \xi + \lambda x_0, \dots, \tau' = \tau + \lambda t_0$, these
satisfying the equation of Ω, we find

$$\lambda = -2[\xi x_0 + \eta y_0 + g(\zeta x_0 + \xi z_0) + f(\eta z_0 + \zeta y_0) + c\zeta z_0 - \zeta t_0 - \tau z_0]F_0^{-1},$$

where F_0 is the value of F for (x_0, y_0, z_0, t_0). Thus $\xi', \eta', \zeta', \tau'$ are
expressible as homogeneous linear functions of ξ, η, ζ, τ. Taking

$$R^2 = [(\xi z_0 - \zeta x_0)^2 + (\eta z_0 - \zeta y_0)^2]z_0^{-2}\zeta^{-2}, \quad k^2 = F_0 z_0^{-2}$$

we easily compute $\zeta'/\zeta = R^2 k^{-2}$ and

$$\frac{\xi' z_0 - \zeta' x_0}{\zeta'} = \frac{k^2}{R^2}\frac{\xi z_0 - \zeta x_0}{\zeta},$$

$$\frac{\eta' z_0 - \zeta' y_0}{\zeta'} = \frac{k^2}{R^2}\frac{\eta z_0 - \zeta y_0}{\zeta};$$

these are the formulae of inversion in the plane ϖ, in which
(ξ, η, ζ) are the coordinates of a point, and (ξ', η', ζ') those of the
inverse point, the circle, $[\lambda]$, of inversion having the equation

$$(xz_0 - x_0 z)^2 + (yz_0 - y_0 z)^2 - k^2 z^2 z_0^2 = 0.$$

Ex. 2. We can (Vol. II, p. 69, Ex. 4) state a relation between
the centres of two circles of the plane ϖ which are inverses of one
another in regard to the circle $[\lambda]$. But the relation in the space
of three dimensions is simpler, these centres being the projections,

from U, of the poles of two plane sections of Ω which lie on a quadric cone of vertex L.

Ex. 3. Two plane sections of a quadric, having two points in common, lie on two quadric cones (Vol. III, p. 10). Projecting these sections upon a plane from any point of the quadric, and at the same time the section by the polar plane of the vertex of one of the containing cones, we see that, if two circles be given in a plane, there are two circles of inversion by means of which one of the given circles is changed into the other.

Ex. 4. The *circle of similitude* of two given circles $[\alpha]$, $[\beta]$, of the plane ϖ (Vol. II, p. 132), may be considered from the present point of view. With a pair of plane sections, α, β, of Ω, intersecting in a line l, and the pair of tangent planes of Ω through the line l, an axial pencil of pairs of planes in involution, can be defined. Thereby, to the plane lU, joining l to the point U of Ω, there is defined a section, γ, passing through l, the two planes, lU, γ, being a pair of the involution. Prove that, if the quadric be projected from U, on to a plane ϖ, whereby the sections α, β become the circles $[\alpha]$, $[\beta]$, then the section γ becomes the circle, coaxial with $[\alpha]$ and $[\beta]$, passing through the centres of similitude of these, that is, the circle of similitude. The centres of similitude are the two intersections of common tangents of $[\alpha]$, $[\beta]$ which lie on the line joining the centres of these circles. The circles are supposed unequal.

Ex. 5. If we have four circles in a plane meeting, in threes, in four points, there are six circles of similitude, one for each pair. It can be shewn that these meet in a point. This follows easily if we assume (see below, Ex. 6) that, if A, B, C, D, U be points of a quadric, Ω, then the quadric, Σ, described to touch the planes BCD, CAD, ABD, ABC and to contain the generators of Ω at U, contains two other generators of Ω, say those at U'. For then, regarding Ω and Σ as defined by their tangent planes, and considering the degenerate tangential quadric which consists of all the planes through the points U and U', we have three quadrics such that all the tangent planes common to any two of them are equally tangent planes of the third, these consisting of all the planes through the generators of Ω (and Σ) at U and U'. The three pairs of tangent planes of these three quadrics which pass through any line are, therefore, three pairs of an axial involution. In particular, taking AB for the line, the pair of planes ABU, ABU' belong to the involution determined by two pairs, namely the pair ABC, ABD, and the pair which consists of the tangent planes to Ω drawn from AB. From this it follows that the sections of Ω by the planes ABC, ABD project from U, on to any plane, into two circles whose circle of similitude passes through the projection of U'. Thus, by a similar argument, the sections of Ω by the four planes BCD, ..., ABC

become, projected from U, four circles whose six circles of similitude all pass through the projection of U'.

Ex. 6. To prove that the quadric, Σ, containing the generators of a quadric Ω at a point U of this, and also touching four planes, α, β, γ, δ, whose four points of intersection lie on Ω, also contains two other generators of Ω, say at U', we use the fact that a quadric which touches the tangent planes of Ω drawn from one point, P, also touches the tangent planes of Ω drawn from another point, P'. If we denote the tangential equation of Ω, referred to the planes α, β, γ, δ, by $\Omega = 0$, and (ξ, η, ζ, τ), $(\xi', \eta', \zeta', \tau')$ be the coordinates of P, P', the equation of the quadric in question is then $\Omega - kPP' = 0$, where $P = l\xi + m\eta + n\zeta + p\tau$, $P' = l\xi' + \ldots + p\tau'$. If this quadric touch the planes α, β, γ, δ, its equation must contain no terms in l^2, m^2, n^2, p^2. Thus, if the coefficients of these terms in Ω be A, B, C, D, the quadric touching α, β, γ, δ and also the tangent cone of Ω drawn from the point (ξ, η, ζ, τ), touches also the tangent cone of Ω drawn from $(A\xi^{-1}, B\eta^{-1}, C\zeta^{-1}, D\tau^{-1})$. Now, in particular, let (ξ, η, ζ, τ) be on Ω; then we know, or it is easily verified, that, as Ω contains the four intersections of the planes α, β, γ, δ, the point $(A\xi^{-1}, \ldots, D\tau^{-1})$ is likewise on Ω (Vol. III, p. 55). Thence, the quadric in question, containing the tangent planes of Ω from a point of itself, contains the generators of Ω at this point; and it is proved that it then contains the generators at another point.

Ex. 7. At the point (ξ, η, ζ, τ) of the quadric

$$fyz + \ldots + uxt + \ldots = 0,$$

let the equation of the tangent plane be written

$$\lambda x + \mu y + \nu z + \varpi t = 0.$$

Using A, B, C, D, respectively, for fvw, gwu, huv, fgh, prove that the polar reciprocal of the quadric, in regard to the quadric

$$A^{-1}\xi\lambda x^2 + B^{-1}\eta\mu y^2 + C^{-1}\zeta\nu z^2 + D^{-1}\tau\varpi t^2 = 0,$$

is a quadric containing the generating lines of the original both at (ξ, η, ζ, τ) and at $(A\xi^{-1}, B\eta^{-1}, C\zeta^{-1}, D\tau^{-1})$.

Ex. 8. Prove that the quadrics $xy + hzt = 0$, $xy + kzt = 0$ are polar reciprocals of one another in regard to any quadric whose equation is of the form $ax^2 + by^2 + hkz^2 + abt^2 = 0$, and that tetrads of points of the first quadric exist which are self-polar in regard to the last, two points of the tetrad, which are conjugate in regard to the last quadric, being arbitrary. Prove, also, that the four points of such a tetrad, taken with the points to which the coordinates are referred, are eight associated points; and that the cubic curve containing the points of the tetrad and two of the reference points whose join is a generator, lies on the first quadric. Hence, also, if AC, AD be two intersecting generators of a quadric, and P, Q, R be any three points of the quadric, the cubic curves on the quadric defined, respectively, by P, Q, R, A, C and by P, Q, R, A, D, have, as their fifth intersection, a point S, such that another quadric can be constructed, touching the four planes QRS, ..., PQR, to contain the generators of the original quadric at C and at D (cf. Vol. III, p. 55).

Ex. 9. If we employ the *cross-ratio* of four lines (cf. Vol. II, p. 166), it is a fundamental property that, if A, B, C, D be four fixed points of a conic, the lines joining these to any point of the conic have an invariable cross-ratio, say $(A, B; C, D)$. It can be shewn that if two circles meet in points A, B, beside the Absolute points I, J, the circle of similitude of these circles has, for its value of $(A, B; I, J)$, the product of the values belonging to this on the two circles. From this it can be shewn at once that, if the circles defined by the triads of coplanar points (B, C, D), (C, A, D), (A, B, D), (A, B, C) be denoted, respectively, by S_1, S_2, S_3, S, then the circles of similitude of the pairs (S, S_1), (S, S_2), (S, S_3), supposed undegenerate, meet in a point. And the theorem of Ex. 5 is deducible.

It can be shewn that the two tangents drawn to one circle of a pair from any point of their circle of similitude are at the same angle as the tangents from this point to the other circle. Hence, from what is proved above, there is a point at which the four circles, drawn through the triads of four points of a plane, subtend equal angles. This result was given by Dr G. T. Bennett.

If V be the vertex of one of the two quadric cones which contain the sections of a quadric by two planes a, β, whose line of intersection is l, and the line joining V to a point, U, of the quadric, meet the quadric again in U_1, the plane lU_1 meets the quadric in a section projecting into the circle of similitude of the two circles into which the sections a, β project, the centre of projection being U.

If W be the vertex of the other quadric cone containing the sections a, β, the planes lU, lU_1 are harmonic conjugates in regard to the planes lV, lW.

If a conic meet the joins BC, CA, AB, of three points A, B, C, respectively in P, P'; Q, Q'; R, R': if QR, $Q'R'$ meet in X; RP, $R'P'$ meet in Y, and PQ, $P'Q'$ meet in Z: if H be any point of the conic, and the lines HX, HY, HZ, meet the conic again in X_1, Y_1, Z_1, respectively: then the lines AX_1, BY_1, CZ_1 meet in a point. We can deduce from this that the three circles of similitude of the three pairs of three arbitrary circles, in a plane, are coaxial.

Ex. 10. For any quadric, given in point coordinates $ax^2 + ... = 0$, let (ξ, η, ζ, τ), $(\xi', \eta', \zeta', \tau')$ be two points of the quadric, let $\phi = a\xi\xi' + ...$ be the polar form, linear and symmetrical in (ξ, η, ζ, τ) and $(\xi', \eta', \zeta', \tau')$, let Δ be the discriminantal determinant, and $\sigma = \phi/2\Delta$. If the tangent planes at (ξ, η, ζ, τ), $(\xi', \eta', \zeta', \tau')$ meet in a line which cuts the quadric in (x, y, z, t) and (x', y', z', t'), prove that $(xx'/A)^{-1}(\sigma - \xi\xi'/A) = (yy'/B)^{-1}(\sigma - \eta\eta'/B) = $ etc., where A, B, ... are the minors of the diagonal elements of Δ.

Ex. 11. Given a quadric, Ω, and five planes, a, β, γ, δ, ϵ, let a quadric be drawn through the four intersections of a, β, γ, δ to contain the section $\Omega\epsilon$; this quadric will have with Ω another plane section, say ϵ_1. Similarly a quadric can be drawn through the four intersections of α, β, γ, ϵ to contain the section $\Omega\delta$; this quadric will have with Ω another plane section, say δ_1. Shew that the sections δ_1, ϵ_1 lie on a quadric cone whose vertex is the intersection of the planes α, β, γ.

Projecting on to a plane, we have five given circles, (α), (β), (γ), (δ), (ϵ), and five other circles, (α_1), ..., (ϵ_1), such that (δ_1) and (ϵ_1) are inverses of one another in regard to the circle which cuts all of (α), (β), (γ) at right angles; and so on. There are ten pairs such as (δ_1), (ϵ_1), and each circle such as (ϵ_1) is the inverse of all the other four with appropriate circles of inversion.

Relation between Wallace's theorem for four circles, and Moebius's figure of two inscribed tetrads. We have given (Vol. II, p. 71) the theorem that, if four arbitrary lines be given in a plane, the four circles, each defined by the intersections of three of the lines, meet in a point. This result is ascribed to Wallace (1806; see *Camb. Phil. Proc.*, xxi, 1923, p. 348). We have also (Vol. I, p. 61) given the theorem, due to Moebius (1828; see *Ges. Werke*, I, 1885, p. 443), that it is possible for two tetrads of points in three dimensions to be so situated that every point of either tetrad lies on the plane determined by three points of the other tetrad. We give here some interesting results of which one incidental consequence is the essential identity of the figures arising in these two theorems. The order in which these results are given is chosen, however, with a further aim.

Let P_1, P_2, P_3, P_4 be any four points in a plane; through each of the six joins of these four points let an arbitrary plane be drawn, the plane through the join of P_i and P_j being denoted by α_{ij}. Every three of these planes, by their intersection, determine a point. Let the point of intersection of the planes, α_{23}, α_{31}, α_{12}, drawn through the joins of the three points P_1, P_2, P_3, be denoted by P_4'. There arise four such points, P_1', P_2', P_3', P_4'. We prove that these lie in a plane. This is, in fact, only a restatement of Moebius's theorem, as may be seen by comparison with p. 61 of Vol. I, where a diagram is given: denote the points P_1, P_2, P_3, P_4, respectively, by A, B, C and D'; let the point of intersection of the arbitrary planes drawn through BC, CA, AB be denoted by D, the point of intersection of the planes drawn through BC, BD', CD' be denoted by A', and similarly for B' and C'; consider then the line of intersection of the planes ABC, $A'B'C'$, denoting the points where this line is met, respectively, by BC, AD', CA, BD', AB, CD', by P, P', Q, Q', R, R'; then, the points B', C' are both on the plane which was drawn through AD', so that $B'C'$ meets AD', and, hence, $B'C'$ passes through P'; in the same way, $C'A'$ passes through Q', and $A'B'$ passes through R'; as P, P'; Q, Q'; R, R' are the points of intersection of a line with the pairs of opposite joins of the four points A, B, C, D', these three pairs of points are in involution; whence it follows (Pappus' theorem being assumed) that the three lines PA', QB', RC' meet in a point, say D_1; the point D, of the theorem, lies on the plane BCA', for both A' and D are on the same arbitrary plane drawn through the line BC; thus we infer that D_1 is the same as D; this, however, shews that the four points A', B', C', D are in one plane; this is what was to be proved.

The dual of the preceding theorem is as follows: From the point O, let there be drawn, in space of three dimensions, four arbitrary planes, $\alpha_1, \alpha_2, \alpha_3, \alpha_4$; upon the line of intersection of any two of these planes, say of α_1 and α_2, let there be taken an arbitrary point, which, for the case of these two planes, may be denoted by P_{12}. The three points so taken upon the lines of intersection of any three of the original planes determine a new plane; for example, the points P_{23}, P_{31}, P_{12} determine a plane which we may denote by α_{123}. There are four such planes; the theorem is that these meet in a point, say O'. The figure contains eight points, $O, P_{23}, ..., P_{34}, O'$, and eight planes, $\alpha_1, ..., \alpha_4, \alpha_{234}, ..., \alpha_{123}$; of these, four planes pass through every point, and four points lie in every plane. Moreover, the points are a system of *associated* points (Vol. III, p. 154), being the intersections of three linearly independent quadrics; for instance, the plane α_1 contains $O, P_{12}, P_{13}, P_{14}$, and the plane α_{234} contains $P_{23}, P_{34}, P_{42}, O'$; thus the plane-pair (α_1, α_{234}) is such a quadric, and two others, independent of this, are formed by the plane-pairs $(\alpha_2, \alpha_{314}), (\alpha_3, \alpha_{124})$. The dual theorem, that the eight planes of two Moebius tetrads touch three independent quadric envelopes, is also easily verified. (Cf. Vol. III, p. 138, Ex. 14.)

Now suppose that we have four lines in a plane, and also two points, I, J, in this plane. We may consider the four conics, all passing through I and J, each determined as also passing through the three intersections of three of the lines. The Wallace theorem is that these four conics have another common point. Take an arbitrary point, O, outside the plane, and describe a non-degenerate quadric containing O and the two lines OI, OJ. Consider the four planes joining O to the four lines of the plane; say, these are $\alpha_1, \alpha_2, \alpha_3, \alpha_4$; let the line of intersection of α_1 and α_2 meet the quadric, beside O, in the point P_{12}; and so on; the plane, α_{123}, containing the points P_{23}, P_{31}, P_{12}, will meet the quadric in a conic which projects from O, on to the plane, into one of the four conics of the Wallace theorem. The point, O', in which the four planes such as α_{123} intersect, will lie on this quadric, by the property of eight associated points. The Wallace theorem thus follows.

The theorem that the four planes $\alpha_{234}, ..., \alpha_{123}$ meet in a point, O', is capable of a suggestive analytical proof, as follows: Denote the equations of the four planes α_1, ..., α_4, drawn through O, respectively by $x_1 = 0$, $x_2 = 0$, $x_3 = 0$, $x_4 = 0$; these will be connected by a linear identity, say, $ax_1 + bx_2 + cx_3 + dx_4 = 0$. Let $t = 0$ be the equation of any chosen plane not passing through O. The equation of any one of the four planes α_{234}, ..., α_{123} may be expressed in terms of t and any three of x_1, x_2, x_3, x_4. We may then suppose the equations of these four planes to be, respectively,

$$k_{12}x_2 + k_{13}x_3 + k_{14}x_4 - at = 0,$$
$$\cdots\cdots\cdots\cdots\cdots$$
$$k_{41}x_1 + k_{42}x_2 + k_{43}x_3 - dt = 0,$$

where x_1 does not appear in the first equation, ..., and x_4 does not appear in the last. Then the point P_{23}, for example, lying on $x_2 = 0$, $x_3 = 0$, lies on α_{234} and on α_{123}; thus it satisfies both the equations $k_{14}x_4 - at = 0$, $k_{41}x_1 - dt = 0$, as well as the identity $ax_1 + dx_4 = 0$. Thence we infer that $k_{14} + k_{41} = 0$. In general, then, in the same way, we have $k_{rs} + k_{sr} = 0$, and, with these identities, the four planes are adequately given by the equations. The condition that these four planes meet in a point is then given, if we add to the four equations the identity $ax_1 + ... + dx_4 = 0$, by the vanishing of a determinant of five rows and columns, in which the elements of the first row are 0, k_{12}, k_{13}, k_{14}, $-a$, and so on, and the elements of the fifth row are a, b, c, d, 0. As this determinant is skew symmetrical, and of odd order, it vanishes identically. See, below, p. 61; and, for another proof of Wallace's theorem, see p. 64.

Ex. 1. Moebius's theorem, above identified with a figure from which Wallace's theorem follows, can be stated thus (cf. Vol. I, p. 61): If a line meet the joins of three points, A', B', C', respectively, in P', Q', R', and the joins of these to three other points, A, B, C, namely the lines $P'A$, $Q'B$, $R'C$, meet in a point, D', then the intersections of the line with the joins of A, B, C, say the points P, Q, R, when joined to A', B', C', will give three lines PA', QB', RC', which also meet in a point, D.

The dual result is that if the joins of a point, O, to three points A', B', C', meet the joins BC, CA, AB, of three points A, B, C, in three points which are in line, then the joins of O to A, B, C meet $B'C'$, $C'A'$, $A'B'$, respectively, in three points also in line. It may easily be found that O lies on a certain cubic curve, whose equation thus appears as capable of two forms.

Ex. 2. We may similarly, in three dimensions, consider the locus of a point whose joins to four given points meet four given planes in points lying on a plane. This locus is a quartic surface, considered by Bauer (*Sitzber. bayer. Ak.*, München, XVIII, 1888, p. 337). When the four given planes determine a tetrad of points in perspective with the four given points, the locus is the Hessian of a cubic surface. (Cf. Vol. III, p. 223.) In general, the reciprocity which holds for the case of Ex. 1 does not hold in three dimensions. If the four given points are in one plane, and are the poles of the four given planes in regard to a conic in this plane, then, with this conic as Absolute conic, the locus is that of a point such that the perpendiculars from it to the given planes have their feet on a plane; this locus is a cubic surface with four double points (see below, p. 25).

Ex. 3. The reader may compare the result of Ex. 1 with the theorem given by Steiner (*Ges. Werke*, I, p. 157): If the perpendiculars drawn from three points, A', B', C', to the joins, BC, CA, AB, of three other points, meet in a point, so do the perpendiculars drawn from A, B, C to the joins $B'C'$, $C'A'$, $A'B'$, of the first three points. This will be found to be true in the more general form : Given two triads of points, A', B', C', and A, B, C, and also an arbitrary conic; if A', B', C' be in perspective with the triad formed by the poles of BC, CA, AB, taken in regard to this conic, then A, B, C are in perspective with the poles of $B'C'$, $C'A'$, $A'B'$. And generalisation is possible to two tetrads of points in three dimensions.

Additions to Wallace's theorem in a plane. Wallace's theorem has additions (Vol. II, p. 72), ascribed to Steiner (1828, cf. *Ges. Werke*, I, p. 197), to the effect that the feet of the perpendiculars, drawn from the Wallace point to the four lines, lie in line, and that the centres of the four circles which meet in the Wallace point, lie on a circle containing the Wallace point. We proceed now to obtain these results by projection from the figure in three dimensions. Let O, A, B, C be four points of a quadric; let P be any point of the section of the quadric by the plane ABC; and let I, J be the points in which the generators of the quadric at O meet the generators at P. Thus the planes OPI, OPJ are the tangent planes of the quadric drawn from the line OP, and OP is the polar line of IJ. Next, let U, V, W be the poles, lying in the tangent plane at O, respectively of the planes OBC, OCA, OAB. We can prove that the six points O, I, J, U, V, W are on a conic. For, if OI, OJ meet the section of the quadric by the plane ABC in I', J', the fact that this section contains the six points P, I', J', A, B, C involves (Vol. II, p. 29) that the planes joining O to the six lines $I'J'$, PI', PJ', BC, CA, AB touch a quadric cone; and the poles of these planes, respectively, are the points O, I, J, U, V, W. Conversely, if conics be taken, in the tangent plane at O, through the four points O, U, V, W, and one of these meet the generators at O in the points I, J, the second tangent plane of the quadric, drawn from the line IJ, touches the quadric in a point, P, of the section by the plane ABC. When we project the figure on to a plane, from the point P, the three sections OBC, OCA, OAB become circles, the projections of I, J being the Absolute points, their centres being the projections of U, V, W; and we have the result that, if three circles in a plane have a common point, their three other intersections being in line, then their centres determine a circle passing through their common point. When we project the figure from the point O, the section of the quadric by the plane OUP becomes a line, through the projection of P, which is at right angles to the line into which the conjugate section OBC projects, the Absolute points being again the projections of I and J. Let H be the point in which the sections OBC, OUP meet, other than O; also let O' be the point which, on the conic $OIJUVW$, is the harmonic of O in regard to I and J; it can be shewn that the line OH lies on the polar plane of O'. To prove this it is sufficient to shew that the polar line of OH, which lies in the tangent plane at O, and contains the pole, U, of the plane OBC, passes through O'. For this, consider the pole of the plane OUP. This pole is the intersection of the polar planes of O, U, P; that is, it lies in the tangent plane at O, in the plane OBC, and in the tangent plane at P; it is thus the intersection of the tangent line, at O, of the section OBC, with the line IJ; the tangent line, at O, of the section OBC, is however harmonic to OU in regard to OI, OJ. Thus the pole of the plane OUP is the point, of IJ, where this is met by the line, from U, which is harmonic to UO in regard to UI and UJ. This line contains O'; as it contains U, it is the polar line of OH. Now the point O' is determined by O, I, J and the conic $OIJUVW$ alone. Thus, as we have proved that the polar plane of O' contains the line (OBC, OUP), it follows that this plane also contains the lines (OCA, OVP) and (OAB, OWP); in other words the points, other than O, in which the sections OBC, OCA, OAB are respectively met by the planes OPU, OPV, OPW, lie in a plane through O. Wherefore, projecting from O, on to a plane, we have the result that if the perpendiculars be drawn to the joins of three points (the projections of A, B, C), from any point (the projection of P) on the circle containing these points, then the feet of these perpendiculars are on a line (Vol. II, p. 71). From this result, and the result obtained in regard to the centres of three circles which meet in a point, the additions to Wallace's theorem are obtained at once.

Ex. 1. If, with any two points, I, J, as Absolute points, the perpendiculars be drawn to the joins of three points, A, B, C, of a plane, from any point, P, the feet of these determine in general a circle, the pedal circle of P (Vol. II, p. 87). We have proved (Vol. II, p. 88) that, if P move on a line passing through the centre of the circle which contains A, B, C, the pedal circles have a common point (Bobillier, *Gergonne's Ann.*, xix, 1828, p. 356; Fon-tené, *Nouv. Ann.*

de Math., v, 1905, p. 504; vi, 1906, p. 55). This point will then be the intersection of the pedal lines of the two points of intersection of the circle *A, B, C* with the line on which *P* moves. It can be shewn that these pedal lines are at right angles, and that their point of intersection is on the nine points circle of *A, B, C.* More generally, if *P* move on an arbitrary fixed line, there is a circle to which all the pedal circles are at right angles; this remark the writer owes to Mr J. H. Grace. It will now be shewn further that this remains true when *P* moves on a certain cubic curve, which in particular may degenerate into the aggregate of a line and a conic.

Let *O, A, B, C* be fixed points of a quadric, the poles of the planes *OBC, OCA, OAB,* lying in the tangent plane at *O,* being denoted by *U, V, W.* Let *P* be any other point of the quadric. Let the plane *OPU* meet the section *OBC* in *H,* beside *O*; similarly let *OK, OL,* where *K, L* are on the quadric, be the lines (*OPV, OCA*) and (*OPW, OAB*). When we project from *O,* on to a plane, and take for Absolute points, *I, J,* the points of the plane lying on the generators of the quadric at *O,* the pedal circle, in regard to the projections of *A, B, C,* of the point which is the projection of *P,* is the circle obtained as the projection of the section *HKL* of the quadric. To prove that the pedal circles, for certain positions of *P,* cut a fixed circle at right angles, we prove that the corresponding planes *HKL* pass through a fixed point of space; when this point is on the quadric the pedal circles meet in a point.

We can suppose the equation of the quadric, referred to *O,A,B,C,* to be

$$fyz + gzx + hxy + t(x + y + z) = 0,$$

where $t = 0$ is the plane *ABC,* and $x = 0$ the plane *OBC,* etc. The point *U* then has coordinates $(-f, w, v, fu)$, where $2u = g + h - f$, $2v = h + f - g$, $2w = f + g - h$. If the point *P,* on the quadric, be (ξ, η, ζ, τ), the point *H,* where the line (*OBC, OUP*) meets the quadric, beside *O,* is then found to be

$$\{0, \; w\xi + f\eta, \; v\xi + f\zeta, \; -(w\xi + f\eta)(v\xi + f\zeta)(\xi + \eta + \zeta)^{-1}\},$$

with similar coordinates for *K* and *L*; from these the plane *HKL* is found, expressed without τ, to be

$$x\xi X + y\eta Y + z\zeta Z + t(\xi + \eta + \zeta)(f\eta\zeta + g\zeta\xi + h\xi\eta) = 0,$$

where

$$X = (u\eta + g\zeta)(u\zeta + h\eta), \quad Y = (v\zeta + h\xi)(v\xi + f\zeta),$$
$$Z = (w\xi + f\eta)(w\eta + g\xi).$$

It will be shewn that when *P* moves on a section of the quadric by a fixed plane containing *O* and the pole of the plane *ABC,* the plane *HKL* passes through a definite point of the quadric; this leads to the theorem above referred to, given in Vol. ii, p. 88. More generally, regarding x, y, z, t, in the above equation of the plane HKL, as fixed, and regarding (ξ, η, ζ) as variable, these being taken as proportional to the coordinates of the projection of P, the equation will represent a locus for P such that the corresponding pedal circles cut a fixed circle at right angles; the Absolute points are given by

$$f\eta\zeta + g\zeta\xi + h\xi\eta = 0, \quad \xi + \eta + \zeta = 0.$$

It can be shewn, in particular, that if l, m, n be arbitrary, and x, y, z, t be given, with $q = vw + wu + uv$, by $ux = mn - t$, $vy = nl - t$, $wz = lm - t$ and

$$u^{-1}v^{-1}w^{-1}q^2t = q(mnu^{-1} + nlv^{-1} + lmw^{-1})$$
$$- u(m-n)^2 - v(n-l)^2 - w(l-m)^2,$$

then the cubic relation connecting ξ, η, ζ is satisfied either by $l\xi + m\eta + n\zeta = 0$ or by $lf\xi^{-1} + mg\eta^{-1} + nh\zeta^{-1} = 0$. More particularly, if l, m, n be such that $lfu + mgv + nhw = 0$, the values for x, y, z, t, namely, $u^{-1}l(m+n)$, $v^{-1}m(n+l)$, $w^{-1}n(l+m)$, and $-(mn+nl+lm)$, correspond to a point lying on the quadric, and the equation $l\xi + m\eta + n\zeta = 0$ corresponds to a line through the centre of the circle ABC; we then have the result of Vol. II, p. 88.

Ex. 2. If four planes, $\alpha, \beta, \gamma, \delta$, be drawn through a point, O, of a quadric, the lines of intersection (β, γ), (γ, α), (α, β), (α, δ), (β, δ), (γ, δ) meeting the quadric again, respectively, in A, B, C, A', B', C', we have proved that the planes $ABC, AB'C', BC'A', CA'B'$ meet in a point, O', and inferred that O' is on the quadric by remarking that $O, A, B, C, A', B', C', O'$ are a set of associated points. This follows from what is said above, p. 21; for let the poles of $\alpha, \beta, \gamma, \delta$ be, respectively, U, V, W, T; let the conic through O, U, V, W, T meet the generators at O again in I and J. Then the second tangent plane to the quadric from IJ has its point of contact on every one of the planes $ABC, AB'C', BC'A', CA'B'$.

Ex. 3. We may consider the generalisation to three dimensions of the pedal properties above given for a plane. A particular result, leading to a cubic surface with four double points, has been mentioned in Vol. III, p. 223. Defining the perpendicular from a point to a plane by means of an Absolute conic, ω, namely as the line, from this point, to the pole of the plane in regard to this conic, ω, the condition that the feet of the perpendiculars, drawn from a point E, to the planes containing four points A, B, C, D, should lie in a plane, is the same as that the quadric which touches these planes, and has the cone $E\omega$ as enveloping cone, should touch the plane of ω. To prove this, consider first, instead of the conic ω, a quadric, Ω, the perpendicular from E to a plane being the join of E to the pole of this plane in regard to Ω. Let the poles of the planes $BCD, \ldots, ABC$ be denoted, respectively, by

A', ..., D', and let the lines EA', ..., ED' meet these planes, respectively, on a plane ϖ, whose pole is P. Let the polar plane of E be ϵ, and denote the planes BCD, ..., ABC by α, ..., δ, respectively. The line EA' then intersects the line (ϖ, α), which is the polar line, in regard to Ω, of the line PA'; in other words, the lines EA', PA' are conjugate. So, likewise, are the pairs (EB', PB'), (EC', PC'), (ED', PD'). The locus of a point whose joins to E and P are conjugate lines in regard to the quadric Ω is, however, that quadric, through the plane sections of Ω by the polar planes ϵ, ϖ, which contains both E and P, as is easy to see. This description of the quadric locus is, however, redundant, since any quadric, through the section of Ω by the polar plane ϵ, which contains also the pole E, will meet Ω in another conic and contain the pole of the plane of this. Thus the condition for E, in relation to A, B, C, D, is that the quadric, which contains A', B', C', D', and also the conic in which Ω is met by the polar plane, ϵ, of E, should also contain E. Or, the condition may be stated in dual form, recalling that the planes ϵ, ϖ meet each of α, β, γ, δ in a pair of lines which are conjugate in regard to Ω, and reciprocating in regard to Ω, by saying that the quadric which touches α, β, γ, δ, and has, for enveloping cone, that drawn from E to Ω, should touch the polar plane, ϵ, of E. When Ω consists of the planes containing the tangent lines of a conic, ω, the polar plane of E becomes the plane of ω, and the condition is as stated above; namely, the quadric touching α, β, γ, δ and the tangent planes of the cone $E\omega$, touches the plane of ω. This quadric has then, also, as generating lines, the two tangent lines of ω which pass through the point of contact of the quadric with this plane.

Ex. 4. If the perpendiculars, in regard to an Absolute conic ω, drawn from a point E to the four planes containing points A, B, C, D, meet these planes in points which lie on a plane, then each of the five points A, B, C, D, E is in the same relation to the other four. For let σ be the conic, in the plane of ω, in regard to which the joining lines and planes of the points A, B, C, D, E determine a polar system. Then, as we have seen (above, p. 11), the enveloping cone from E to any quadric touching the four planes BCD, ..., ABC, and touching also the plane of ω, meets this plane in a cone which is inpolar to σ. Thus, from Ex. 3 preceding, when the perpendiculars, in regard to ω, drawn from E to the four planes BCD, ..., ABC, meet these in points of a plane, the conic ω is inpolar to σ, and conversely. As ω is given, and σ is symmetrical in regard to A, B, C, D, E, the property in question is likewise symmetrical.

Ex. 5. The theorem of the symmetry of the points A, B, C, D, E in Ex. 4 is due to W. Mantel, *Wiskundige Apgaven*, 1899–1902, p. 396, No. 199. See H. W. Richmond, *Camb. Phil. Proc.*, Vol. xxii,

Part I, 1924. In this paper Mr Richmond proves, for the rational curve of order n in space of n dimensions, in extension of the Wallace theorem for a circle in a plane, a result which, for the cubic curve in three dimensions, may be stated : The necessary and sufficient condition that a cubic curve should be such that, in regard to an Absolute conic ω, the perpendiculars from any point of the curve to the four planes containing any other four points of the curve, should have their feet on a plane, is that the conic ω should touch the three chords of the curve which lie in its plane. Let the three points of the curve which lie in this plane be P, Q, R; let σ be the conic, of the plane PQR, in regard to which the joining points and planes of five points, A, B, C, D, E, of the curve, form a self-polar system. We have shewn above (p. 12) that the only condition for σ is that P, Q, R should form a self-polar triad in regard thereto; also (Ex. 4, preceding) that the condition for A, B, C, D, E is that ω should be inpolar to σ. A conic, ω, inpolar to all conics, σ, for which P, Q, R form a self-polar triad, is given by any conic which touches PQ, QR, RP; and, conversely, only by such a conic. This proves the result.

Ex. 6. Prove that, if Ω be any quadric, the necessary and sufficient condition that the lines joining a point E to the poles, A', B', C', D', respectively, of the planes BCD, CAD, ABD, ABC, in regard to Ω, should meet these planes, respectively, in points lying on a plane, is that the lines $D'E$, $D'E_1$ should be conjugate in regard to Ω, where E_1 is the definite point such that the enveloping cones from E and E_1 to Ω are also enveloping cones of a quadric touching the four planes $BCD, ..., ABC$. It can be shewn (cf. Ex. 11 above) that, if D_1 be defined from D, and the four planes containing A, B, C, E, just as E_1 is here defined, then the line D_1E_1 contains D'. Hence the condition for E in regard to A, B, C, D, is that the line $D_1D'E_1$ should intersect the polar line of $D'E$. The condition that the lines joining D to the poles of the planes BCE, CAE, ABE, ABC, in regard to Ω, should meet these planes in points of a plane, is that the line $D_1D'E_1$ should intersect the polar line of $D'D$. One condition does not involve the other, in general; this will be so, however, when the points D', D, E are in line. It will also be so when Ω reduces to a conic.

Ex. 7. The results may be obtained analytically. As before, let A, B, C, D, E be any five points, and E_1 be determined from E by the condition that the quadric which touches the planes $BCD, ...,$ ABC, and the tangent cone from E to the given quadric Ω, has E_1 as the vertex of the further common enveloping cone of the two quadrics, while D_1 is determined from D and Ω similarly, by a quadric touching the four planes joining A, B, C, E and the enveloping cone to Ω drawn from D; let coordinates be taken with

respect to A, B, C and a further arbitrary point; and the tangential
equation of Ω be $(A,B,C,D,F,G,H,U,V,W \delta l,m,n,p)^2 = 0$. Then
prove that the line D_1E_1 contains the pole, D', of the plane ABC,
and meets the plane ABC in the point whose coordinates are

$$\frac{A\tau\tau' - U\,(\xi\tau' + \xi'\tau) + D\xi\xi'}{\xi\tau' - \xi'\tau}, \qquad \frac{B\tau\tau' - V\,(\eta\tau' + \eta'\tau) + D\eta\eta'}{\eta\tau' - \eta'\tau},$$

$$\frac{C\tau\tau' - W\,(\zeta\tau' + \zeta'\tau) + D\zeta\zeta'}{\zeta\tau' - \zeta'\tau}, \quad 0,$$

where (ξ, η, ζ, τ), $(\xi', \eta', \zeta', \tau')$ are the coordinates of D and E,
respectively. In order that the perpendiculars from E, in regard to
Ω, drawn to the planes $BCD, \ldots, ABC$, should meet these in the
points of a plane, we have to express that the lines $D'E$, $D'E_1$
should be conjugate in regard to Ω; that is that the polar plane
of E should contain the point in which $D'E_1$ meets the plane ABC.
If the point equation of Ω be $(a, b, c, d, f, g, h, u, v, w \delta x, y, z, t)^2 = 0$,
this condition is that E should lie on the quartic surface whose
equation, with x, y, z, t as current coordinates, is

$$\frac{A\tau t - U\,(\xi t + \tau x) + D\xi x}{\xi t - \tau x}\,(ax + hy + gz + ut) + \ldots$$

$$+ \frac{C\tau t - W\,(\zeta t + \tau z) + D\zeta z}{\zeta t - \tau z}\,(gx + \ldots + wt) = 0.$$

When coordinates are referred to A, B, C, D, so that $\xi = 0$, $\eta = 0$,
$\zeta = 0$, this equation is

$$\frac{At - Ux}{x}\,(ax + hy + gz + ut) + \ldots + \frac{Ct - Wz}{z}\,(gx + \ldots + wt) = 0,$$

or

$$\frac{A\,(ax+hy+gz+ut)}{x} + \ldots + \frac{C\,(gx+\ldots+wt)}{z} + \frac{D\,(ux+\ldots+dt)}{t} - \Delta = 0,$$

where Δ is the discriminantal determinant of the point equation
of Ω.

If Ω be a conic, lying in the plane $\lambda x + \mu y + \nu z + \varpi t = 0$, so that
there are four equations such as $A\lambda + H\mu + G\nu + U\varpi = 0$, the point
equation of Ω is $(\lambda x + \ldots + \varpi t)^2 = 0$. The equation then divides by
$\lambda x + \ldots + \varpi t = 0$. The remaining cubic surface is given by

$$\lambda\,(\xi t - \tau x)^{-1}\,[A\tau t - U\,(\xi t + \tau x) + D\xi x] + \ldots + t = 0$$

or, referred to A, B, C, D, by

$$A\lambda x^{-1} + B\mu y^{-1} + C\nu z^{-1} + D\varpi t^{-1} = 0,$$

and is a cubic surface with four double points. The equation expresses that the point E_1, derived from E as in the preceding
Example, is on the plane of the Absolute conic.

Ex. 8. In regard to the Absolute conic given by

$$\lambda x + \mu y + \nu z + \varpi t = 0, \quad Ax^2 + By^2 + Cz^2 + Dt^2 = 0,$$

the four points of reference for the coordinates are such that their opposite joins are at right angles; and they have an orthocentre, the intersection of the four perpendiculars from the points to the opposite planes, which is $(\lambda A^{-1}, \mu B^{-1}, \nu C^{-1}, \varpi D^{-1})$.

Ex. 9. The quadrics touching $x = 0$, $y = 0$, $z = 0$, $t = 0$ in regard to which the point (ξ, η, ζ, τ) is the pole of the plane $\lambda x + \mu y + \nu z + \varpi t = 0$, give, in this plane, ∞^2 conics. The quadrics of the form $Ax^2 + By^2 + Cz^2 + Dt^2 = 0$, which pass through (ξ, η, ζ, τ), give, in this plane, other ∞^2 conics. Prove that every conic of the latter system is outpolar to every conic of the former system.

Ex. 10. The surface $x^{-1} + y^{-1} + z^{-1} + t^{-1} = 0$ is the locus of a point from which the perpendiculars to $x = 0, y = 0, z = 0, t = 0$ have their feet on a plane, when a proper conic is taken as Absolute conic. Prove that, of such conics, there are ∞^2 in an arbitrary plane, $\lambda x + \mu y + \nu z + \varpi t = 0$, namely the intersections of this plane with quadrics touching $x = 0$, $y = 0$, $z = 0$, $t = 0$ which are such that the pole of this plane, in regard to the quadric, is

$$(\lambda^{-\frac{1}{2}}, \mu^{-\frac{1}{2}}, \nu^{-\frac{1}{2}}, \varpi^{-\frac{1}{2}}).$$

Ex. 11. It should be remarked in connexion with preceding theorems for a cubic curve, that, if three tetrads of points of the cubic be taken, say, A, B, C, D; A', B', C', D'; A'', B'', C'', D'', then the twelve planes, of which each contains three points of one tetrad, are all touched by a quadric. Further, taking any two other arbitrary points of the curve, the two tangent planes of the quadric, through the chord joining these two points, determine two further points of the curve, and hence a tetrad, of which the two other connecting planes also touch the quadric.

Let the planes $BCD, \ldots, ABC$ be denoted, respectively, by $\alpha, \beta, \gamma, \delta$, with a similar notation for the planes of the two other given tetrads, namely $\alpha', \beta', \gamma', \delta'$, and $\alpha'', \beta'', \gamma'', \delta''$. Since the quadrics touching the planes $\alpha, \beta, \gamma, \delta, \alpha', \beta', \gamma'$ likewise touch δ', these quadrics, expressed tangentially, are ∞^2, and such a quadric exists touching also the planes joining D to the chords $A''B''$, $A''C''$. The enveloping cone from D to this quadric then touches the five planes $\alpha, \beta, \gamma, DA''B'', DA''C''$. But the lines DA, DB, DC, DA'', DB'', DC'' lie on a quadric cone (of vertex D, containing the curve). Hence the quadric cone touching the five planes α, β, γ, $DA''B''$, $DA''C''$ also touches the plane $DB''C''$. Wherefore, the quadric constructed, touching the planes $\alpha, \beta, \gamma, \delta, \alpha', \beta', \gamma', \delta'$ and $DA''B'', DA''C''$, equally touches the plane $DB''C''$. As touching $\alpha', \beta', \gamma', \delta'$ and $DB''C'', DC''A'', DA''B''$, this quadric also touches the eighth plane, $A''B''C''$, or δ'', these eight planes being associated, Vol. III, p. 139. Similarly it touches the planes α'', β'', γ''.

Now, let A_3 be any further point of the curve, and let the tangent planes of the quadric from DA_3 meet the curve again in B_3 and C_3. Then, as above, the quadric also touches the plane DB_3C_3, and thence, also, the plane $A_3B_3C_3$. If, then, D_3 be taken arbitrarily, on the curve, and the tangent planes be drawn from A_3D_3 to this quadric, determining, by their further intersections, the points B_3', C_3' of the curve, it follows, as before, that this quadric also touches $A_3B_3'C_3'$, and, thence, also the plane $D_3B_3'C_3'$. This establishes the statement made, the two arbitrary points taken being A_3 and D_3.

Ex. 12. If the coordinates of a point of the cubic curve be written $(\theta^3, \theta^2, \theta, 1)$, and the points A, B, C, D be given by $a_0\theta^4 + a_1\theta^3 + \ldots + a_4 = 0$, or say $f = 0$, the points A', B', C', D' being given by $\phi = 0$, where $\phi = b_0\theta^4 + b_1\theta^3 +$ etc., and the points A'', B'', C'', D'' by $\psi = 0$, where $\psi = c_0\theta^4 + c_1\theta^3 +$ etc. $= 0$, shew that the planes of every tetrad of points of the cubic, given by an equation of the form $\xi f + \eta \phi + \zeta \psi = 0$, touch the quadric whose tangential equation is

$$\begin{vmatrix} a_0, & a_1, & a_2, & a_3, & a_4 \\ b_0, & b_1, & b_2, & b_3, & b_4 \\ c_0, & c_1, & c_2, & c_3, & c_4 \\ l, & m, & n, & p, & 0 \\ 0, & l, & m, & n, & p \end{vmatrix} = 0.$$

Theorems for the circumscribed circles when an indefinite number of lines is given in a plane. Recurring now to the theorem in which four arbitrary points were taken in a plane, and arbitrary planes were drawn through the joins of each two, we were able, with the use of Pappus' theorem, to shew that the four points, obtained as intersections of the planes through the joins of each triad of points, lie in a plane (above, p. 18). If we start with five arbitrary points, it can be shewn that the five planes, each obtained as above from four of these, all meet in a point; and for the proof of this the Propositions of Incidence only are sufficient. If we begin with six points, the six points so obtainable, one from each five of these, lie in a plane. And so on, indefinitely.

In a plane, α, let five points P_1, P_2, $\ldots$, P_5 be arbitrarily taken; and through the join of every two let an arbitrary plane be drawn, for instance the plane α_{12} through the join P_1P_2. The planes $\alpha_{23}, \alpha_{31}, \alpha_{12}$ determine a point, P_{123}, and so on. The four points P_{123}, P_{234}, P_{314}, P_{124} lie in a plane; this is what was proved above (p. 18). When we start with five points there arise five such planes. Denoting these by α_{1234}, $\ldots$, α_{2345}, we now prove that these five planes meet in a point.

In the plane α_{12} are five points, namely P_1, P_2, P_{123}, P_{124}, P_{125};

and the process followed gives another plane than α_{12} passing
through the line joining any two of these points. Of these five
points consider, at first, only the four P_1, P_{123}, P_{124}, P_{125}, and apply
to these the process which was followed when we started with four
points, P_1, P_2, P_3, P_4, lying in a plane α. Through the lines joining
the respective six pairs, of the four points,

$$(P_{124}, P_{125}), \; (P_{125}, P_{123}), \; (P_{123}, P_{124}), \; (P_1, P_{123}), \; (P_1, P_{124}), \; (P_1, P_{125})$$

there pass, beside the plane α_{12}, the planes

$$\alpha_{1245}, \quad \alpha_{1235}, \quad \alpha_{1234}, \quad \alpha_{13}, \quad \alpha_{14}, \quad \alpha_{15} \; ;$$

of these, however, the planes α_{1245}, α_{14}, α_{15} meet in P_{145}; the planes
α_{1235}, α_{13}, α_{15} meet in P_{135}; and α_{1234}, α_{13}, α_{14} meet in P_{134}. Thus, the
plane α_{1345}, which is the plane of the points P_{145}, P_{135}, P_{134}, contains
the point of intersection of the planes α_{1245}, α_{1235}, α_{1234}. If we next
consider the four points P_2, P_{123}, P_{124}, P_{125}, we similarly prove that
the same point of intersection lies on the plane α_{2345}. That is, the
five planes, $\alpha_{p,q,r,s}$, where p, q, r, s are every four of the numbers
1, 2, 3, 4, 5, meet in a point. This is what we desired to prove.
The final point may be denoted by P_{12345}.

If we now begin with six points, $P_1, \ldots, P_6$, of the original plane,
every five of these, by the same process, will give rise to a point.
It can be shewn that the six points so obtained lie in a plane. The
process gives rise, with others, to the four points P_{123}, P_{124}, P_{125}, P_{126},
all lying in the plane α_{12}. Beside this plane there pass, through the
six joins of pairs of these points, namely

$$(P_{124}, P_{125}), \; (P_{125}, P_{123}), \; (P_{123}, P_{124}), \; (P_{123}, P_{126}), \; (P_{124}, P_{126}), \; (P_{125}, P_{126}),$$

respectively the planes

$$\alpha_{1245}, \quad \alpha_{1235}, \quad \alpha_{1234}, \quad \alpha_{1236}, \quad \alpha_{1246}, \quad \alpha_{1256},$$

which, for brevity, we may denote, momentarily, by (45), (53), (34),
(36), (46), (56), respectively. By what we have shewn, the points
of intersection of threes of these planes, which are

$$(45, 53, 34), \quad (45, 46, 56), \quad (53, 56, 36), \quad (34, 36, 46),$$

lie in one plane. These points are, however, respectively, P_{12345},
P_{12456}, P_{12356}, P_{12346}. By a similar argument it follows that every four
of the six points in question, whose symbols have two digits in
common, are in one plane. From this, the fact that all six points
are in one plane is clear.

If we start with seven points, $P_1, \ldots, P_7$, in the original plane,
there will be seven planes, of which one arises, as in the case considered, from every six of these points. It can be shewn that these
seven planes meet in a point. The proof may begin by considering
the five coplanar points P_{12r}, where $r = 3, 4, \ldots, 7$. And so on.

We may consider the dual of these theorems, obtained by taking planes all passing through a point, and an arbitrary point on the line of intersection of any two of these planes. The case when four planes are drawn through the point was considered above. In particular, the points which are taken on the lines of intersection of the pairs of planes, may be the further intersections of these lines with a quadric drawn through the original point. When the number of planes originally taken is even, the theorem leads finally to a point; it can be shewn, as in the case of four planes, that this point is also on the quadric. When we begin by considering five planes through the original point, we obtain, by projection on to a plane, the part of the theorem proved above (p. 1) which is expressed by saying that the foci of the five parabolas, which touch five lines in fours, lie on a circle. When we begin by considering six planes through the original point, the result is that the six circles, obtained from every five parabolas, have a point in common. And so on. These theorems were given by Clifford (1870), *Math. Papers* (1882), p. 38, whose method of proof (in dual statement) is by a curve of order n with a fixed $(n-1)$-fold point.

Ex. 1. Another representation of the preceding results may be referred to. On a quadric surface, any generator of one system may be associated with a value of a parameter, say θ, and any point of the surface with the values of a pair of parameters, say θ and ϕ. Now take two fixed lines of space, say x and y, both meeting a line h; and associate the points of the line x with the values of θ, the point (h, x) corresponding to $\theta = 0$, say, and two other assigned points of x corresponding to assigned values of θ; similarly, associate the points of the line y with the values of ϕ. Next, take a further line p, meeting h but not x nor y. Then any transversal of the three lines x, y, p, as meeting x and y in points, say θ and ϕ, corresponds to a definite point (θ, ϕ) of the original quadric; and the aggregate of all such transversals, which meet x and y in related ranges, corresponds to the points of a plane section of the original quadric; upon this plane section there is a point, say H, independent of the line p, which corresponds to the transversal h. Conversely, any plane section of the quadric passing through the point H, corresponds to such an aggregate of transversals of the fixed lines x, y and another line p, meeting h but not x nor y, wherein however the line p is not unique. Two such plane sections, α_1 and α_2, have, beside H, another common point, say P_{12}; this corresponds to the fact that two such reguli of lines meeting x and y, both containing h, have in common another line. If three such sections of the quadric be taken through H, and the three further intersections of pairs of these be P_{23}, P_{31}, P_{12}, the points of the plane section of the quadric, which contains these points P_{23}, P_{31}, P_{12}, have parameters

θ, ϕ connected by an equation, linear in both these; the points of this plane section thus correspond to a regulus of lines meeting x and y, in points forming two related ranges; this regulus does not contain h, but contains three lines corresponding, respectively, to P_{23}, P_{31}, P_{12}. The correspondence may be pursued when more than three sections of the quadric are taken through H. In particular we have the following result: Let h be a given transversal of two given non-intersecting lines, x and y. Consider five reguli of lines, each consisting of lines meeting x and y in related ranges, each regulus having h as one line. Two of these reguli have common a line beside h, say q_{ij}; the lines q_{12}, q_{23}, q_{31}, common to the pairs of three of the reguli, determine a further regulus of lines meeting x and y, say q_{123}, not containing h; the four reguli, q_{234}, q_{314}, q_{124}, q_{123}, so arising from four of the original reguli, have then a line in common, say q_{1234}; and the five lines, q_{1234}, ..., q_{2345}, all belong to a regulus of lines meeting x and y.

From this result, taking a section by an arbitrary plane, we obtain the figure considered above (p. 31), in which we have considered the circles through the intersections of the triads selected from five arbitrary lines of a plane; there is the slight generalisation, however, unless the plane of section be taken to contain the line h, that here, instead of five lines in a plane, we obtain five circles with a common point.

Ex. 2. Let S be the conic drawn to touch five given lines of a plane; let S_1, ..., S_5 be conics each touching four of these lines; suppose the six conics are all triangularly inscribed to another conic. Prove that S_1, ..., S_5 have a common tangent line (Wakeford, *Proc. Lond. Math. Soc.*, xv, 1916, p. 340, who uses the result to establish the theorem of a double six of lines).

SECTION II. THEOREMS OF THREE AND FOUR DIMENSIONS

Tetrahedral complex as determined by planes meeting three lines in four dimensions. We have considered in Vol. III (p. 99) the system of lines which are normals of the quadrics of a confocal system. These constitute a tetrahedral complex; this is an aggregate (∞^3) of lines, in space of three dimensions, determined by the single condition that the range of points, on every one of the lines of the complex, in which the line meets four given fixed planes, is related to a given range of four points. Or, what is the same thing (Vol. I, p. 30), by the condition that the planes, joining every one of the lines of the complex to the four points of intersection of the four given planes, form an axial pencil of four planes which is related to the same given range of four points.

We assume the elementary properties of space of four dimensions (cf. Vol. I, pp. 33, 37, etc.); that a general line meets a threefold space in a point; that a general plane meets a threefold space in a line; that two general planes have a point in common; and two general threefold spaces have a plane in common; that two non-intersecting lines determine a threefold space; that three non-intersecting lines have a common transversal line, which meets each of the three lines in the point in which this meets the threefold space determined by the other two. We also assume that through a plane, ϖ, there pass ∞^1 threefold spaces which meet any two lines in related ranges of points. When the two lines lie in a plane, say α, if P, P' be the points in which these lines meet one of the three-folds through ϖ, the line PP' meets ϖ, and lies in α, and so passes through the point common to ϖ and α; the ranges determined on the two lines by the threefolds are thus in perspective. From this it is easy to deduce the general statement.

Now suppose that, in the fourfold space, we have three non-intersecting lines, and also a (flat) threefold space, Ω, not containing any of the lines, nor their transversal. As the join of three arbitrary points, one on each of the lines, there is determined a plane, and there are ∞^3 such planes. We prove that the lines, in which the threefold Ω is met by these planes, constitute a tetrahedral complex in Ω. And, conversely, we shew that any tetrahedral complex in Ω can be determined in such a way.

Let the lines be a, b, c; let the points in which they meet Ω be A, B, C, these being supposed not to lie in line. Then the common transversal of a, b, c does not lie in Ω; let this transversal meet Ω in D. Also let this transversal meet a, b, c in A', B', C'. Consider a plane, ϖ, not lying in Ω, meeting a, b, c, in P, Q, R, respectively. The threefold space, ϖA, containing ϖ and A, as containing P, contains the line a, and hence also the point A'. Thus the four threefolds $\varpi A, \varpi B, \varpi C, \varpi D$ meet the transversal line of a, b, c, respectively, in the points A', B', C', D; these points are independent of the plane ϖ. Next, let l be the line in which the plane ϖ meets Ω; the threefold space ϖA then meets Ω in the plane lA; and any line of Ω meets the plane lA in the point in which this line meets the threefold ϖA. Thus the planes lA, lB, lC, lD are met by any line of Ω in the range in which this line meets the threefold spaces $\varpi A, \varpi B, \varpi C, \varpi D$; and this range is therefore related to the fixed range A', B', C', D, by what we have seen. Wherefore, the lines l, in which Ω is met by the planes ϖ, constitute a tetrahedral complex, relative to A, B, C, D.

Conversely, given in Ω any tetrahedral complex relative to four points, A, B, C, D, of Ω, we can draw, in a fourfold space containing Ω, an arbitrary line through D, not lying in Ω; we can then take on this line three points, A', B', C', in such a way that the range A', B', C', D is related to the range determined, on any line of Ω, by the planes joining A, B, C, D to a line of the complex. Then the tetrahedral complex is that constituted by the lines of Ω which lie in planes meeting the three lines AA', BB', CC', of the fourfold space.

Ex. Let the fourfold space be referred to the five points A, B, C, D and C', and, in terms of the symbols of D and C', let the symbols of A' and B' be, respectively, given by $A' = D + C'$, $B' = D + \lambda C'$. A plane, ϖ, meeting the lines AA', BB', CC', contains three points with symbols, respectively,

$$A + p(D + C'), \quad B + q(D + \lambda C'), \quad C + rC',$$

so that a general point of this plane has a symbol

$$\xi[A + p(D + C')] + \eta[B + q(D + \lambda C')] + \zeta(C + rC');$$

in particular, the points of this plane which lie in the space (A, B, C, D) are such that $\xi p + \eta q \lambda + \zeta r = 0$; such points then have symbols $r\xi[A + pD] + r\eta[B + qD] - (\xi p + \eta q \lambda)C$, and are those of the line joining the two points

$$r(A + pD) - pC, \quad r(B + qD) - q\lambda C.$$

The coordinates of these points, relative to A, B, C, D, are $(r, 0, -p, rp)$ and $(0, r, -q\lambda, rq)$; and the line joining them has the coordinates $(-rq, rp, qp - qp\lambda, p, q\lambda, r)$; denoting these by $(l, ..., n')$, they satisfy $\lambda ll' + mm' = 0$. This is the equation for the tetrahedral complex. It may be readily verified that the planes joining any line of it to A, B, C, D meet an arbitrary line in a range related to A', B', C', D.

Analogous generation of a linear complex. In the preceding section it was assumed that the common transversal of the three lines a, b, c was not in the threefold space Ω. We now shew that, if this be so, the planes meeting a, b, c meet Ω in the lines of a linear complex. The definition of a linear complex has been given in Vol. III, pp. 56 ff., and it has been shewn that this can be stated without the use of the algebraic symbols (Vol. III, p. 64): suppose we have, in the space of three dimensions, two planes, intersecting in a line TU; and, in one of the planes, a flat pencil of lines of vertex T, and in the other plane a flat pencil of lines of vertex U; these pencils being related to one another, and the ray TU, of the one, corresponding to the ray UT of the other. Then the aggregate of the lines joining a point, of one ray of one pencil, to any point of the corresponding ray of the other pencil, is a linear complex; and, conversely, any linear complex can be so determined.

Let c, t, u be three non-intersecting lines of a fourfold space; let k be the common transversal line of these, meeting them, respectively, in C, T, U. Let Ω be a threefold space containing the line k, but not the lines c, t, u. Take fixed points, E, F, G, arbitrarily upon c, t, u, and let EF, which does not lie in Ω, because c and t do not, meet Ω in A; similarly let EG meet Ω in B. Thus AB is a particular fixed line of Ω lying in a plane which meets the lines c, t, u, and we may regard Ω as determined by the line k and the points A, B. If now L be any point of the line c, the line LF, lying in the plane $CEFA$, will meet the fixed line CA, say in P; and the line LG, lying in the plane $CEGB$, will meet the fixed line CB, say in Q; and as L varies on the line c, the ranges $(P), (Q)$, on CA and CB, will be related. In particular, when L is at C, both P and Q are also at C. Thus, the pencil $T(P)$, in the plane kA, is

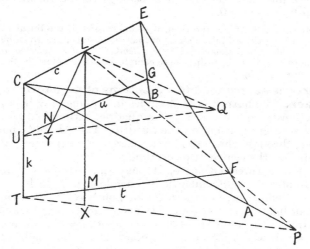

related to the pencil $U(Q)$, in the plane kB; and these pencils have the ray TU in common. If now M be any point of the line t, the line LM, which lies in the plane TLF, meets the line TP, say in X; and as T, C, A, P are in Ω, so is X; again if N be any point of the line u, the line LN, which lies in the plane ULG, meets the line UQ, say in Y; and as U, C, B, Q are in Ω, so is Y. Thus XY is a line of Ω, in which this is met by a plane LMN which meets the given lines c, t, u; and X, Y are, respectively, upon the corresponding rays TP, UQ, of the two related pencils spoken of. Thus, as L, M, N vary, the line XY describes a linear complex in Ω, of which AB and PQ are particular lines.

Ex. Let the fourfold space be referred to the five points A, B, T, U, E, the space Ω being referred to the first four of these. We may then, without loss of generality, suppose the symbols of C, F, G, in terms of those of the fundamental points, to be $C = T + U$, $F = A + E$, $G = B + E$. If then the symbol of L be written $C + xE$, or $T + U + xE$, the point P, lying on LF and CA, has the symbol $P = T + U - xA$, and the point Q, lying on LG and CB, has the symbol $Q = T + U - xB$. Whence, if the symbols of M and N be written, respectively, $M = A + E + yT$, $N = B + E + zU$, the point X, lying on LM and TP, has the symbol $X = T(1 - xy) + U - xA$; and the point Y, lying on LN and UQ, has the symbol

$$Y = U(1 - xz) + T - xB.$$

That is, the coordinates of X and Y relatively to A, B, T, U are, respectively, $(-x, 0, 1 - xy, 1)$ and $(0, -x, 1, 1 - xz)$. From these we at once find that the line XY belongs to the linear complex expressed by $m + m' = 0$.

The determination of the tetrahedral complex, and of the linear complex, by planes meeting three lines in space of four dimensions, are given by Segre, "Alcune considerazioni elementari," etc., *Rend. Circ. Mat. Palermo*, II, 1888, p. 45. Another way of regarding these complexes arises below, in Section III of this Chapter (p. 40).

Spheres as determined from sections of a quadric in fourfold space. We have seen that circles in a plane may be regarded as projections of plane sections of a quadric in threefold space, the centre of projection being on the quadric. The Absolute points of the plane, through which the circles pass, are on the generators of the quadric at the centre of projection.

Spheres in a threefold space, Π, may similarly be obtained by projection of sections of a quadric in fourfold space. Such a quadric, Ω, may be defined here as the locus of points (∞^3 in multiplicity) whose coordinates, in the fourfold space, satisfy a general homogeneous quadratic equation. Let $x = 0$, $y = 0$, $z = 0$, $t = 0$ represent four threefolds which intersect in a point, O, of the quadric Ω, and $v = 0$ be another threefold. The equation of Ω will then have a form

$$ax^2 + by^2 + cz^2 + dt^2 + 2fyz + 2gzx + 2hxy + 2pxt + 2qyt$$
$$+ 2rzt + 2(Ax + By + Cz + Dt)v = 0.$$

Any line drawn through O will meet Ω in one point at O. Lines can be drawn through O which meet the quadric in two coincident points at O; these, it is easily proved, lie in and generate a threefold space; for the given equation of Ω this is given by

$$Ax + By + Cz + Dt = 0;$$

it is called the *tangent threefold* of Ω at O. If P be any point, other than O common to Ω and its tangent threefold at O, the line OP must lie entirely on Ω, because it meets Ω in two coincident points at O and in a further point P, while a line not lying on Ω meets this only in two points. Thus the common points of Ω and its tangent threefold at O are the points of an aggregate of lines through O. Now, an arbitrary threefold meets Ω in a quadric surface, lying in this threefold; an arbitrary plane, which may be regarded as the intersection of two threefolds, meets Ω in a conic; and, if any line, in the tangent threefold at O, but not passing through O, meet the quadric Ω in P and Q, the conic, in which the plane OPQ meets Ω, consists of the two lines OP, OQ. Thus, the intersection of Ω with the tangent threefold at O consists of a quadric cone, of vertex O, lying in this threefold, the ∞^1 generators of this cone being the intersections of Ω with the tangent threefold, constituting all the lines of Ω which pass through O. With the equation taken for Ω, this cone is the intersection of the threefold given by $Ax + By + Cz + Dt = 0$, with the locus represented by

$$(a,\ b,\ c,\ d, f,\ g,\ h,\ p,\ q,\ r \varochi x,\ y,\ z,\ t)^2 = 0.$$

This last consists of an ∞^2 of lines, joining O to the points of a quadric surface lying in the threefold $v = 0$.

If we project the points of Ω from O, on to an arbitrary threefold, Π, the quadric cone of ∞^1 lines of Ω, passing through O, will evidently give rise to a conic, σ, in Π, lying in the plane in which Π is met by the tangent threefold of Ω at O. Now consider an arbitrary threefold, Σ, not passing through O; this meets Ω in a quadric surface; and it is intersected by every line of Ω which passes through O. The section of Ω by Σ will thus project, from O, into a quadric surface in the threefold Π, with the property that it passes through the conic σ. Or, if we regard σ as the Absolute conic of the threefold space Π, the sections of Ω, by all threefold spaces not passing through O, project into spheres in Π.

The tangent plane of such a sphere, at any point, is the projection of the tangent plane of the quadric section of Ω at the corresponding point, say P, of Ω; this tangent plane at P is the intersection of the tangent threefold of Ω, at P, with the threefold, Σ, whose section of Ω is under consideration. But, in particular, if we consider the tangent plane of the sphere, in the space Π, at a point of the conic σ, the corresponding point, P, of Ω, lies on the tangent threefold of Ω at O, and the tangent threefold of Ω at P then passes through O. This tangent threefold of Ω at P contains also the pole point, say S, of the threefold Σ, in regard to Ω, and thus contains the line OS. The point in which OS meets the threefold Σ thus lies on the tangent plane at P of the quadric section (Ω, Σ). Wherefore, the tangent planes of the sphere, in the space Π, at all points of the conic σ, pass through the point in which the line OS, from O to the pole of the cutting threefold Σ, meets the space Π. In other words, the centre of the sphere is the projection of the pole of the cutting threefold Σ, in regard to Ω. In this description, we have assumed that the notion of a polar point, and polar threefold, in regard to the quadric Ω, is clear, from the analogy of the cases in two and three dimensions, without detailed explanation; the equation of the polar threefold of a point (x', y', z', t', v') is formed from the equation of Ω by the operator $x'\partial/\partial x + \dots + v'\partial/\partial v$; and if the polar threefold of a point P contain a point Q, the polar threefold of Q contains P.

Consider two points of Ω, say O and Q; the tangent threefolds of Ω at these points will meet in a plane, and this plane will meet Ω in a conic. Every line, lying on Ω, which passes through O, will therefore meet this conic, and will intersect a particular line, lying on Ω, of those passing through Q, at a point of this conic. Through this point, which is on Ω, there pass ∞^1 lines, lying on Ω, to points such as Q. Thus every line of Ω meets ∞^2 others, and there are ∞^3 lines lying upon Ω. When the points O and Q are such that each lies on the tangent threefold of Ω at the other, the plane common to these two tangent threefolds passes through both O and Q, and contains the line OQ. This line, meeting Ω in two coincident points both at O and Q, lies entirely on Ω. It can be shewn that, in this case, the plane common to the two tangent threefolds meets Ω only in this line, OQ, taken twice over, and touches the cones, of lines of Ω, drawn from O and Q, along this line. Further, a tangent plane of the cone of lines of Ω, through O, is characterised by the fact that every threefold which contains this plane is a tangent three-fold of Ω.

Ex. 1. Given the six equations

$$x^2 + y^2 + z^2 + t^2 + v^2 = 0, \qquad x_0{}^2 + \dots + v_0{}^2 = 0, \qquad x_1{}^2 + \dots + v_1{}^2 = 0,$$

$$x_0 x_1 + y_0 y_1 + \dots + v_0 v_1 = 0, \qquad x x_0 + \dots + v v_0 = 0, \qquad x x_1 + \dots + v v_1 = 0,$$

deduce that

$$[x\,(y_1 z_0 - y_0 z_1) + y\,(z_1 x_0 - z_0 x_1) + z\,(x_1 y_0 - x_0 y_1)]^2 = 0.$$

Ex. 2. If the equation of Ω be written, as above,

$$\phi\,(x, y, z, t) + 2v\,(Ax + By + Cz + Dt) = 0,$$

the cone of lines of Ω from $(0, 0, 0, 0, 1)$ is given by $\phi\,(x, y, z, t) = 0$, $Ax + \dots + Dt = 0$. If (x', y', z', t') be such that $Ax' + \dots + Dt' = 0$, and ϕ' denote $\phi\,(x', y', z', t')$, this cone is touched by the plane given by $Ax + \dots + Dt = 0$ and $x\partial\phi'/\partial x' + \dots + t\partial\phi'/\partial t' = 0$.

Ex. 3. It follows from what is said that a sphere may be represented by the threefold whose section with Ω projects into the sphere. The sphere may, therefore, equally be represented by the point which is the pole of this threefold in regard to Ω. A circle in the space Π, through which an infinite aggregate of spheres can be drawn, may thus be represented, in the fourfold space, by a plane, through which an infinite aggregate of threefolds can be drawn; or the circle can equally be represented by the line of the fourfold space which is the pole, in regard to Ω, of the plane; this being the locus of the poles, in regard to Ω, of the threefolds drawn through the plane, as we easily verify. A consequence of this representation will be that two lines of the fourfold space which have a point in common, being the polars in regard to Ω of two planes which lie in the same threefold space, will represent two circles of the threefold space which lie on the same sphere, that is, two circles which have two points in common. In particular, from the theorem that three lines in fourfold space have a common transversal, if no two of the lines meet, we infer that three circles in threefold space, of which no two lie on a sphere, are all met in two points by a properly chosen circle. (Darboux, *Compt. Rend.*, xcii, 1881, p. 447.)

Ex. 4. As an exercise in regard to the derivation of a sphere from a threefold section of a quadric in four dimensions, we may prove that two spheres cut everywhere at the same angle. By this we understand that the four planes, passing through the tangent line of the circle of intersection of the two spheres, at any point, P, of this circle, which consist of the two tangent planes of the spheres at P, together with the two tangent planes of the Absolute conic drawn from this tangent line, form an axial pencil which is related to that similarly arising for any other point of this circle of intersection. We shall assume that if four threefolds, in space of four dimensions, have a common plane, and an arbitrary threefold be taken, to meet this plane in a line, and each of the four threefolds in a plane, then the axial pencil of four planes so obtained is related to that obtained similarly by another arbitrary threefold; in fact, a line can be drawn (in many ways) meeting the four threefolds in points lying on the four planes of the axial pencil.

Consider two quadric sections of Ω, by two threefolds, Σ and Σ'; let ϖ be the plane common to these threefolds, and P a point of the conic in which this plane meets Ω; let T be the tangent threefold of Ω at P, and l the line in which T meets the plane ϖ. The range of threefolds which can be drawn through ϖ meets T in a range, or axial pencil, of planes, all passing through l; of these threefolds Σ and Σ' are two, and these meet T in the tangent planes at P of the quadric sections of Ω by Σ and Σ'. There are also, however, through ϖ, as through any other plane, two threefolds which touch Ω. By what is said above, these meet the tangent threefold T in planes which touch the cone of lines, from P, lying on Ω. Thus these planes touch the conic in which the cone of lines of Ω from P is met by the cone of lines of Ω drawn from any other point, O, of Ω. These planes thus project from O, on to any threefold Π, into planes touching the Absolute conic of the space Π. These planes are, however, by definition, the intersections of the tangent threefold T, at P, with the two tangent threefolds drawn to Ω from the plane ϖ, and these two threefolds are independent of the position of P upon the conic in the plane ϖ.

Thus it appears that the two spheres obtained by projection from O of the two sections $\Omega\Sigma$, $\Omega\Sigma'$, have the same angle of intersection at all their common points. This angle is a right angle when Σ, Σ' are conjugate in regard to Ω.

As many detailed applications of the geometry of space of four dimensions occur in subsequent pages of this Volume, we content ourselves here with these indications.

SECTION III. USE OF SPACE OF FIVE DIMENSIONS

The representation of the lines of space of three dimensions by points in space of five dimensions. A line, in space of three dimensions, depends on *four* parameters; it may be given, for example, by the two points in which it meets two specified planes, though this would fail if the line were in one of the two planes. It appears that the only representation which does not fail for any line is that given by Cayley* (see above, Vol. III, p. 56), wherein a line is represented by the ratios of six coordinates, l, m, n, l', m', n', connected by an equation $ll' + mm' + nn' = 0$. If, then, these six coordinates are taken to be the coordinates of a point in space of five dimensions, the lines of the original threefold space are represented by the points of a quadric fourfold lying in the fivefold space. In the representation, considered above, in Section I, of the points of a plane by the points of a quadric surface in three dimensions, there are two points of the plane to which there correspond all the points of two lines of the quadric surface; in the representation (Vol. III, p. 189) of the points of a plane upon a cubic surface, there are, similarly, six points of the plane each corresponding to a line on the cubic surface. It is remarkable that no such singular elements occur in the correspondence between the lines of space of three dimensions, and the points of the fourfold quadric in space of five dimensions; to every element in either figure corresponds a definite element in the other. Moreover, the quadric fourfold is quite general; for, as in preceding cases (Vol. III, p. 15), any general quadratic relation in six homogeneously entering variables is reducible to the form $u^2 + v^2 + w^2 - x^2 - y^2 - z^2 = 0$, and, by putting $l = u + x$, $l' = u - x$, $m = v + y$, $m' = v - y$, $n = w + z$, $n' = w - z$, this is of the form $ll' + mm' + nn' = 0$.

Through any point of such a quadric fourfold, Ω, there pass ∞^3 lines which meet Ω in two points coinciding at this point; these lines, therefore, have no other point of intersection with Ω,

---

* Cf., also, Klein, *Ges. Math. Abh.*, I, pp. 107 ff.

unless they lie entirely thereon. These lines generate a tangent fourfold of Ω at this point. If $\lambda, \ldots, \lambda', \ldots$ be current coordinates, the equation of this tangent fourfold is found by acting upon the equation of Ω with an operator $\lambda \partial/\partial l + \ldots + \lambda' \partial/\partial l' + \ldots$. There are, however, ∞^2 lines, through any point of Ω, which lie entirely on Ω. The points of these lines constitute the aggregate of the points common to Ω and the tangent fourfold at the point considered. These lines meet any (flat) fourfold space in the points of a quadric surface, which lies in the threefold space in which this fourfold is met by the tangent fourfold at the point considered. If one point of Ω lie in the tangent fourfold at another point of Ω, this latter lies in the tangent fourfold at the former; two such points of Ω may be spoken of as *conjugate* to one another; they are such that the line joining them lies entirely on Ω. More generally, to any point of the fivefold space there corresponds a fourfold, given by operating on the equation of Ω by $\lambda \partial/\partial l + \ldots + \lambda' \partial/\partial l' + \ldots$, where $(l, \ldots, l', \ldots)$ are the coordinates of the point, and $(\lambda, \ldots, \lambda', \ldots)$ are current coordinates. This is called the *polar* fourfold of the point. Thus to any line of the fivefold space there corresponds a *polar threefold*, which is common to the polar fourfolds of all points of the line.

When the equation of Ω is $ll' + mm' + nn' = 0$, the condition that two points of Ω, $(l, m, \ldots)$, $(\lambda, \mu, \ldots)$, should be conjugate is $\lambda l' + \lambda' l + \ldots = 0$. This is the condition that the two corresponding lines of the original threefold space should intersect one another. The coordinates of any point of the line joining the two conjugate points are of the forms $(\sigma l + \lambda, \sigma m + \mu, \ldots)$, for a proper value of σ; such points evidently represent the lines of the original threefold space which lie in the plane of the two intersecting lines, and pass through their point of intersection; they describe a line lying on Ω. The aggregate of all the lines of the threefold space, which meet a given line $(l, m, \ldots)$, is represented by the points of Ω lying on the tangent fourfold at the point $(l, m, \ldots)$.

Consider the lines of a linear complex in the original space, satisfying an equation $al' + \ldots + a'l + \ldots = 0$ (cf. Vol. III, p. 61). These are represented by the points of Ω lying on the fourfold represented by this linear equation; these points constitute a threefold quadric in this fourfold space. The lines, of a linear congruence, common to two linear complexes, are then represented by the points of Ω lying on a threefold given by two such equations, say $al' + \ldots + a'l + \ldots = 0$ and $Al' + \ldots + A'l + \ldots = 0$; these points constitute a quadric surface lying in this threefold space. This space is the intersection of the polar fourfolds of two points $(a, b, \ldots)$, $(A, B, \ldots)$; it is equally given by the intersection of the tangent fourfolds of Ω at the two points where Ω is met by the line joining these two points. Thus, the lines of a linear congruence, in the original space, consist of the lines which meet two properly chosen lines of this space. Incidentally we thus again reach the correspondence, above referred to (p. 31), between the points of a quadric surface, and the lines meeting two arbitrary skew lines. In particular, the lines of a linear complex, $(a, b, ...)$, of the original space, which meet an arbitrary line, $(l, m, ...)$, where $ll' + mm' + nn' = 0$, are represented by the points of Ω lying on the threefold intersection of the tangent fourfold of Ω, at the point $(l, m, ...)$, with the polar fourfold of the point $(a, b, ...)$; this threefold lies equally on the tangent fourfold of Ω at the point, other than $(l, m, ...)$, in which Ω is met by the join of the two points $(l, m, ...)$, $(a, b, ...)$. Thus, all the lines of a linear complex, $(a, b, ...)$, which meet an arbitrary line $(l, m, ...)$, equally meet another line, say $(\lambda, \mu, ...)$. This other line is that known as the *polar line* of the first in regard to the focal system associated with the linear complex (Vol. iii, pp. 61, 64); its coordinates are of the forms $l - \rho a, m - \rho b, ...$, where ρ is such that these satisfy the equation of Ω, namely such that $(l - \rho a)(l' - \rho a') + ... = 0$, so that $\rho = (al' + a'l + ...)/(aa' + bb' + cc')$. In the space of five dimensions, the condition for the two points of Ω, that correspond to a pair of polar lines in regard to a linear complex of the original space, is that the line joining these points should pass through the pole, in regard to Ω, of the fourfold which defines the linear complex; or, that the tangent fourfolds of Ω at these two points should meet on this fourfold.

Ex. 1. Consider two linear complexes $(a, b, ...)$, $(A, B, ...)$, for which the linear invariant $aA' + a'A + ...$ is zero. In the space of five dimensions, they correspond to two fourfolds of which each contains the pole of the other, in regard to Ω; so that, in place of being spoken of as *apolar*, they may be spoken of as *conjugate*. A line from the pole of one of these fourfolds, to a point of Ω which lies on the other fourfold, evidently meets Ω again in a point also on this other fourfold. Thus, in the original space, the polar line, in regard to one of the complexes, of a line belonging to the other, is equally a line of the other (cf. Vol. iii, p. 65).

Ex. 2. Prove that every two of the six linear complexes expressed by $l + l' = 0$, $l - l' = 0$, $m + m' = 0$, $m - m' = 0$, $n + n' = 0$, $n - n' = 0$ are conjugate. Also, that the poles of an arbitrary plane, in the original threefold space, in regard to the six focal systems determined by these, lie on a conic; and, that the six polar planes of an arbitrary point, in these focal systems, touch a quadric cone.

Ex. 3. If, instead of the coordinates $l, m, ...$, we use coordinates, x, y, z, u, v, w, such that $l = x + iu, l' = x - iu, m = y + iv, m' = y - iv, n = z + iw, n' = z - iw$, the equation of Ω takes the form

$$x^2 + y^2 + z^2 + u^2 + v^2 + w^2 = 0.$$

Two linear complexes may then be represented by
$$ax + by + cz + fu + gv + hw = 0 \text{ and } a'x + \ldots + h'w = 0.$$
With these forms the condition that these complexes should be conjugate is easily found to be $aa' + bb' + \ldots + hh' = 0$. Suppose now we have six linear complexes, $X_1 = 0$, ..., $X_6 = 0$, where $X_r = a_r x + b_r y + \ldots + h_r w$, of which every two are conjugate, so that $a_r a_s + \ldots + h_r h_s = 0$; without loss of generality we can suppose that also $a_r^2 + b_r^2 + \ldots + h_r^2 = 1$, $(r = 1, \ldots, 6)$. From these twenty-one conditions there can be inferred (see below) the twenty-one conditions $a_1 b_1 + a_2 b_2 + \ldots + a_6 b_6 = 0$, ..., $g_1 h_1 + g_2 h_2 + \ldots + g_6 h_6 = 0$, $a_1^2 + a_2^2 + \ldots + a_6^2 = 1$, ..., $h_1^2 + h_2^2 + \ldots + h_6^2 = 1$. Hence the equation of Ω can equally be written $X_1^2 + X_2^2 + \ldots + X_6^2 = 0$. Thus the property enunciated in Ex. 2 holds for any six linear complexes of which every two are conjugate.

To shew that the second form of the conditions follows from the first, consider the matrix, M, of six rows and columns, in which the r-th row consists of the elements a_r, b_r, ..., h_r. Using $\overline{M}$ for the transposed matrix, obtained from M by interchanging rows and columns, the first form of the conditions is expressed by $M\overline{M} = 1$ (cf. Vol. I, p. 67, and Vol. III, p. 71). This, however, gives $\overline{M} = M^{-1}$, and hence $\overline{M}M = 1$. This expresses the second form of the conditions.

Consider, now, *the lines common to three linear complexes* of the original threefold space. These are represented, in the fivefold space, by the points of Ω lying on the plane which is common to the three fourfolds which represent these linear complexes in the fivefold space. This plane meets Ω in the points of a conic. If $(l_1, m_1, \ldots)$, $(l_2, m_2, \ldots)$, $(l_3, m_3, \ldots)$, be three points of this conic, any other point of the conic has coordinates $(\sigma_1 l_1 + \sigma_2 l_2 + \sigma_3 l_3, \sigma_1 m_1 + \sigma_2 m_2 + \sigma_3 m_3, \ldots)$, provided σ_1, σ_2, σ_3 are subject to the quadratic condition which expresses that this point lies on Ω. In the original space the corresponding lines are, in general, those of one system of generators of a quadric surface (cf. Vol. III, pp. 58, 60; Exx. 4, 9). To any plane of the fivefold space there corresponds another plane; for the polar fourfolds, in regard to Ω, of any three points of the plane intersect in another plane, through which there passes the polar fourfold of every point of the original plane; the polar fourfold of any point of the second plane equally contains the first plane; the planes may be spoken of as polars of one another. Thus, to a conic determined on Ω by its intersection with a plane, there corresponds another conic also on Ω, of which every point is conjugate to every point of the former conic; so that the line joining any point of one conic to any point of the other conic lies entirely on Ω. The points of these two conics correspond, in the original threefold space, to the two systems of generators of a quadric surface. The two lines, of either of these systems of generators, which meet an arbitrary line of the threefold space, correspond to the two points of Ω, on one of these conics, which lie on the tangent fourfold of Ω at the point corresponding to the arbitrary line.

If we take four linear complexes, they will have two lines in common, in general; these correspond, in the fivefold space, to the two points in which Ω is met by the line common to the fourfolds representing the complexes. In particular, the two transversals of four skew lines of the threefold space require the consideration of the line common to four tangent fourfolds of Ω.

If we take three lines in the original space of which every two intersect, say $(l_1, m_1, \ldots)$, $(l_2, m_2, \ldots)$, $(l_3, m_3, \ldots)$, the coordinates $(\sigma_1 l_1 + \sigma_2 l_2 + \sigma_3 l_3, \sigma_1 m_1 + \sigma_2 m_2 + \sigma_3 m_3, \ldots)$ are those of a line whatever $\sigma_1, \sigma_2, \sigma_3$ may be. Thus, in the fivefold space, if three points of Ω be taken of which every two are conjugate, *the plane containing these points lies entirely upon* Ω. Two lines $(l_1, m_1, \ldots)$, $(l_2, m_2, \ldots)$, of the threefold space, which intersect, give rise to ∞^1 other lines, with coordinates of the forms $(l_1 + \sigma l_2, m_1 + \sigma m_2, \ldots)$, which pass through the point of intersection and lie in the common plane of the lines; these correspond to the points of Ω lying on the line joining the points $(l_1, m_1, \ldots)$, $(l_2, m_2, \ldots)$. There are, now, in the threefold space, two systems of lines, each ∞^2 in aggregate, all of which intersect both the given lines; namely, first, the lines through the point of intersection of the two given lines, and, second, the lines in the common plane of the two given lines. If $(l_3, m_3, \ldots)$ be a line of the first system, not lying in the plane of the two given lines, all lines of the first system have coordinates of the forms $(\rho_1 l_1 + \rho_2 l_2 + \rho_3 l_3, \ldots)$; and, if $(\lambda_3, \mu_3, \ldots)$ be a line of the second system, that is, a line in the plane of the two given lines, not passing through their point of intersection, all lines of the second system have coordinates of the forms $(\sigma_1 l_1 + \sigma_2 l_2 + \sigma_3 \lambda_3, \ldots)$. To these two systems of lines there evidently correspond, in the space of five dimensions, two planes lying entirely on Ω, both passing through the line of Ω which joins the two given conjugate points, $(l_1, m_1, \ldots)$, and $(l_2, m_2, \ldots)$. Considering the matter more generally, in the fivefold space, let x, y, z, u, t, v denote coordinates therein, of which $t = 0$ represents the tangent fourfold of Ω at the point $(0, 0, 0, 0, 0, 1)$, lying on Ω, and $v = 0$ represents the tangent fourfold of Ω at the point $(0, 0, 0, 0, 1, 0)$, also lying on Ω, the joining line of these two points being, therefore, $x = 0$, $y = 0$, $z = 0$, $u = 0$. These points are supposed not to be conjugate. The equation of Ω will then be of the form $tv - \phi = 0$, where ϕ is a homogeneous quadratic polynomial in x, y, z, u. As $\phi = 0$ is satisfied by an ∞^2 of values of the form $x/x' = y/y' = z/z' = t/t'$, this equation, $\phi = 0$, represents an aggregate of ∞^2 planes, all passing through the line $x = 0$, $y = 0$, $z = 0$, $u = 0$. The intersection of Ω with the tangent fourfold $t = 0$ consists of the ∞^2 lines in which this fourfold meets these planes. The points common to Ω and the tangent fourfolds $t = 0$, $v = 0$, consist of the ∞^2 points of a quadric surface, lying in the threefold given by $t = 0$, $v = 0$; this surface is the intersection of Ω with the polar threefold of the line $x = 0$, $y = 0$, $z = 0$, $u = 0$. Now suppose we take coordinates so that the points $(0, 0, 0, 0, 0, 1)$ and $(0, 0, 0, 0, 1, 0)$, lying on Ω, are conjugate to one another. The equation of Ω then takes a form $tV + vT - \psi = 0$, where ψ is a quadratic polynomial in x, y, z, u, but T, V are linear forms in x, y, z, u. The tangent fourfold at $(0, 0, 0, 0, 0, 1)$ is now $T = 0$, and that at $(0, 0, 0, 0, 1, 0)$ is $V = 0$. The points of Ω which are conjugate to both these points, which are given by $T = 0$, $V = 0$, $\psi = 0$, then lie on two planes; the equations of these planes are, respectively, of the forms $x/x_1 = y/y_1 = z/z_1 = t/t_1$, and $x/x_2 = y/y_2 = z/z_2 = t/t_2$; they are the intersection of Ω with the threefold space common to the tangent fourfolds at the two given conjugate points of Ω. This threefold space is the polar of the line joining these points, and the planes in which it intersects Ω contain this line. In the original threefold space there are ∞^3 points, through each of which pass ∞^2 lines; and there are ∞^3 planes, in each of which there lie ∞^2 lines. There are thus, on Ω, in the fivefold space, two different systems of planes, each ∞^3. Through two points of the original threefold space there passes one line, and two planes meet in one line; but there is not, generally, a line through a given point which lies in a given plane. Thus, in the fivefold space, two planes of Ω, of the same system, have a point in common, but two planes of different systems do not, in general, intersect. In the original space, if a point lie in a plane, there is a range (pencil) of lines through this point which lie in this plane. Thus, in the fivefold space, if two planes on Ω, of different systems, have a point in common, they intersect in a line, lying on Ω, and every two points of the line are conjugate to one another.

The planes of the quadric fourfold treated with the symbols. The existence of the planes of the quadric fourfold Ω is immediately clear from the equations. The equation of Ω can be supposed to be $\lambda'^2 + \mu'^2 + \nu'^2 = \lambda^2 + \mu^2 + \nu^2$, where, in terms of the original line coordinates, $\lambda = l - l'$, $\lambda' = l + l'$, etc. If a_1, b_1, c_1, a_2, b_2, c_2, a_3, b_3, c_3 be nine quantities subject to the six equations $a_r^2 + b_r^2 + c_r^2 = 1$, $a_r a_s + b_r b_s + c_r c_s = 0$ ($r, s = 1, 2, 3$), so that, as will appear in a moment, they are expressible by three parameters, the equation of Ω is identically satisfied by supposing

$$\lambda' = a_1\lambda + a_2\mu + a_3\nu, \qquad \mu' = b_1\lambda + b_2\mu + b_3\nu, \qquad \nu' = c_1\lambda + c_2\mu + c_3\nu;$$

these three equations connecting the six coordinates represent a plane. The coefficients, a_r, b_s, c_t, in these equations, are such that the square of the determinant (a_1, b_2, c_3) has the value unity, so that there are two cases, according as this determinant is $+1$ or -1. We can easily see that the three equations of the plane express, either that a line, in the original threefold space, passes through a definite point, or lies in a definite plane; this point, or plane, is determined by the coefficients a_r, b_s, c_t, that is, by the three parameters upon which these depend. For, the conditions that a line, $(l, m, \ldots)$, of the original threefold space, contains the point (ξ, η, ζ, τ), may be taken to be the three, $l'\tau + m\zeta - n\eta = 0$, $m'\tau + n\xi - l\zeta = 0$, $n'\tau + l\eta - m\xi = 0$, which, together, unless $\tau = 0$, involve

$$ll' + mm' + nn' = 0;$$

replacing $l, l', \ldots$ by $\frac{1}{2}(\lambda' + \lambda)$, $\frac{1}{2}(\lambda' - \lambda)$, $\ldots$, we obtain three linear equations by which λ', μ', ν' may be expressed in terms of λ, μ, ν; thereby, explicit expressions, in terms of ξ, η, ζ, τ, are obtained, in place of the coefficients a_r, b_s, c_t, in the three equations above. The conditions that a line, $(l, m, \ldots)$, in the original threefold space, lies in a plane, of coordinates (u, v, w, p), are three similar equations, with ξ, η, ζ, τ replaced by u, v, w, p, respectively, but $l, l', \ldots$ respectively replaced by $l', l, \ldots$ (Vol. III, p. 57); thus, if we solve the three latter equations for λ', μ', ν', we obtain the same linear functions of λ, μ, ν as before, with two differences: the unimportant difference that ξ, η, ζ, τ are replaced by u, v, w, p, and the important difference, arising from the interchange of l, m, n with l', m', n', that the signs of λ', μ', ν' are all changed. Explicitly, with the notation of matrices (Vol. I, p. 67), the formulae in the former case are found to be

$$(\lambda', \mu', \nu') = (\tau - \omega)^{-1}(\tau + \omega)(\lambda, \mu, \nu), = (\tau + \omega)(\tau - \omega)^{-1}(\lambda, \mu, \nu),$$

where
$$\omega = \begin{pmatrix} 0, & -\zeta, & \eta \\ \zeta, & 0, & -\xi \\ -\eta, & \xi, & 0 \end{pmatrix},$$

while, in the latter case,

$$(\lambda', \mu', \nu') = -(p - \psi)^{-1}(p + \psi)(\lambda, \mu, \nu),$$
$$= -(p + \psi)(p - \psi)^{-1}(\lambda, \mu, \nu),$$

where ψ is obtained from ω by replacing ξ, η, ζ, respectively, by u, v, w. It follows that the aggregate of the lines through a point, in the original space of three dimensions, corresponds to the aggregate of the points on a plane on Ω, given by such a set of equations as the former, in which the determinant, (a_1, b_2, c_3), of the coefficients, has the value $+1$; but the lines in a plane of the original space correspond to the points of a plane of Ω for which the corresponding determinant has the value -1. Conversely, when the nine coefficients

a_r, b_s, c_t, in the equations of a plane lying on Ω, in the fivefold space, are given, there are unique values for $\xi/\tau, \eta/\tau, \zeta/\tau$, or $u/p, v/p, w/p$, necessary to put the equations in the forms above (p. 205, below).

The properties of the planes of Ω, in regard to intersection, which we have deduced from consideration of the original threefold space, can be deduced directly from the above equations. If $(\lambda', \mu', \nu') = D(\lambda, \mu, \nu)$ denote a plane on Ω of the first system, depending on (ξ, η, ζ, τ), and the matrix replacing D for another plane of this system, depending on $(\xi_1, \eta_1, \zeta_1, \tau_1)$, be D_1, it is easy to verify that the matrix $D_1^{-1}D$ is also of the same form, say D_2, in terms of suitable parameters, $(\xi_2, \eta_2, \zeta_2, \tau_2)$. Now, the condition that the planes (D), (D_1) should intersect is, that we should be able to solve, for λ, μ, ν, the three equations $(D - D_1)(\lambda, \mu, \nu) = 0$, or $(D_1^{-1}D - 1)(\lambda, \mu, \nu) = 0$, or $(D_2 - 1)(\lambda, \mu, \nu) = 0$; namely is, that the determinantal equation for ρ, $|D_2 - \rho| = 0$, should have the root $\rho = 1$. In fact, it is easy to prove that the determinant $|D_2 - \rho|$ has the form $(1 - \rho)(\rho - e^{i\theta})(\rho - e^{-i\theta})$, with a proper value for θ. Thus two planes of the first system, or, similarly, two planes of the second system, lying on Ω, in the fivefold space, have a point in common. For an intersection of a plane, (D), of the first system, with a plane, $(-D_1)$, of the second system, we should, similarly, require the determinantal equation $|D_2 + 1| = 0$. With the notation above, this can only be satisfied if $\theta = \pi$; when this is so, however, all second minors of the determinant $|D_2 + 1|$ vanish, as is easily seen; the three linear equations for λ, μ, ν are then satisfied by an infinite aggregate (∞^1) of values. Thus, if a plane of the first system, on Ω, in the fivefold space, have a point common with a plane of the second system, it has a line of common points. In general, the value of θ is $2 \tan^{-1}[(\xi_2^2 + \eta_2^2 + \zeta_2^2)^{\frac{1}{2}} \tau_2^{-1}]$; this is π when $\tau_2 = 0$.

The aggregate of the planes lying on the quadric Ω in fivefold space. We have introduced a plane on Ω as that containing three points of Ω of which every two are conjugate. Such a plane is, therefore, its own polar plane in regard to Ω. And the converse is true. In general, the plane given by the three equations

$$\lambda' = a_1\lambda + a_2\mu + a_3\nu, \quad \mu' = b_1\lambda + b_2\mu + b_3\nu, \quad \nu' = c_1\lambda + c_2\mu + c_3\nu$$

contains the three points $(1, 0, 0, a_1, b_1, c_1)$, $(0, 1, 0, a_2, b_2, c_2)$, $(0, 0, 1, a_3, b_3, c_3)$; the polar fourfolds of these points, in regard to Ω, whose equation is $\lambda'^2 + \mu'^2 + \nu'^2 = \lambda^2 + \mu^2 + \nu^2$, are, respectively, given by

$$\lambda = a_1\lambda' + b_1\mu' + c_1\nu', \quad \mu = a_2\lambda' + b_2\mu' + c_2\nu', \quad \nu = a_3\lambda' + b_3\mu' + c_3\nu';$$

the plane given by the aggregate of the three latter equations agrees with the former if, and only if, the coefficients a_r, b_s, c_t satisfy the six equations which characterise a plane that lies on Ω.

Through any line lying in a plane on Ω, there passes another plane on Ω, of the other system; the two planes are, together, the complete intersection of Ω with the polar threefold, in regard to Ω, of the line. Conversely, if any threefold be drawn through a plane on Ω, this threefold has for its polar line, in regard to Ω, a line lying on the plane; and the threefold meets Ω in a further plane. Thus, any plane on Ω meets, each in a line, ∞^2 planes of the other system. The ∞^2 lines on Ω, that can be drawn through any point, O, of Ω, arrange themselves in ∞^1 planes, on Ω, passing through this point, there being one plane of each system through every such line. These planes meet any arbitrarily taken fourfold, Π, each in a line; these lines lie in the threefold in which Π is met by the tangent fourfold of Ω at O. Two of the planes of the same system, through O, have no other common point, and give lines in Π which do not intersect; but two planes of different systems, through O, have a line in common, and meet Π in intersecting lines. Thus, in the threefold intersection of Π with the tangent fourfold of Ω at O, the lines obtained by the planes on Ω, of the two systems, which pass through O, are the generators of a quadric surface. If ϖ_1, ϖ_2 be two planes of the first system, from these, whose only common point is O, the plane, ϖ', of the second system, which passes through any line, l, lying in ϖ_2 but not passing through O, will not meet ϖ_1; for the line of meeting would intersect l in a point lying both on ϖ_1 and ϖ_2. Conversely, if ϖ' be any plane of the second system which does not meet ϖ_1, the plane, ϖ_2, of the first system, which contains a line, l, lying on ϖ', will meet ϖ_1 in a point, say O; and, then, in an infinite number of ways, a plane of the second system can be drawn through O to meet ϖ_1 and ϖ_2 in a line. We thus see how to pass from any one plane on Ω to every other plane on Ω; namely, by suitably repeating the process of drawing, through a plane of either system, a threefold which determines a plane of the other system as its residual intersection with Ω.

Ex. 1. Let Λ, Λ' be two conjugate fourfold sections of Ω. Consider the line in which Λ meets a plane, ϖ_1, of the first system on Ω; through this line there passes a plane, ϖ', of the second system; consider the line in which ϖ' meets Λ'; through this line there passes a plane, ϖ_2, of the first system. Prove that ϖ_2 is also obtained from ϖ_1 if the same construction is made with the interchange of Λ and Λ'. Two planes on Ω, of different systems, which meet in a line lying on Λ, correspond, in fact, to a point and its polar plane, in the original threefold space, taken in regard to the focal system given by Λ (cf. Vol. III, p. 66, Ex. 10).

Ex. 2. The representation, in the space of five dimensions, of the polar line of a line, l, of the original threefold space, in regard to a given quadric surface of this threefold space, is also interesting.

Let one point where l meets the quadric be the intersection of the generators p_1, q_1 of the quadric, and the other point, where l meets the quadric, the intersection of the generators p_2, q_2. The generators p_1, q_2, of opposite systems, meet in a point, and the generators p_2, q_1 meet in a point, the join of these two points being the polar line, l', of l, in regard to the quadric. Correspondingly, in the five-fold space, let L be any point of Ω; let the tangent fourfold of Ω at L meet the conic, on Ω, which represents one system of generators of the original quadric, in P_1 and P_2; to this conic there is a conjugate conic on Ω, as we have seen above; let this be met by the tangent fourfold of Ω at L in Q_1 and Q_2. Thus the lines P_1Q_1, P_2Q_2 lie on Ω, but do not intersect; and the planes LP_1Q_1, LP_2Q_2 lie on Ω, intersecting in L. These are of the same system. Through each of P_1Q_1, P_2Q_2 there pass also planes, on Ω, of the other system; their point of intersection, L', represents the polar line, l'.

We may remark, further, that the lines P_1Q_2, P_2Q_1 lie on Ω, and there are planes, on Ω, of the same system, through these, which intersect in L', and planes of the other system, on Ω, through these, which intersect in L. In other words, the construction, as stated, is valid without distinction of P_1 and P_2, or of Q_1 and Q_2. There are in the figure, six points on Ω, $(P_1, P_2, Q_1, Q_2, L, L')$; twelve lines on Ω, $(P_1Q_1, P_2Q_2, P_1Q_2, P_2Q_1, LP_1, LP_2, LQ_1, LQ_2, L'P_1, L'P_2, L'Q_1, L'Q_2)$; four planes on Ω of one system, $(P_1Q_1L, P_2Q_2L, P_1Q_2L', P_2Q_1L')$, and four others of the other system, $(P_1Q_1L', P_2Q_2L', P_1Q_2L', P_2Q_1L')$.

The representation of congruences of lines of the original space. The lines of the original threefold space, whose coordinates are subject to two (rational algebraic) conditions, are ∞^2 in aggregate; they are said to form a *congruence* of lines. We have, in particular, spoken of the linear congruence, of lines common to two linear complexes, which are the transversals of two fixed lines; they are such that one of the lines passes through any arbitrarily taken point, of general position, and one lies in any arbitrarily taken plane, of general position. The lines of a linear congruence are represented, in the fivefold space, by the points of a quadric surface lying on Ω, having, clearly, one point of meeting with every plane on Ω. A more general congruence of lines, in the original threefold space, is that formed by the chords of a (not plane) curve of this space; this, also, consisting of ∞^2 lines, is represented, in the five-fold space, by the points of a surface lying on Ω. We proceed, in illustration of general ideas, to find the order of this surface. In general, we mean by the *order of a locus*, of k dimensions, lying in space of n dimensions, the number of points of the locus which lie on a planar manifold of dimension $n - k$, when this number is the same for every such general manifold. Thus, the order of a surface, in space of five dimensions, will be the number of its points of intersection with an arbitrary planar threefold. We have seen that there are threefolds meeting Ω in two planes, of different systems; to determine the order of a surface lying entirely on Ω it is convenient to take such a threefold. This will meet the surface only on the two planes. Thus the order of a surface lying on Ω is the sum of the numbers of points in which this surface meets two planes of different systems on Ω, say h and h', respectively; that these numbers are both unaltered whatever two planes of the two systems be taken, appears from the possibility we have remarked, of passing from any plane on Ω to any other by means of intersecting threefolds. In the case of a surface, on Ω, which represents the chords of a curve in the original threefold space, the number, h, of points common to the surface and a plane of the first system on Ω, is the number of chords of the curve, in the original space, which pass through an arbitrary point of general position in this space; the number, h', of points of the surface on a plane of the second system, is the number of chords of the curve which lie in an arbitrary plane of general position. In the fivefold space there are two tangent fourfolds of Ω passing through a given threefold; their points of contact with Ω are the intersections of this with the polar line of the threefold in regard to Ω. Thus the order of a surface, lying on Ω, is also the number of points of the surface which are conjugate to two arbitrarily taken points of Ω; in the original threefold space, this number is that of the lines of the congruence, which corresponds to the surface on Ω, which belong to an arbitrary linear congruence. For the congruence consisting of the chords of a curve of order m, in the threefold space, the number, by what we have seen, is $h + \tfrac{1}{2}m(m-1)$, where h is the number of chords of the curve passing through an arbitrary point. If $p = \tfrac{1}{2}(m-1)(m-2)-h$, this number is $(m-1)^2 - p$.

Number of lines common to two congruences in three-fold space. It can be shewn that two surfaces on Ω, of which the first meets general planes on Ω, of the first and second systems, respectively, in h and h' points, and the second meets general planes on Ω, of the first and second systems, respectively, in k and k' points, have in common $hk + h'k'$ points. In the original threefold space this is the statement that two congruences of lines, of which the first has h lines through an arbitrary point and h' lines in an arbitrary plane, and the second has k lines through an arbitrary point and k' lines in an arbitrary plane, have in common a number of lines given by $hk + h'k'$. There is a theorem, for algebraic curves lying on a quadric surface in space of three dimensions, that such a curve meets all the generators of the quadric, of the same system, in the same number of points; that, if these numbers, for the two systems of generators, be r and r', respectively, the order of the curve is $r + r'$; and that such a curve, say (r, r'), meets another curve, (s, s'), of the quadric, in $rs' + r's$ points; the theorem may be proved by projecting the curves into plane curves, from a point of the quadric, so obtaining a curve with two multiple points of orders r, r', and a curve with two multiple points of orders s, s'. We shall assume this result, in order to obtain the theorem above enunciated for the number of intersections of two surfaces lying on Ω, in the space of five dimensions. We may denote these surfaces, respectively, by (h, h') and (k, k'). Let O be a point of Ω, not lying on either surface, and Π be an arbitrary fourfold space. When we project the points of Ω, from O, upon Π, the planes on Ω which pass through O determine, in Π, the generators of a quadric surface, ω, as we have seen; this quadric lies in the threefold space, (Π, T), in which Π is met by the tangent fourfold, T, of Ω at O. The surface (h, h'), on Ω, being met by an arbitrary threefold in $h + h'$ points, will project into a surface in the fourfold space Π; and this, being met by an arbitrary plane in Π in $h + h'$ points, will be of order $h + h'$. The surface (h, h') will be met by the tangent fourfold T in a curve; this projects into a curve lying on the quadric ω, having the property of meeting the generators of the two systems of ω, respectively, each in h and h' points. We assume that two surfaces in fourfold space, of respective orders M and N, which have not an infinite number of common points (as those of a line, etc.), have MN common points. Thus the two surfaces, in Π, obtained by projection from O of two surfaces, (h, h') and (k, k'), upon Ω, have $(h + h')(k + k')$ common points. These, however, arise in part from only apparent intersections of the two surfaces (h, h'), (k, k') upon Ω, namely by lines through O meeting both these surfaces, but not in the same point. Every such line, meeting Ω in three distinct points, will lie entirely on Ω, and be, therefore, in the tangent fourfold T; such a line will give an intersection of the curves lying on the quadric ω, which are the intersections of ω with the surfaces, in the fourfold space Π, arising from projection of the surfaces (h, h'), (k, k'), on Ω. The number of such intersections, we have remarked, is $hk' + h'k$. The number of intersections on Ω of the surfaces (h, h'), (k, k') is thus $(h + h')(k + k') - hk' - h'k$, or $hk + h'k'$; as we desired to prove.

Ex. If h, k be the respective numbers of chords that can be drawn from an arbitrary point, to two given curves of orders m, n, in threefold space, the number of common chords of these curves is $hk + \frac{1}{4}(m - 1)(n - 1)mn$.

It may be remarked that the surface on Ω, corresponding to the chords of a curve, in space of three dimensions, meets a plane on Ω which corresponds to a point of the curve, in an *infinite* aggregate of points; these points form a curve of order $m-1$, if the original curve be of order m.

The chords of a cubic curve in the original threefold space. Veronese's surface. An important particular case of the preceding is that of the chords of a cubic curve in the original threefold space. Of these chords, one passes through an arbitrary point, and three lie in an arbitrary plane. The corresponding surface on Ω thus meets the planes on Ω, of the first and second systems, respectively, in one and three points, and is of order four, having four points of intersection with an arbitrary threefold. The axes of a cubic developable, in the original space, similarly give a surface on Ω for which the corresponding numbers are three and one. We may denote the former surface by V, and the latter by V'. That the surface V is of order four is equivalent, we have seen, to the fact that there are four chords, of a cubic curve in threefold space, which meet two arbitrary lines of this space. This may be verified directly, for example, from the fact that the chords of the curve which meet one line are projected from a point of the curve by the tangent planes of a quadric cone (Vol. III, p. 135, Ex. 8). The surface V meets ∞^1 planes of the first system on Ω, those corresponding to the points of the cubic curve in the original space, each in a conic; and two such conics intersect in the common point of their planes. But there are, in fact, ∞^2 conics lying on the surface V, whose planes do not lie on Ω. For, if an arbitrary line be drawn through a point of the cubic curve in the original space, there is an infinite aggregate of chords of the cubic curve meeting this line; these chords form one system of generators of a quadric surface, and meet the curve in the pairs of points of an involution on the curve. (Cf. Vol. III, p. 128.) We have seen that the lines of one system of generators, of a quadric surface, in the original space, correspond to the points of Ω lying on a conic; further, an involution on the cubic curve is determined by two pairs of points, that is by two chords of the curve; and two such involutions have a common pair of points. Hence there are, as stated, ∞^2 conics of Ω which lie on the surface V; one of these conics passes through any two chosen points of V, and any two of these conics have a common point. The conics on V, previously remarked, whose planes lie on Ω, arise among these.

In more detail, every quadric surface containing the original cubic curve, is determined by the chords of this curve which meet a line drawn through an arbitrary fixed point, K, of the curve. Thus, if we take any fixed plane, κ, lying on Ω, from among the ∞^1 which meet V in a conic, and take, in this plane, a point, P, not generally lying on V, then, the tangent fourfold of Ω, at this point, not only contains the plane κ, but also meets V in a conic; and, as

P varies in the plane κ, all the ∞^2 conics of V are obtained. When
P is taken on V, that is, on the conic in which the plane κ meets
V, the tangent fourfold of Ω, at P, contains, beside the plane κ, a
conic whose plane lies on Ω; and all the ∞^1 such conics are obtainable in this way. Conversely, the plane of any one of the conics of
V meets the plane κ in a point, and lies in the tangent fourfold of
Ω at this point, which also contains the plane κ. If a line be taken
in the plane κ, the tangent fourfolds of Ω, at the points of this
line, will meet V in conics having a common point; and the tangent
fourfold of Ω at this point will meet the plane κ in this line. This
follows from what has been said, but is obvious by recurring to the
original threefold space, wherein, corresponding to the line of κ,
there is a flat pencil of lines passing through the point K of the
cubic curve. Thus the points of V are in unique correspondence
with the lines of the plane κ. Further, it can be shewn that the
points of V which lie on an arbitrary (planar) fourfold give rise, in
this correspondence, to lines of the plane κ which touch a conic;
for this fourfold will meet the conic, on V, which corresponds to
any point, P, of the plane κ, in two points; the tangent fourfolds
of Ω at these two points of V will meet the plane κ in two lines
passing through P. Thus, the coordinates of any point of V are
proportional to quadratic polynomials in three parameters, the
coordinates of any line of the plane κ. (Cf. Vol. III, p. 223, Ex. 5.)
We may obtain the explicit expressions by reference to the original
threefold space. The cubic curve being given by points $(\theta^3, \theta^2, \theta, 1)$,
the point K being for $\theta = k$, a plane, through K and the chord (θ, ϕ)
of the curve, has the equation, in terms of coordinates x, y, z, t in
the threefold space, $u(x - yk) + v(y - zk) + w(z - tk) = 0$, where
$\theta + \phi = -v/u$, $\theta\phi = w/u$. Thus the coordinates of the chord (θ, ϕ)
are $(wu - v^2, uv, -u^2, wu, wv, w^2)$; these are the coordinates of a
point of the surface V in terms of the parameters u, v, w. The
tangent fourfold of Ω at this point of V, with $l, m, \dots$ as current
coordinates, has the equation

$$lwu + l'(wu - v^2) + mwv + m'vu + nw^2 - n'u^2 = 0.$$

The line which is the intersection of this with the plane κ may be
found by substituting herein for l', m', n' from the equations of the
plane κ; these are $l'\tau + m\zeta - n\eta = 0$, $m'\tau + n\xi - l\zeta = 0$, $n'\tau + l\eta - m\xi = 0$,
where $\xi = k^3$, $\eta = k^2$, $\zeta = k$, $\tau = 1$. The result of the substitution is
found to be

$$(k^2u + kv + w)[lu + m(v - ku) + n(w - kv)] = 0;$$

the first factor of this vanishes only when the point (u, v, w), of the
surface V, is on the plane κ; the vanishing of the second factor
determines uniquely the point (u, v, w), of V, when the line in the
plane κ is given. The lines of the plane κ which correspond to the

points of the surface V which lie on the polar fourfold, in regard to Ω, of the point (a, b, c, a', b', c'), touch the conic given tangentially by $awu + a'(wu - v^2) + bwv + b'vu + cw^2 - c'u^2 = 0$. As has already been remarked (Ex. 4, p. 6, above), when the point $(a, b, \ldots)$ is on Ω, this conic is triangularly inscribed to $xz - y^2 = 0$.

Ex. 1. The section of the surface V by any (planar) fourfold is a rational quartic curve; the conic just obtained arises from the tangent fourfolds of Ω at all the points of this curve. The values of u, v, w corresponding to a point of this curve can be expressed, in terms of a parameter θ, in the forms

$$u = a'\theta^2 - b\theta - c, \quad v = a'\eta\theta^2 + (a + a')\theta - c\xi,$$
$$w = -a'(\xi - \eta)\theta + b\xi + a + a',$$

where, for convenience, in place of b' and c', we have ξ, η, such that $b' = a'(\xi + \eta)$, $c' = a'\xi\eta$. From these values of u, v, w the coordinates of a point of the quartic curve can be expressed as quartic polynomials in θ.

Ex. 2. The conics of a plane may evidently be represented by the points of a fivefold space, the coefficients of the equation of the conic being taken as coordinates. More generally, if any conic in a plane be expressed as a linear function of six given conics in that plane (themselves not linearly connected), the coefficients in the expression may be interpreted as coordinates of a point in space of five dimensions. The condition that such a conic, supposed given in point coordinates, should be a pair of lines, is, then, a single equation of the third order in the coordinates; this represents a fourfold of the third order in the fivefold space, say M_4^3. The conditions that these two lines coincide reduce the number of independent coordinates further to two, the surface in the fivefold space representing this being essentially the surface V, of the fourth order, just discussed. This lies upon the M_4^3, and is a double surface thereon. See Segre, *Atti d. r. Acc. d. Sc. d. Torino*, xx (1884–85), pp. 367–384.

Ex. 3. The surface V was briefly considered by Cayley, *Papers*, vi, p. 198 (1868); at length by Veronese, *Mem. d. Acc. d. Lincei*, xix (1883–84), and by Segre (see the reference in Ex. 2). Bertini, *Geom. d. iperspazi* (Pisa, 1907), gives, with others, the following results in regard thereto: A surface, in space of any number of dimensions, which contains ∞^2 conics, is the surface of Veronese, or one of its projections. (Cf. Darboux, *Bull. d. sc. math.*, iv, 1880: A surface, in threefold space, containing ∞^2 conics, is a quadric, a ruled cubic surface, or the quartic Steiner surface.) A surface, in space of any number, r, of dimensions, for which $r > 4$, with the property that every two tangent planes have a common point, is a cone, or is the surface of Veronese (Del Pezzo, *Rend. d. Cir. Mat. d. Palermo*, i,

1887). A surface, in space of any number of dimensions, which is not a cone, and is such that no line can be drawn through an arbitrary point of the space to meet the surface in two points, is the surface of Veronese. (Cf. Severi, *Rend. d. Palermo*, xv, 1901 ; and, *Mem. d. Acc. d. Torino*, lii, 1902.) A surface of order $r - 1$, in space of r dimensions, is a rational ruled surface, or, for $r = 5$, is the surface of Veronese (Del Pezzo, *Rend. d. Acc. d. Napoli*, 1885 and 1886). A surface, in space of r dimensions, which meets any (planar) $(r - 1)$-fold in a rational curve, unless it is a rational ruled surface, is the Veronese surface, or one of its projections. (Cf. Picard, *Théorie des fonctions algébriques*, ii, 1900, p. 59: A surface, in space of three dimensions, whose plane sections are rational curves, is a rational ruled surface, or the quartic Steiner surface.) Compare the theorem of Kronecker and Castelnuovo quoted in Vol. iii, p. 223.

Lie's correspondence between lines and spheres in space of three dimensions. Sophus Lie has enunciated, with the help of the equations, a relation whereby, to a line of threefold space, there corresponds a sphere (or plane) ; and, conversely, to a sphere, there corresponds a pair of lines ; with the property that, to two lines which intersect, there correspond two spheres which touch (1869. See Lie u. Scheffers, *Berührgstransftn.*, 1896, pp. 453 ff.; Klein, *Ges. Math. Abh.*, i, p. 96). The correspondence will be obtained here in a geometrical manner. We have seen above (Section ii, p. 36) that, in a fourfold space, a section of a threefold quadric, by a planar threefold, projects, from a point of this quadric, on to a threefold, into a sphere. We may regard such a fourfold space as lying in a fivefold space, wherein the lines of a threefold space are represented by the points of a quadric fourfold.

Let Ω be a fourfold quadric, of fivefold space, whose points represent the lines of the original threefold space; let Π be an arbitrary (planar) fourfold space; this meets Ω in a threefold quadric, U. The tangent fourfold of Ω, at a point P, meets Π in a threefold, ϖ ; the section of U by this threefold, ϖ, is a surface, which, projected from any point, O, of U, upon any threefold lying in Π, gives a sphere therein. It is sufficient, then, to regard the sphere as determined by the section of U by ϖ. This sphere is determined by the point P, that is, by the line of the original threefold space which is represented by P. From the threefold, ϖ, there can be drawn another tangent fourfold of Ω, touching this, say, in P' ; thus, to the sphere arising from P, there corresponds also P'. The lines, of the original threefold space, which correspond to P and P', are, as we have seen above (p. 42), polars of one another in regard to the focal system, or linear complex, which is represented by the section of Ω by the fourfold Π. We have shewn that, to a line, of the threefold space, there can be made to correspond a sphere; a sphere, conversely, corresponding to a pair of lines, polars of one another in a certain focal system.

Now consider two points, P, Q, of Ω, which are conjugate to one another, and so represent intersecting lines of the original threefold space. We desire to see that the spheres, which correspond to these lines, touch one another. The line PQ lies wholly on Ω; it meets Π in a point, say H, also lying on Ω; this point is, therefore, on the quadric, U, given by Ω and Π. Let the tangent fourfolds of Ω at the points P, Q, H be, respectively, denoted by t_P, t_Q, t_H; the tangent planes, at H, of the sections (U, t_P) and (U, t_Q), are, respectively, the planes (Π, t_P, t_H) and (Π, t_Q, t_H). Now, it is easily seen that the tangent fourfolds of Ω, at the points of a line which lies thereon, form a range (pencil) of fourfolds all passing through a threefold; this threefold is then determined as the intersection of any two of the fourfolds. The threefolds (t_P, t_H) and (t_Q, t_H) are thus identical. Wherefore, the planes (Π, t_P, t_H) and (Π, t_Q, t_H) are identical. This shews that the spheres obtained by projection of the sections (U, t_P) and (U, t_Q), from any point, O, of U, touch one another at the point which is the projection of H. As a particular case, the point O may lie on t_Q; the projection of the section (U, t_Q) will then be, not a sphere but, a plane. This plane will touch the sphere obtained by projection of (U, t_P).

Ex. 1. We may take coordinates, x, y, z, t, u, v, in the fivefold space, such that $u = 0$ is the fourfold, Π, of the above statement; while the point, O, from which the projection is to be made, is given by $x = y = z = t = u = 0$; the tangent fourfold of Ω, at O, may be supposed to be $t = 0$; and the tangent fourfold of Ω at the point, other than O, in which the line $x = y = z = u = 0$ meets Ω, to be $v = 0$. Then, the equation of Ω will contain no term in v^2, and the coefficient of v in this equation will be a constant multiple of t; similarly, there will be no term in t^2, and the terms involving t will reduce to tv. Thus the equation of Ω will be of the form $-2vt + u^2 + uL + Q = 0$, where L and Q are respectively linear and quadratic polynomials in x, y, z only; by proper choice of x, y, z we may then suppose the equation to be

$$-2vt + u^2 + 2auz + x^2 + y^2 + z^2 = 0,$$

in which a is a constant. The surface obtained by the intersection of Ω with the fourfold Π, or $u = 0$, and the tangent fourfold of Ω at a point (x', y', z', t', u', v'), is then to be found from

$$u = 0, \quad -2vt + x^2 + y^2 + z^2 = 0, \quad -vt' - v't + au'z + xx' + yy' + zz' = 0.$$

The projection of this surface from O, by elimination of v from the two latter equations, satisfies the equation

$$-2t(xx' + yy' + zz' + au'z - v't) + t'(x^2 + y^2 + z^2) = 0;$$

this is of the form

$$x^2 + y^2 + z^2 - 2\,(fx + gy + hz)\,t + ct^2 = 0,$$

appropriate to a sphere, with

$$ft' = x', \quad gt' = y', \quad ht' = z' + au', \quad ct' = 2v'.$$

If f, g, h and c be given, and we seek (x', y', z', t', u', v'), expressing that this point is on Ω we first obtain

$$t'^2\,(f^2 + g^2 + h^2 - c) = u'^2\,(a^2 - 1),$$

and then, to each of the two solutions of this, a unique set of ratios $x' : y' : z' : t' : u' : v'$. Thus the two points of Ω, or the two lines of the original threefold space which correspond to a given sphere, are obtainable.

The condition that two spheres (f_1, g_1, h_1, c_1) and (f_2, g_2, h_2, c_2) should touch one another, which is (Vol. III, pp. 77, 78)

$$f_1 f_2 + g_1 g_2 + h_1 h_2 \pm (f_1^2 + g_1^2 + h_1^2 - c_1)^{\frac{1}{2}} (f_2^2 + g_2^2 + h_2^2 - c_2)^{\frac{1}{2}} = \tfrac{1}{2}(c_1 + c_2),$$

leads, for the corresponding points $(x_1, y_1, \ldots)$ and $(x_2, y_2, \ldots)$ of Ω, to the condition

$$x_1 x_2 + y_1 y_2 + z_1 z_2 + (z_1 + au_1)(z_2 + au_2) \pm u_1 u_2 (a^2 - 1) = v_1 t_2 + v_2 t_1;$$

if the lower of the ambiguous sign be taken this is the same as

$$x_1 x_2 + y_1 y_2 + z_1 z_2 + a\,(u_1 z_2 + u_2 z_1) + u_1 u_2 = v_1 t_2 + v_2 t_1;$$

this is the condition that the two points of Ω should be conjugate, or that the corresponding lines of the original threefold space should intersect. In words, if two spheres touch, then either of the lines corresponding to one of the spheres intersects one of the lines corresponding to the other.

Ex. 2. In a preceding section (II, p. 36), we have regarded a sphere in threefold space as arising by projection of a section, by a threefold, of a quadric threefold, U, in fourfold space. Two spheres will then touch if the corresponding quadric sections touch, that is, if the poles of the threefolds which determine these sections, taken in regard to U, lie on a line which touches U. With the figure now under consideration, let M be the pole, in regard to Ω, of the fourfold Π; then prove that, if P, Q be two points of Ω which are conjugate to one another, the plane MPQ meets Π in a line touching U, or (Ω, Π), at the point H, where the line PQ meets Π.

Ex. 3. Given four spheres, we can prove, either from the point of view of the preceding section (II, p. 36), or from the point of view of the present section, that there are eight pairs of spheres touching all of them. From the former point of view, we consider four threefold sections of a quadric U in fourfold space, and the enveloping cones of U at the points of each of these sections. The polar threefold of a point which is common to these four cones, in regard to U, will give a section touching the four given sections (cf. below, Chap. II). There are however sixteen common points of these four cones. From the present point of view, we assume that four lines in a threefold space, of general positions, have two common transversals; this is only equivalent to the statement that the tangent fourfolds at four points of the fourfold quadric Ω, in the space of five dimensions, meet in a line, which then intersects Ω in two points. To the four given spheres there correspond, in the manner explained, four pairs of lines. Every four lines, chosen from the available eight, which are so taken that no two belong to the same pair of lines, have two transversals; these give two spheres touching the four original spheres.

Ex. 4. The special case of the preceding example, that there are eight spheres touching four given planes, of a threefold space, may likewise be examined from both points of view. It will be sufficient to take the point of view of the present section. To a point of Ω lying in the tangent fourfold of Ω at O, there corresponds, not a sphere, but a plane. To four points of Ω lying in this tangent fourfold at O will then correspond four planes. These four points determine a threefold, from which only one tangent fourfold can be drawn to touch Ω, beside the tangent fourfold at O. This gives rise to one sphere touching the four planes, in the original threefold space. Conversely, a given plane of this space leads to two points of Ω, lying in the tangent fourfold at O, whose join passes through the pole of Π in regard to Ω. From this the eight spheres touching four given planes can be inferred.

Ex. 5. The figure of a double six of lines in threefold space is obtained by starting from five lines having a common transversal (Vol. III, p. 159). Thus, if five planes be taken, and a properly chosen one of the spheres touching every four of these planes, the five spheres so obtained touch another sphere (J. H. Grace, *Camb. Phil. Transactions*, xvi, 1898, p. 167).

Generalisation of Wallace's theorem. The theorem of the double six. A theorem for six lines with a common transversal. We consider now a set of six connected theorems. We state them, in the first instance, for space of five dimensions, remarking, later, on their meaning in ordinary threefold space.

1. In fivefold space, let six arbitrary fourfolds be given passing through a point which lies on a quadric, Ω. Every four of these meet in a line, having a further intersection with Ω. Thus, from five of these fourfolds we obtain five further points of Ω; and these five points determine a fourfold, passing through them. In all, then,

we have six such fourfolds. The theorem is that these six fourfolds meet in a point, and that this point lies in Ω.

2. Dually, let six arbitrary points be taken on the tangent fourfold at any point of Ω. Four of these points determine a threefold, from which can be drawn, to Ω, a further tangent fourfold, beside the original. Thus, from five of the points we obtain five further tangent fourfolds of Ω; and these meet in a point, not generally on Ω. In all, then, we have six such points. These six points lie on a fourfold; and this touches Ω.

3. Through a point of Ω let five fourfolds be taken, each of which touches Ω, at a point other than that through which they all pass. The line of intersection of every four of these meets Ω in a further point. In all there are five such points, lying on Ω. The theorem is that the fourfold determined by these five points touches Ω. Thus, also, the tangent fourfolds of Ω at these five points have for their intersection a point which lies on Ω.

4. Dually, let five points be taken, on a tangent fourfold of Ω, which are also on Ω itself; any four of these determine a threefold, from which another tangent fourfold can be drawn to Ω. In all there are five such new tangent fourfolds. The point of intersection of these is on Ω. Thus, also, the points of contact of the five fourfolds are in another tangent fourfold of Ω.

5. Through a point of Ω let six fourfolds be taken each of which touches Ω, at a point other than that through which they all pass. By theorem (3) above, every five of these fourfolds determine a further tangent fourfold of Ω. In all there will be six such further tangent fourfolds. The theorem is that these meet in a point, and that this point is on Ω. Thus, their six points of contact are in another tangent fourfold of Ω.

6. Dually, let six points be taken on a tangent fourfold of Ω which are also on Ω itself. By theorem (4) above, every five of these determine a new point of Ω. The six new points of Ω so found, lie on a fourfold; and this touches Ω.

In regard to theorem (1), we recall the proof above given (p. 19) of the theorem of Wallace, that the four circles containing the triads of intersections, of the sets of three out of four arbitrary lines, in a plane, meet in a point. A corresponding theorem, for the loci which correspond to circles, holds in space of any *even* number of dimensions (J. H. Grace, *Camb. Phil. Trans.*, xvi, July 1897, p. 163; Kühne, *Crelle*, cxix, 1898, p. 186; cf. *Camb. Phil. Proc.*, xxii, Part i, 1924). That this is so will be clear from the proof of theorem (1). In regard to theorem (4), we notice that, when points of a quadric Ω, in space of five dimensions, represent lines of a threefold space, this is equivalent to the theorem of a double six of lines; the theorem (4) is equivalent to (3). The theorem

(6), in virtue of (4), is a particular case of (2); and the equivalent theorem (5), in virtue of (3), is a particular case of (1). This theorem (6), however, is equivalent to the following, for lines in space of three dimensions: Consider six lines, in space of three dimensions, which have a common transversal, no two of the lines intersecting; from every five of these, by the theorem of the double six, let there be found another line. Then the six new lines so found have a common transversal. Let the six lines be denoted here by a, b, c, d, e, f, their common transversal being t. It can be shewn that the locus of a point, such that the planes joining it to the seven lines all touch a quadric cone, is a cubic curve, say ϑ, having the six lines a, b, c, d, e, f as chords; dually, the planes meeting the seven given lines in the points of a conic constitute a cubic developable, say Θ, of which the six lines a, b, c, d, e, f are axes (Vol. III, p. 195, Ex. 7). Further, if, with the fixed lines a, b, c, d, e, and their common transversal, t, we define a double six of lines, and denote the completing line of this by f_1 (Vol. III, p. 159), then f_1 is both a chord of the curve ϑ and an axis of the developable Θ. In fact, the points where ϑ meets f_1 are on the planes of Θ which meet in f, and the points where ϑ meets f are on the planes of Θ which meet in f_1. Thus we have a tetrad of points of ϑ determined by a tetrad of planes of Θ. We may, however, select any five of the lines a, b, c, d, e, f, and with these, and t, form a double six. Thereby we shall obtain six such tetrads of points and planes, associated with ϑ and Θ. But any two tetrads of points of a cubic curve are both self-polar in regard to a proper quadric (Vol. III, p. 148). We can thus infer that there exists a quadric in regard to which Θ is the polar reciprocal of ϑ, and the six pairs of lines such as f and f_1 are mutually polar lines. Then, from the fact that the lines a, b, c, d, e, f have a common transversal, it follows that the six lines $a_1, \ldots, f_1$ have also a common transversal. The theorem was given by Mr J. H. Grace, and this proof by Mr E. K. Wakeford (cf. *Proc. Lond. Math. Soc.*, XXI, 1922, p. 127).

In order to prove the six theorems of fivefold space enunciated above, it is clearly sufficient, after what has been said, to prove (1) and (3). Proofs are given in the examples following.

Ex. 1. Let $x_1 = 0, \ldots, x_6 = 0$ be any six fourfolds, of fivefold space, passing through the same point, so that there exists an identity of the form $a_1 x_1 + \ldots + a_6 x_6 = 0$. On the line of intersection of any four of these, say on $x_1 = 0, x_2 = 0, x_3 = 0, x_4 = 0$, let another arbitrary point be taken, say, for this case, P_{1234}. The five points so arising, by taking the fours from $x_1, \ldots, x_5$ only, determine a fourfold, say Π_6. We first prove, as in a preceding case, that the six fourfolds thus obtained, $\Pi_1, \Pi_2, \ldots, \Pi_6$, meet in a point.

Let $v = 0$ be any fourfold not passing through the point common to the original six fourfolds. We can suppose the equation of Π_6 to be expressed in terms of $x_1, \ldots, x_5$ only, and v, say $\xi_6 = 0$, where

$$\xi_6 = a_{61}x_1 + \ldots + a_{65}x_5 - a_6v.$$

The point P_{1234}, for which $x_1 = x_2 = x_3 = x_4 = 0$, will then be such that $a_{65}x_5 - a_6v = 0$, as well as $a_5x_5 + a_6x_6 = 0$. If we similarly express the equation of Π_5, without use of x_5, in the form $\xi_5 = 0$, where

$$\xi_5 = a_{51}x_1 + \ldots + a_{54}x_4 + a_{56}x_6 - a_5v,$$

the point P_{1234} will also be given by $a_{56}x_6 - a_5v = 0$. By comparison of these two forms for P_{1234} we infer that $a_{56} = -a_{65}$. With similar forms for the equations of $\Pi_1, \ldots, \Pi_4$ we prove, in the same way, that in general $a_{rs} = -a_{sr}$ $(r, s = 1, \ldots, 6)$. If then we add to the six equations $\xi_1 = 0, \ldots, \xi_6 = 0$ the identical equation

$$a_1x_1 + \ldots + a_6x_6 = 0,$$

we have seven equations in $x_1, \ldots, x_6$, v whose coefficients form a skew-symmetrical system; the determinant of these, being of odd order, vanishes identically. This proves that $\Pi_1, \ldots, \Pi_6$ meet in a point; and there is, evidently, a corresponding theorem in space of any *odd* number of dimensions.

Ex. 2. In space of n dimensions, all $(n-1)$-dimensional quadrics which pass through $\frac{1}{2}n(n+1)+1$ general points are expressible linearly, when their equations are in point-coordinates, by n linearly independent quadrics passing through these points; all such quadrics, therefore, pass through $2^n - 1 - \frac{1}{2}n(n+1)$ other points. In particular, in space of five dimensions, all quadrics (of dimension four) through sixteen given general points are expressible linearly by five such quadrics. In the preceding example, the original point, through which the fourfolds $x_1 = 0, \ldots, x_6 = 0$ pass, and the fifteen points such as P_{1234}, lie on the degenerate quadric expressed by $x_6\xi_6 = 0$. For $x_6 = 0$ contains, beside the original point, the ten points P_{rst6}, in which r, s, t are any triad from $1, 2, 3, 4, 5$; and $\xi_6 = 0$ contains the five points P_{rstk}, in which r, s, t, k are any set of four from $1, 2, 3, 4, 5$. Similarly, each of the quadrics $x_r\xi_r = 0$ contains the same sixteen points. These six quadrics are equivalent to five in virtue of the identity $\Sigma x_r\xi_r = 0$, but, in general, are otherwise linearly independent. Hence, any quadric through the original point and the fifteen points such as P_{1234} is capable of being written in the form $\lambda_1 x_1\xi_1 + \ldots + \lambda_6 x_6\xi_6 = 0$, wherein, in virtue of $\Sigma x_r\xi_r = 0$, only the differences $\lambda_r - \lambda_s$ are definite. Every quadric through the sixteen points named thus passes through the point common to $\xi_1 = 0, \ldots, \xi_6 = 0$. Thus, conversely, if we suppose the original point, $x_1 = 0, \ldots, x_6 = 0$, to be upon an arbitrary given quadric, Ω, and determine the fifteen points such as P_{1234} as the points of intersection, other than the original point, of the lines such as

$x_1 = x_2 = x_3 = x_4 = 0$ with Ω, then the final point $\xi_1 = \xi_2 = \ldots = \xi_6 = 0$ will also lie on Ω. The $2^5 - 16 - 1$, or fifteen, further intersections of the quadrics $x_r \xi_r = 0$ are the intersections of the fifteen lines, such as $\xi_1 = \xi_2 = \xi_3 = \xi_4 = 0$, each with a fourfold, such as $x_5 = 0$ or $x_6 = 0$, these giving the same point. There is in fact exact reciprocity, which we can put in evidence by the equations: Let the minors of the elements of the last row of the vanishing skew-symmetrical determinant, of seven rows and columns, which we have considered above, be denoted, respectively, by $A_1, \ldots, A_6, \Delta$; thus Δ is the skew-symmetrical determinant of six rows and columns whose general element is a_{rs}. And let A_{rs} denote the minor of a_{rs} in Δ itself. Then it is easy to see that $A_{rs} = -A_{sr}$, and

$$A_s = a_1 A_{1s} + a_2 A_{2s} + \ldots + a_6 A_{6s}, \; (r, s = 1, 2, \ldots, 6).$$

Hence we find $\Delta x_s = A_{1s} \xi_1 + \ldots + A_{6s} \xi_6 + A_s v$, and $A_1 \xi_1 + \ldots + A_6 \xi_6 = 0$. Further, the final point, $\xi_1 = \ldots = \xi_6 = 0$, is given by

$$x_r / v = A_r / \Delta \; (r = 1, \ldots, 6).$$

Ex. 3. If desired, the proof of the form $\lambda_1 x_1 \xi_1 + \ldots + \lambda_6 x_6 \xi_6 = 0$, for the quadric Ω, may be obtained directly. The quadric, passing through $x_1 = x_2 = \ldots = x_6 = 0$, is necessarily capable of an equation of the form

$$\Sigma \, (g_r x_r^2 + 2g_{rs} x_r x_s) + 2v \, (\lambda_1 a_1 x_1 + \ldots + \lambda_6 a_6 x_6) = 0, \; (r, s = 1, 2, \ldots, 6);$$

here, for symmetry, we allow terms in a redundant coordinate, say x_6, and $a_1, \ldots, a_6$ are as taken above, in the equations of $\xi_1 = 0, \ldots, \xi_6 = 0$. The condition that this quadric contain the point P_{1234}, for which $x_1 = x_2 = x_3 = x_4 = 0, \, -x_5/a_6 = x_6/a_5 = v/a_{56}$, is one of fifteen equations, which are all of the form

$$2g_{rs} = g_r a_s a_r^{-1} + g_s a_r a_s^{-1} - 2a_{rs} (\lambda_r - \lambda_s);$$

if we regard these equations as determining the coefficients g_{rs}, and notice the identities

$$\overset{1 \ldots 6}{\underset{r, s}{\Sigma}} \, [g_r x_r^2 + (g_r a_s a_r^{-1} + g_s a_r a_s^{-1}) \, x_r x_s], = (\Sigma g_r a_r^{-1} x_r) \, (\Sigma a_s x_s), = 0,$$

$$\underset{r, s}{\Sigma} \, a_{rs} (\lambda_r - \lambda_s) \, x_r x_s + v \Sigma \lambda_r a_r x_r = -\underset{r}{\Sigma} \lambda_r x_r \underset{s}{\Sigma} \, (a_{rs} x_s - a_r v),$$

the equation of the quadric is obtained in the form specified.

Ex. 4. With a loss of symmetry the algebra may be simplified. We may suppose the fourfold $v = 0$ to be taken so as to be the same as $\xi_6 = 0$, so that $a_{16}, a_{26}, \ldots, a_{56}$ are zero; and may take $\lambda_6 = 0$. That is, we may begin by supposing Ω to have an equation of the form

$$\overset{1 \ldots 5}{\underset{r, s}{\Sigma}} \, c_{rs} x_r x_s - (x_1 + x_2 + \ldots + x_5) \, v = 0,$$

the sixth plane drawn through the original point of Ω being of equation such as $c_1x_1 + \ldots + c_5x_5 = 0$. It is then easy to compute the forms of $\xi_1, \ldots, \xi_5$; and to obtain the result of Ex. 2 as the result of the vanishing of a skew-symmetric determinant of five rows and columns.

Ex. 5. The theorems (3), (4) above, equivalent to the double six theorem in threefold space, may be proved independently of this, as follows: Let X, Y, Z, U, T be points of the quadric Ω, lying in the tangent fourfold of Ω at a point V'. Let T' be the other point of Ω, beside V', which is conjugate to the four points X, Y, Z, U. Take, for the fundamental points of the fivefold space, the six points X, Y, Z, U, V', T', and for coordinates x, y, z, u, v, t; so that $x = 0$ contains all these points except X, and so for $y = 0, z = 0$, $u = 0$, the last containing all these points except U; while $v = 0$ contains all these points except V', and is the tangent fourfold of Ω at T'; and $t = 0$ contains all these points except T', and is the tangent fourfold of Ω at V'. The equation of Ω will then be of the form $\phi - \Delta tv = 0$, where ϕ is of the form

$$2fyz + 2gzx + 2hxy + 2pxu + 2qyu + 2rzu,$$

and Δ is the discriminantal determinant of ϕ. The equation $\phi = 0$ represents the quadric, lying on Ω, in the threefold $t = 0, v = 0$; this quadric we denote by H. Let the minors in Δ be A, B, C, D, F, G, H, P, Q, R.

The coordinates of the point T, lying in the tangent fourfold, $t = 0$, of Ω, at V', we denote by $(\xi, \eta, \zeta, \omega, 2, 0)$. The quadric H contains then the point $(\xi, \eta, \zeta, \omega, 0, 0)$; thus this quadric also contains (Vol. III, p. 55) the point $(\xi_1, \eta_1, \zeta_1, \omega_1, 0, 0)$, where

$$\xi_1 = A\xi^{-1}, \quad \eta_1 = B\eta^{-1}, \quad \zeta_1 = C\zeta^{-1}, \quad \omega_1 = D\omega^{-1}.$$

Therefore Ω contains the point $(\xi_1, \eta_1, \zeta_1, \omega_1, 0, 2)$. We prove that this is, in fact, the final point of the theorem under consideration.

Let X' be the point of Ω, beside V', which is conjugate to Y, Z, U, T; so Y' the point of Ω, beside V', conjugate to Z, X, U, T; and Z' the point of Ω, beside V', conjugate to X, Y, U, T; and U', the point of Ω, beside V', conjugate to X, Y, Z, T. We can shew that the coordinates of these points are, respectively,

$$X'(A, H, G, P, \xi_1, \xi); \quad Y'(H, B, F, Q, \eta_1, \eta);$$
$$Z'(G, F, C, R, \zeta_1, \zeta); \quad U'(P, Q, R, D, \omega_1, \omega).$$

We have defined T' as the point of Ω, beside V', conjugate to X, Y, Z, U. We then shew that the point $(\xi_1, \eta_1, \zeta_1, \omega_1, 0, 2)$ is conjugate to X', Y', Z', U', T'.

The point of coordinates (A, H, G, P, ξ_1, ξ) is at once verified to lie on Ω. The tangent fourfold at this point is found to be

$2x - t\xi_1 - v\xi = 0$. This contains the points Y, Z, U, all lying on $x = v = t = 0$; and it contains T, or $(\xi, \eta, \zeta, \omega, 2, 0)$. A precisely similar verification is possible for Y', Z', U'. Again, the tangent fourfold at $(\xi_1, \eta_1, \zeta_1, \omega_1, 0, 2)$ is expressed by

$$\xi_1(hy + gz + pu) + \eta_1(hx + fz + qu)$$
$$+ \zeta_1(gx + fy + ru) + \omega_1(px + qy + rz) = \Delta v;$$

it is at once seen that this contains the points X', Y', Z', U', T'. This proves what has been stated.

Note to p. 20. Wallace's theorem, that the four conics, through two points, I, J, of a plane, each containing the intersections of three out of four lines of the plane, all have another common point, may also be proved by remarking that, if four arbitrary planes be drawn, one through each line, meeting in threes in X, Y, Z, T, then the cubic curve through I, J, X, Y, Z, T meets the plane in another point. For the four conics are the projections of this cubic, from X, Y, Z, T, respectively.

More generally if, in n-fold space, there be given two points I, J, and also $(n+2)$ *primes* (spaces of $(n-1)$-dimensions), and we consider the $(n+2)$ rational curves of order n, each passing through I, J and the $(n+1)$ intersections of $(n+1)$ of these primes, then these $(n+2)$ curves have $(n-1)$ further common points. For these curves are all projections of a curve of order $(n+1)$, in space of $(n+1)$ dimensions, defined as passing through I, J and through the intersections of $(n+2)$ spaces of dimension n, arbitrarily drawn through the given $(n+2)$ primes.

For generalisation of this to the proof of Clifford's chain of theorems (above, p. 31), see F. P. White, *Camb. Phil. Proc.*, xxii, 1925. For instance, that the foci of the five parabolas which touch the fours of five given lines, lie on a circle, arises from the fact that the chords of a rational quartic curve, in fourfold space, which meet a plane meeting the curve in two points, meet this plane in the points of a conic passing through these two points.

Wallace's theorem is also a particular case of the theorem that plane cubic curves through eight points have another common point (Vol. iii, p. 217).

CHAPTER II

HART'S THEOREM, FOR CIRCLES IN A PLANE, OR FOR SECTIONS OF A QUADRIC

Given three lines in a plane, there are four circles touching them; these circles, we know, are all touched by another circle, the nine-points circle (Feuerbach's theorem; see Vol. II). In other words, given three lines, we can add to them a circle such that the four, these lines and the circle, are all touched by four other circles.

In the present chapter we shew how, given any three circles in a plane, we can add to them another circle, which we call the Hart circle, such that the four circles are all touched by four other circles (Hart, *Quart. J. of Maths.*, IV (1861), p. 260).

The three original circles are in fact touched by eight other circles, as we shall prove. There are fourteen ways of choosing, from these eight, four circles which all touch another circle. In six of these ways, the four circles chosen have a common orthogonal circle; and the four circles consisting of the original circles, and their Hart circle, have also a common orthogonal circle.

We have shewn that circles in a plane may be regarded as projections of plane sections of a quadric. We prove the results enunciated as theorems for such plane sections. This appears greatly to increase the interest and clearness of the matter.

The sections of a quadric which touch three given sections. Let the lines of intersection of the pairs of the planes of three sections of a quadric be OX, OY, OZ. Let the sections in the planes YOZ, ZOX, XOY be named, respectively, α, β, γ; let the common points of β, γ, on the line OX, be A and A'; the common points of γ, α, on OY, be B and B'; and the common points of α, β, on OZ, be C and C'. Let the poles of the planes YOZ, ZOX, XOY, in regard to the quadric, be, respectively, P, Q, R, so that the plane PQR is the polar plane of O. The conics β, γ, having two common points, lie on two quadric cones; the vertices of these cones lie on the line QR, and are harmonic conjugates in regard to Q and R; the polar plane of either vertex, in regard to the quadric, contains the line of intersection of the planes of the sections, and also contains the other vertex. Any plane section of the quadric, which touches both the sections β, γ, is a tangent plane of one of the two cones. When the sections are projected on to any plane, from any point of the quadric, they become two curves which we have agreed to call circles, the projections of Q and R being the centres of the circles; the projections of the vertices of the two cones which contain β and γ are the centres of similitude of these circles. Each centre of similitude is the centre of a circle, coaxial with the two circles, in regard to which these two circles are inverses of one another.

Consider, then, the six quadric cones, two containing every pair of the three sections α, β, γ; by Pascal's theorem, for the section of the quadric by the plane PQR, we see that these vertices lie, in threes, on four lines in this plane. Let l be one of these lines, containing the vertices of three of these cones, each cone containing two of the sections α, β, γ. Consider two of these three cones; as they have one common plane section (one of α, β, γ), they will have another common plane section; they will therefore have two common tangent planes, which must intersect in the line l containing the vertices of the two cones. This shews that a tangent plane drawn from l to one of the three cones whose vertices are on l, is equally a tangent plane of the other two cones. This plane will then meet the quadric in a section touching the three conics α, β, γ. As there are four lines l, we thus construct eight sections of the quadric touching the three sections α, β, γ.

Ex. 1. If two fixed sections α, β, of a quadric, which meet in A, B, be both touched by two other sections ξ, η, which also touch one another, and the planes of ξ, η meet the line AB in P and Q, the range A, B, P, Q is related to the range A, B, P', Q', similarly obtained from two other such sections, ξ', η'. It is understood that the sections ξ, η touch the same one of the cones which contain α and β; if V be the vertex of this cone, the theorem is obtained by considering, on the plane of the polar of V, the section both by the quadric and by the cone. We then have two conics which touch at two points, and a pair of tangent lines, to one of these, drawn from a variable point of the other.

Ex. 2. The problem of drawing a circle to touch three given circles in a plane is sometimes called Apollonius's problem. Deduce from what is said above the following solution of this, given by Gergonne (*Ann. d. Math.*, vii, 1817, p. 289): Find O, the centre of the common orthogonal circle of the three given circles; take one of the four lines which contain three of the centres of similitude of pairs of the circles, and let U, V, W be the respective poles of this line in regard to the three given circles. Then the joining lines OU, OV, OW contain the points of contact with the given circles of two of the circles which touch them all. (Cf. Poncelet, *Prop. Proj.*, i, 1865, p. 138.)

Ex. 3. Prove that there are four circles touching three lines in a plane. Also, that the circle through the centres of three of these circles, and the circle containing the three intersections of the lines, are two circles which have two common tangents, meeting in the centre of the fourth of the original circles.

Ex. 4. Regarding two circles in a plane as projections of plane sections, α, β, of a quadric, prove that a circle cutting these circles at equal angles is the projection of a section of the quadric by a plane passing through the vertex of one of the two quadric cones containing the sections α, β. Hence, given three circles in a plane, prove that a circle cutting these three circles all at equal angles belongs to one of four systems of coaxial circles, of which the radical axis contains three of the centres of similitude of pairs of the three given circles.

Ex. 5. Consider four plane sections of a quadric, and the vertices of cones containing pairs of these sections, twelve points in all. The plane containing the poles of three of the four given planes contains, we have seen, six of these vertices. Omitting the four such planes, prove that the twelve points, with their joining lines and planes, form a figure of structure 12 ($\cdot$, 4, 4) 16 (3, $\cdot$, 2) 8 (6, 4, $\cdot$), arising from two of three desmic tetrads (Vol. II, p. 213). The eight planes of this figure enable us to construct eight circles meeting four circles, given in a plane, all at equal angles.

Ex. 6. Regarding two circles in a plane as projections of plane sections of a quadric, the condition that the two circles should cut at a given angle is that the range, of four points, consisting of the two poles of the corresponding plane sections of the quadric, and the two points where the join of these meets the quadric, should be related to a given range. Prove that a circle cutting two given circles at given angles cuts any circle coaxial with the given circles at a constant angle, and touches two fixed circles. Construct a circle cutting three given circles at given angles (Steiner, *Ges. Werke*, I, p. 21, 1826).

Ex. 7. If three circles have three tangents, each touching two of the circles, which meet in a point, they have three other tangents, also each touching two of the circles, which likewise meet in a point.

Let α, β, γ be three plane sections of a quadric, U a quadric cone containing β and γ, V a quadric cone containing γ and α, and W a quadric cone containing α and β. The pair of cones V, W, as both containing α, have two common tangent planes; so W and U have two common tangent planes, and U and V have two common tangent planes. There is thus a quadric, say Σ, of which U, V, W are all enveloping cones (Vol. III, p. 10). Any section of the original quadric touching β and γ lies in a plane touching one of the two cones containing β and γ; if this cone be U, this plane touches the quadric Σ. Consider three sections of the original quadric whose planes meet in a line, l, the planes of these sections touching, respectively, the cones U, V, W; these planes, therefore, all touch the quadric Σ. Of this quadric, then, the line l is a generator. Let m be any generator of Σ which meets l; this lies in a particular tangent plane of every enveloping cone of Σ. There is therefore a plane through m touching the sections β, γ; another plane through m touching the sections γ, α; and another plane through m touching the sections α, β.

Regard, now, three circles in a plane as projections of three sections of a quadric, the centre of projection being a point, H, of this quadric. A common tangent line of two of the circles is the projection of a section of the quadric, touching the two given sections, by a plane which passes through H. Three concurrent tangents, each of two of the given circles, are thus projections of three sections of the quadric by planes which intersect in a line, l, passing through H. By taking for m the other generator of the quadric Σ which passes through H, the theorem above stated is obtained.

Ex. 8. Given three circles in a plane, construct three other circles each to touch two of the given circles, each of the constructed circles, moreover, touching the other two. For the case when the given circles are lines, the problem was put by Malfatti, *Mem. d. Soc. Ital. d. Sc.*, Modena, 1803, Vol. x. For this case, and the general one, a solution was enunciated, without proof, by Steiner, *Ges. Werke*, i, pp. 35–39. Steiner's construction was justified by Hart, *Quarterly J.*, i, 1857, p. 219. An analytical investigation of the corresponding theorem for sections of a quadric was given by Cayley, *Papers*, ii, p. 57 (1852).

The theorem is as follows: Let α, β, γ be three plane sections of a quadric, whose planes meet in O, the poles of these planes being P, Q, R, respectively: let U be the vertex of a cone containing the sections β, γ, the vertex of a cone containing γ and α being V, and of a cone containing α and β being W, these points U, V, W being in a line in the plane PQR. Next, let θ, ϕ, ψ be, respectively, the sections by the polar planes of U, V, W; the planes of these will meet in a line passing through O, and the sections will have two points in common: let I be one of these common points. The section θ, for example, contains the two points common to the sections β, γ, and the plane of θ contains the vertex, other than U, of a cone containing the sections β, γ. Now, let λ be a section touching ϕ, ψ and α, the point of contact with α being A; similarly, let μ be a section touching ψ, ϑ, and touching β, in the point B; and let ν be a section touching ϑ, ϕ, and touching γ, in the point C. The sections λ, μ, ν, being touched in pairs by sections, θ, ϕ, ψ, whose planes meet in the line OI, are also (Ex. 7) touched in pairs by sections, say θ', ϕ', ψ', whose planes meet in another line OI'; it can in fact be shewn, the sections λ, μ, ν being properly taken, that there is a section, with plane passing through O, which passes through A, and touches μ and ν; let this be θ'; and that there is a section, ϕ', through O and B, touching ν and λ, and a section, ψ', through O and C, touching λ and μ. There is then a section touching β, γ, and ϕ', ψ', say ξ; also a section, η, touching γ, α and ψ', θ'; and a section, ζ, touching α, β and θ', ϕ'. Further, the sections η, ζ touch one another, touching θ' at the same point; similarly ζ and ξ touch one another, and ξ and η touch one another. The sections ξ, η, ζ are those to be constructed.

Ex. 9. We may easily investigate the equations of the sections touching the three given sections α, β, γ. Let the planes of these be, respectively, $x = 0$, $y = 0$, $z = 0$, the polar plane of the point, O, in which these meet, being $t_1 = 0$, so that the equation of the quadric is of the form

$$ax^2 + by^2 + cz^2 + 2fyz + 2gzx + 2hxy = t_1^2.$$

Let $A = bc - f^2$, $F = gh - af$, etc., these being the minors of a, f, etc. in the determinant of the quadratic form, Δ, whose value is

$$abc + 2fgh - af^2 - bg^2 - ch^2.$$

The pole, P, of the plane α has the coordinates $(A, H, G, 0)$, the poles of Q and R being, respectively, $(H, B, F, 0)$ and $(G, F, C, 0)$. The equation of a quadric cone containing the sections β, γ is of the form $(a, b, c, f, g, h \gamma x, y, z)^2 + 2\lambda yz = t_1^2$, and λ is such that the three derivatives of the left side all vanish at the vertex, $(x', y', z', 0)$, of this cone. With $\lambda = \Delta [(BC)^{\frac{1}{2}} - F]^{-1}$, these are satisfied by taking

$$(x', y', z') = B^{-\frac{1}{2}}(H, B, F) - C^{-\frac{1}{2}}(G, F, C),$$

that is, $x' = B^{-\frac{1}{2}}.H - C^{-\frac{1}{2}}.G$, $y' = B^{-\frac{1}{2}}.B - C^{-\frac{1}{2}}.F$, etc. Herein the signs of $A^{\frac{1}{2}}$, $B^{\frac{1}{2}}$, $C^{\frac{1}{2}}$ are arbitrary, and $(BC)^{\frac{1}{2}}$ means $B^{\frac{1}{2}}.C^{\frac{1}{2}}$, etc. If, for the sake of brevity, we use X to denote the set of four coordinates $A^{-\frac{1}{2}}(A, H, G, 0)$, with, similarly, Y and Z for the sets $B^{-\frac{1}{2}}(H, B, F, 0)$, $C^{-\frac{1}{2}}(G, F, C, 0)$, then three cones, each containing a pair of the sections α, β, γ, with vertices on a line, l, are those whose vertices have coordinates, respectively, $Y - Z$, $Z - X$, $X - Y$. It is then easy to verify that the plane whose equation is

$$(ax + hy + gz)A^{\frac{1}{2}} + (hx + by + fz)B^{\frac{1}{2}} + (gx + fy + cz)C^{\frac{1}{2}} = t_1 M^{\frac{1}{2}}$$

contains this line l, and touches the three sections α, β, γ, provided M satisfy

$$M = aA + bB + cC + 2f(BC)^{\frac{1}{2}} + 2g(CA)^{\frac{1}{2}} + 2h(AB)^{\frac{1}{2}} - \Delta;$$

for the condition that two sections of the quadric touch one another is that the enveloping cone from the pole of the plane of either section should contain the pole of the plane of the other section.

The possible signs of $A^{\frac{1}{2}}$, $B^{\frac{1}{2}}$, $C^{\frac{1}{2}}$, $M^{\frac{1}{2}}$ give the eight sections touching α, β, γ.

Ex. 10. Through any one of the sections α, in Ex. 9, pass two of the cones used in the construction. These cones have then another plane section; shew that this lies on the plane whose equation is

$$y\left[(CA)^{\frac{1}{2}} - G\right] = z\left[(AB)^{\frac{1}{2}} - H\right].$$

The three such planes thus meet in a line, passing through the point O in which the planes α, β, γ intersect.

Ex. 11. If from the vertex $(x', y', z', 0)$, above taken, of a cone containing the sections β, γ, a cone be drawn to contain the section α, prove that this cone meets the quadric again in a section lying on a plane, through O, whose equation is

$$\left[(BC)^{\frac{1}{2}} - F\right]x + (GB^{\frac{1}{2}} - HC^{\frac{1}{2}})(B^{-\frac{1}{2}}y - C^{-\frac{1}{2}}z) = 0.$$

Ex. 12. Prove that the polar planes of the vertices of the three cones used in Ex. 9, which are on the line l, are respectively

$$B^{-\frac{1}{2}}y - C^{-\frac{1}{2}}z = 0, \quad C^{-\frac{1}{2}}z - A^{-\frac{1}{2}}x = 0, \quad A^{-\frac{1}{2}}x - B^{-\frac{1}{2}}y = 0.$$

These are the sections θ, ϕ, ψ of the enunciation in Ex. 8; the point I is then on the line $A^{-\frac{1}{2}}x = B^{-\frac{1}{2}}y = C^{-\frac{1}{2}}z$.

Ex. 13. Shew that if we introduce λ, μ, ν such that

$$\cos\lambda = (bc)^{-\frac{1}{2}}f, \ \sin\lambda = (bc)^{-\frac{1}{2}}A^{\frac{1}{2}}, \text{ etc.,}$$

and take $s = \frac{1}{2}(\lambda + \mu + \nu)$, the section above found, touching α, β, γ, lies in the plane

$$xa^{\frac{1}{2}}\cos(s-\lambda) + yb^{\frac{1}{2}}\cos(s-\mu) + zc^{\frac{1}{2}}\cos(s-\nu) - t_1 = 0.$$

Herein the signs of $a^{\frac{1}{2}}$, $b^{\frac{1}{2}}$, $c^{\frac{1}{2}}$ are arbitrary, and $(bc)^{\frac{1}{2}}$ means $b^{\frac{1}{2}} . c^{\frac{1}{2}}$; a change of $a^{\frac{1}{2}}$ into $-a^{\frac{1}{2}}$, for example, leads to the substitution of $\pi + \mu$, $\pi + \nu$, $\pi + s$ respectively for μ, ν and s. Four of the eight tangent sections are obtained by substituting $\pm\lambda$, $\pm\mu$, $\pm\nu$, respectively, for λ, μ, ν, retaining the signs of $a^{\frac{1}{2}}$, $b^{\frac{1}{2}}$, $c^{\frac{1}{2}}$; the other four are then obtained by changing $a^{\frac{1}{2}}$, $b^{\frac{1}{2}}$, $c^{\frac{1}{2}}$ into $-a^{\frac{1}{2}}, -b^{\frac{1}{2}}, -c^{\frac{1}{2}}$, in the first four.

Ex. 14. With the substitution of Ex. 13, the plane of Ex. 11 becomes

$$xa^{\frac{1}{2}} + (yb^{\frac{1}{2}}\sin\nu - zc^{\frac{1}{2}}\sin\mu)\sin(\mu - \nu) = 0.$$

Ex. 15. Taking p, q, r such that

$$\tfrac{1}{2}p = 1 - F(BC)^{-\frac{1}{2}}, \quad \tfrac{1}{2}q = 1 - G(CA)^{-\frac{1}{2}}, \quad \tfrac{1}{2}r = 1 - H(AB)^{-\frac{1}{2}},$$

prove that a section touching the sections by $x=0$, $zC^{-\frac{1}{2}}-xA^{-\frac{1}{2}}=0$, $xA^{-\frac{1}{2}}-yB^{-\frac{1}{2}}=0$, namely the section named λ in the enunciation of Ex. 8, lies in the polar plane of the point (ξ, η, ζ, τ), where $\xi = A^{\frac{1}{2}}$, $\eta = B^{\frac{1}{2}}(1-r^{\frac{1}{2}})$, $\zeta = C^{\frac{1}{2}}(1+q^{\frac{1}{2}})$, and τ is given by

$$\tau^2 = (a, b, c, f, g, h \Updownarrow \xi, \eta, \zeta)^2 - \Delta.$$

If we put $-F(BC)^{-\frac{1}{2}} = \cos \lambda'$ (as in Vol. II, p. 205), we have $p^{\frac{1}{2}} = 2 \cos \frac{1}{2} \lambda'$, etc.

The Hart circles of three circles when the four have a common orthogonal circle. We consider, as before, three sections α, β, γ, of a quadric, whose planes meet in O. Let U be the vertex of one of the two quadric cones containing the sections β and γ. The polar plane of U, as has been remarked, contains the line of intersection of the planes of β and γ; in fact, if these sections meet in A and A', the tangent planes of the quadric, at A and A', touch these sections, and pass through U. If we consider, also, the quadric cone obtained by joining U to the points of the section α, this cone will meet the quadric again in a plane section, say α'; thence, the line in which the plane of α' meets the plane of α, lying equally in the polar plane of U, will meet the line of intersection of the planes β, γ; in other words, the plane of α' passes through the point O in which the planes of α, β, γ intersect. But also the two cones of vertex U, namely $(U : \beta, \gamma)$ and $(U : \alpha, \alpha')$, have four common tangent planes; and these will touch all the four sections α, β, γ, α'.

Thus α' is a section which, added to α, β, γ, gives four sections which have four common tangent sections. Thus, on projection of the quadric, the section α' gives a circle, orthogonal to the common orthogonal circle of α, β, γ; such that the four are all touched by four circles; these likewise have a common orthogonal circle, whose centre is the projection of U. The circle given by α' is a particular kind of Hart circle, of those belonging to α, β, γ.

Since any chord of the quadric passing through O is divided harmonically by O and the polar plane of O, it is clear that the eight sections, which touch the sections α, β, γ, consist of four pairs, such that the planes of a pair intersect on the polar plane of O. Also, it is clear that there are six such sets of four sections as those found above, touching β, γ and α, α'; for there are two cones containing every two sections such as β and γ. Each of these sets consists, then, of two of the four pairs into which the eight tangent sections of α, β, γ can be divided.

The equation of the plane of the section α' has been found above, in Exx. 11 and 14. Four planes touching the sections α, β, γ which all pass through the point U, whose coordinates are

$$(x', y', z', 0) = B^{-\frac{1}{2}}(H, B, F, 0) - C^{-\frac{1}{2}}(G, F, C, 0),$$

are found from

$$(ax + hy + gz)\, A^{\frac{1}{2}} + (hx + by + fz)\, B^{\frac{1}{2}} + (gx + fy + cz)\, C^{\frac{1}{2}} = t_1 M$$

by giving to $A^{\frac{1}{2}}$, $B^{\frac{1}{2}}$, $C^{\frac{1}{2}}$ the signs $(+, +, +)$, $(+, -, -)$, $(-, +, +)$, $(-, -, -)$, respectively.

The Hart circles of three circles in general. Given, as before, three sections α, β, γ of a quadric, we now investigate a fourth section δ, whose plane does not pass through the intersection of the planes, α, β, γ, such that α, β, γ, δ are all touched by four other sections of the quadric.

It will appear that the necessary and sufficient condition that four sections of a quadric, by planes which do not meet in a point, should all be touched by four other planes, is very simple. It may add to the interest of the work to state at once what this condition is. Let the planes of the four sections, α, β, γ, δ meet in threes in the points O, X, Y, Z, the sections β, γ meeting OX in A and A', the sections γ, α meeting OY in B and B', the sections α, β meeting OZ in C and C', the sections α, δ meeting YZ in L and L', the sections β, δ meeting ZX in M and M', and the sections γ, δ meeting XY in N and N'. Through each of the six joins of O, X, Y, Z draw the two planes to the points in which the quadric meets the opposite join; as, for instance, draw through OX the planes OXL and OXL'. Thereby twelve planes are found. In general these twelve planes touch another quadric (Vol. III, p. 54, Ex. 16). It may happen, however, that this quadric degenerates into two points, the planes meeting, in sets of six, in these two points. In such case, choosing the notation properly, we may suppose that the planes OXL, OYM, OZN, YZA, ZXB, XYC meet in a point, say S, the other six planes, OXL', ..., YZA', ..., meeting in a point S'. Then the line AL, as lying in the planes OXL, YZA, passes through S, and so on; namely, the three lines AL, BM, CN meet, in the point S. Conversely, if these three lines meet in a point, say S, the six planes OXL, ..., XYC are easily seen to meet in S; and then, similarly, the three lines $A'L'$, $B'M'$, $C'N'$ also meet in a point, say S'. The condition that this be so may be expressed otherwise: Beginning, as before, with the general case, let points, say A_1 and A_1', be taken on OX, so as to be the respective harmonic conjugates of A and A' in regard to O and X; and let a pair of points be similarly defined on each of the six joins of O, X, Y, Z. It is then easy to see that the twelve new points so obtained lie upon another quadric; we may call this the *harmonic conjugate* of the original quadric in regard to the tetrad O, X, Y, Z. In the particular case when AL, BM, CN meet in a point, say S, the harmonically conjugate quadric breaks up into two planes; and conversely. These two planes are, in fact, the polars of the points S and S', in regard to the original quadric.

What we prove is: If the harmonically conjugate quadric of the original quadric, in regard to O, X, Y, Z, break up into two planes, then the sections of the original quadric by the four planes XYZ, OYZ, OZX, OXY are all touched by four planes. We afterwards prove, conversely, that, for such tangency, it is necessary that the harmonically conjugate quadric should so degenerate.

As before, let $x = 0, y = 0, z = 0$ be the planes of the sections α, β, γ, the equation of the quadric being

$$(a, b, c, f, g, h \gimel x, y, z)^2 = t_1^2,$$

where $t_1 = 0$ is the polar plane of $(0, 0, 0, 1)$. For $t_1 = 0$ we can substitute a plane, $t = 0$, in eight ways, so that the condition referred to, that AL, BM, CN meet in a point, is satisfied. For, taking arbitrary signs for the radicals $a^{\frac{1}{2}}, b^{\frac{1}{2}}, c^{\frac{1}{2}}$, and denoting $b^{\frac{1}{2}} c^{\frac{1}{2}}$ by $(bc)^{\frac{1}{2}}$, etc., let

$$u = \tfrac{1}{2} [f + (bc)^{\frac{1}{2}}], \quad v = \tfrac{1}{2} [g + (ca)^{\frac{1}{2}}], \quad w = \tfrac{1}{2} [h + (ab)^{\frac{1}{2}}],$$

and $\quad l = (uvw)^{\frac{1}{2}} u^{-1}, \quad m = (uvw)^{\frac{1}{2}} v^{-1}, \quad n = (uvw)^{\frac{1}{2}} w^{-1},$

so that $mn = u$, $nl = v$, $lm = w$. As the product of $(bc)^{\frac{1}{2}}, (ca)^{\frac{1}{2}}, (ab)^{\frac{1}{2}}$ is without ambiguity there are thus eight possible sets of values for l, m, n. Putting $t = lx + my + nz - t_1$, the equation of the quadric referred to $x = 0, y = 0, z = 0, t = 0$, is

$$ax^2 + by^2 + cz^2 + 2fyz + 2gzx + 2hxy = (lx + my + nz - t)^2;$$

this meets OX in the points A, A' given by $(l \pm a^{\frac{1}{2}}) x = t$. Taking for X, Y, Z the points where OX, OY, OZ meet $t = 0$, the points, L, L', where the quadric meets YZ, are given by

$$(m^2 - b) \dot{y}^2 + 2 (mn - f) yz + (n^2 - c) z^2 = 0;$$

since $f = 2mn - (bc)^{\frac{1}{2}}$, these points are given by

$$(m \pm b^{\frac{1}{2}}) y = (n \pm c^{\frac{1}{2}}) z.$$

Hence, with suitable choice of notation, the lines AL, BM, CN meet in the point, S, given by $(l - a^{\frac{1}{2}}) x = (m - b^{\frac{1}{2}}) y = (n - c^{\frac{1}{2}}) z = t$, and the lines $A'L', B'M', C'N'$ meet in the point, S', given by $(l + a^{\frac{1}{2}}) x = \dots = t$. Conversely, it can be shewn that these intersections require the plane $t = 0$ to be one of the eight we have taken. In general, the harmonically conjugate quadric of that given by the equation $(a, b, c, d, f, g, h, p, q, r \gimel x, y, z, t)^2 = 0$ has an equation obtained from this by change of f, g, h, p, q, r, respectively, into $-f, -g, -h, -p, -q, -r$. In the present case, since $f = 2mn - (bc)^{\frac{1}{2}}$, $g = 2nl - (ca)^{\frac{1}{2}}$, $h = 2lm - (ab)^{\frac{1}{2}}$, the harmonically conjugate quadric is found to be

$$(xa^{\frac{1}{2}} + yb^{\frac{1}{2}} + zc^{\frac{1}{2}})^2 - (lx + my + nz + t)^2 = 0.$$

The point S, or $(l-a^{\frac{1}{2}})\,x = \ldots = t$, is then easily verified to be the pole of the plane $(l-a^{\frac{1}{2}})\,x + (m-b^{\frac{1}{2}})\,y + (n-c^{\frac{1}{2}})\,z + t = 0$, in regard to the original quadric. This plane we denote by σ, the plane which is the polar of S' being denoted by σ'.

To prove, now, that the sections α, β, γ, and the section, say δ, of the quadric, by $t = 0$, are all touched by four planes, consider the section by the plane

$$xa^{\frac{1}{2}}\cos(s-\lambda) + yb^{\frac{1}{2}}\cos(s-\mu) + zc^{\frac{1}{2}}\cos(s-\nu) - t_1 = 0,$$

above found (p. 70, Ex. 13) as touching the sections α, β, γ; here $\cos\lambda = (bc)^{-\frac{1}{2}}f$, etc., and $s = \frac{1}{2}(\lambda + \mu + \nu)$. The cone joining O to the section of the quadric by this plane is given by the equation

$$ax^2 + \ldots + cz^2 + 2yz\,(bc)^{\frac{1}{2}}\cos\lambda + \ldots + 2xy\,(ab)^{\frac{1}{2}}\cos\nu - [xa^{\frac{1}{2}}\cos(s-\lambda) + \ldots + zc^{\frac{1}{2}}\cos(s-\nu)]^2 = 0$$

This is easily found to be the same as is obtained by rationalising the equation

$$[xa^{\frac{1}{2}}\sin(s-\lambda)]^{\frac{1}{2}} + [yb^{\frac{1}{2}}\sin(s-\mu)]^{\frac{1}{2}} + [zc^{\frac{1}{2}}\sin(s-\nu)]^{\frac{1}{2}} = 0.$$

Putting p, q, r as abbreviations for $\sin(s-\lambda)$, $\sin(s-\mu)$, $\sin(s-\nu)$, respectively, any generator of the cone is given, in terms of a parameter, θ, by

$$(xa^{\frac{1}{2}}p)^{\frac{1}{2}} : (yb^{\frac{1}{2}}q)^{\frac{1}{2}} : (zc^{\frac{1}{2}}r)^{\frac{1}{2}} = -\theta : \theta - 1 : 1\,;$$

in particular, one generator is given by

$$xa^{\frac{1}{2}} : yb^{\frac{1}{2}} : zc^{\frac{1}{2}} = p\,(q-r)^2 : q\,(r-p)^2 : r\,(p-q)^2,$$

and the tangent plane of the cone along this generator has the equation

$$xa^{\frac{1}{2}}\,(q-r)^{-1} + yb^{\frac{1}{2}}\,(r-p)^{-1} + zc^{\frac{1}{2}}\,(p-q)^{-1} = 0 \quad \ldots(A).$$

But, in virtue of $f = (bc)^{\frac{1}{2}}\cos\lambda$, and $2u = f + (bc)^{\frac{1}{2}}$, we have $u = (bc)^{\frac{1}{2}}\cos^2\frac{1}{2}\lambda$; and, thence, by $l = (uvw)^{\frac{1}{2}}u^{-1}$, etc., we may suppose the plane $t = 0$ to be that given by

$$xa^{\frac{1}{2}}\,\frac{\cos\frac{1}{2}\mu\cos\frac{1}{2}\nu}{\cos\frac{1}{2}\lambda} + yb^{\frac{1}{2}}\,\frac{\cos\frac{1}{2}\nu\cos\frac{1}{2}\lambda}{\cos\frac{1}{2}\mu} + zc^{\frac{1}{2}}\,\frac{\cos\frac{1}{2}\lambda\cos\frac{1}{2}\mu}{\cos\frac{1}{2}\nu} - t_1 = 0$$
$$\ldots\ldots\ldots(H).$$

There is ambiguity here in the sign of t_1; we can however suppose that the plane

$$xa^{\frac{1}{2}}\cos(s-\lambda) + yb^{\frac{1}{2}}\cos(s-\mu) + zc^{\frac{1}{2}}\cos(s-\nu) - t_1 = 0 \ \ldots(I),$$

above taken from among the tangent planes of the sections α, β, γ, is chosen to have the same sign for t_1. Then the planes, (I), (H), meet in a line which lies in the plane

$$xa^{\frac{1}{2}} \left[\cos\left(s - \lambda\right) - \frac{\cos\frac{1}{2}\mu \cos\frac{1}{2}\nu}{\cos\frac{1}{2}\lambda} \right]$$

$$+ \dots + zc^{\frac{1}{2}} \left[\cos\left(s - \nu\right) - \frac{\cos\frac{1}{2}\lambda \cos\frac{1}{2}\mu}{\cos\frac{1}{2}\nu} \right] = 0;$$

this, however, is the same, as we easily see, as the tangent plane (A), found above, of the cone which joins O to the section of the quadric by the plane (I). As the line of intersection of the planes (I), (H) is thus a tangent line of this cone, and is, therefore, a tangent line of the section of the quadric by the plane (I), it is a tangent line of the quadric; and, as the plane (H) contains this line, the section of the quadric by the plane (H) has this line for a tangent. The sections of the quadric by the planes (I) and (H) thus touch one another; or the plane (I), beside touching the sections α, β, γ, touches also the section, δ, by the plane (H). This section δ is one of the Hart sections to be associated with α, β, γ. For a definite plane (H) we have four possible planes (I), obtainable, namely, by the changes of λ, μ, ν respectively into λ, $\pm \mu$, $\pm \nu$; all these touch the four sections α, β, γ, δ.

In all there are eight possibilities for the plane (H). There are four in which, in place of $a^{\frac{1}{2}}, b^{\frac{1}{2}}, c^{\frac{1}{2}}, \lambda, \mu, \nu$, we have: (1), these; (2), $-a^{\frac{1}{2}}, b^{\frac{1}{2}}, c^{\frac{1}{2}}, \lambda, \pi + \mu, \pi + \nu$, respectively; (3), $a^{\frac{1}{2}}, -b^{\frac{1}{2}}, c^{\frac{1}{2}}, \pi + \lambda, \mu, \pi + \nu$, respectively; and (4), $a^{\frac{1}{2}}, b^{\frac{1}{2}}, -c^{\frac{1}{2}}, \pi + \lambda, \pi + \mu, \nu$. Then there are four others obtainable from these by changing the signs of all of $a^{\frac{1}{2}}, b^{\frac{1}{2}}, c^{\frac{1}{2}}$. Thus, given three plane sections of a quadric, it is possible, in eight ways, to add to these another section, by a plane not passing through the common point of the three given planes, so that the aggregate four sections shall have four common tangent planes. Of this result another proof is contained in the work given below to shew that the conditions we have here verified to be sufficient, are also necessary.

Ex. 1. Make an explicit table of the equations of the eight planes (H), and, in each case, of the four planes touching the sections determined by these.

Ex. 2. The planes ABC, $A'B'C'$, where A, ..., C' are as above, may be taken to be

$$(l - a^{\frac{1}{2}})\, x + (m - b^{\frac{1}{2}})\, y + (n - c^{\frac{1}{2}})\, z - t = 0,$$

and

$$(l + a^{\frac{1}{2}})\, x + (m + b^{\frac{1}{2}})\, y + (n + c^{\frac{1}{2}})\, z - t = 0,$$

respectively. The line of intersection of these planes is the intersection of the polar plane $t_1 = 0$ with the plane $a^{\frac{1}{2}}x + b^{\frac{1}{2}}y + c^{\frac{1}{2}}z = 0$. On the section of the quadric by the plane (I) we have two triads of points: (1), the points of contact of this section with $x = 0$, $y = 0$, $z = 0$; if the points of the section be given, as above, in terms of a parameter θ, by the fact that $(a^{\frac{1}{2}}px)^{\frac{1}{2}}$, $(b^{\frac{1}{2}}qy)^{\frac{1}{2}}$, $(c^{\frac{1}{2}}rz)^{\frac{1}{2}}$ are in the ratios of $-\theta$, $\theta-1$, 1, then these points of contact are given by $\theta = 0, 1, \infty$, respectively; (2), the point of contact of the section (I) with the plane $t = 0$, taken with the two points in which the section (I) is met by the above plane $a^{\frac{1}{2}}x + b^{\frac{1}{2}}y + c^{\frac{1}{2}}z = 0$. The first point of this triad (2) is given, we have seen, by the value $-p(q-r)/r(p-q)$ of the parameter θ; the parameters for the other two points, in which the section (I) is met by the plane we have named, are the roots of the equation

$$p^{-1}\theta^2 + q^{-1}(\theta-1)^2 + r^{-1} = 0.$$

Hence it is easily seen that the points of the first triad are apolar with the points of the second triad, in the sense explained in Vol. II, p. 114. For the condition that θ, θ_1, θ_2 should be apolar with $0, 1, \infty$ is at once seen to be that $\theta = (\theta_1\theta_2 - \theta_1 - \theta_2)/(1 - \theta_1 - \theta_2)$; and, when θ_1, θ_2 are the roots of the quadratic equation just given, this leads to $\theta = -p(q-r)/r(p-q)$. Hence we may say: Upon the section (I), touching the three sections α, β, γ, the point of contact with the Hart section (H) is the apolar complement of the two points in which (I) is met by the plane $a^{\frac{1}{2}}x + b^{\frac{1}{2}}y + c^{\frac{1}{2}}z = 0$, with reference to the three points of contact of (I) with α, β, γ. The plane $a^{\frac{1}{2}}x + $ etc. $= 0$ is the harmonic conjugate of the point O, in which the planes α, β, γ meet, with respect to the planes ABC and $A'B'C'$; it will receive a better definition below (Ex. 5).

Ex. 3. As before, denote by S the point of intersection of the lines AL, BM, CN, and by S' the point of intersection of the lines $A'L'$, $B'M'$, $C'N'$. Also, denote by σ, σ' the respective polar planes of S and S'. Further, denote the planes ABC, $A'B'C'$ by μ, μ', respectively, and the plane of the section (H) by ϖ; let G, G' and H be the poles of the planes μ, μ' and ϖ, respectively. As we have seen, the equations of the planes σ, μ and ϖ are, respectively,

$$(l - a^{\frac{1}{2}})x + \ldots + t = 0, \quad (l - a^{\frac{1}{2}})x + \ldots - t = 0, \quad t = 0.$$

It can be shewn:

(*a*), that the planes μ, μ' meet in a line lying on the polar plane of O; the plane, $a^{\frac{1}{2}}x + b^{\frac{1}{2}}y + c^{\frac{1}{2}}z = 0$, which joins O to this line, contains also the line of intersection of the planes σ, σ'.

(*b*), that the two quadric cones which contain the sections by the planes μ and ϖ have their vertices at S, and at the point where OS' meets the plane σ. The former is clear because AL, BM, CN meet in S. For the latter, since O, the point of intersection of AA', BB', CC', is the vertex of one of the cones containing the sections μ, μ', and S' is the vertex of one of the cones containing the sections μ', ϖ, therefore the line OS' contains the vertex of one of the cones containing the sections μ, ϖ, by what we have seen above; and OS' does not contain S. The vertex of the second cone containing μ and ϖ is therefore on OS'. That this second vertex lies on the polar plane, σ, of the vertex, S, of the former cone, is a general property. And the planes μ, ϖ meet on σ, which is harmonically separated from S by these two planes.

(*c*), that the lines OS, OS' lie in a plane containing the lines OG, OH, and are harmonic conjugates in regard to these. This follows from (*b*), because the vertices of the two cones containing the sections μ, ϖ lie on the line GH, and are harmonically conjugate in regard to G and H. To point out the meaning of this result, consider the quadric as a sphere of which O is the centre; then the Hart circle of a spherical triangle, ABC, meets BC in L, L'; CA in M, M'; AB in N, N', such that the arcs AL, BM, CN meet in a point S, and the arcs AL', BM', CN' in a point S', these points, S and S', being the centres of similitude of the circle, (μ), circumscribed to ABC, and the Hart circle (ϖ).

It may also be remarked that, from any point, P, and four given points O, X, Y, Z, we can define a plane thus: Let OP, XP, YP, ZP meet the planes XYZ, OYZ, OZX, OXY, respectively, in O', X', Y', Z'. The planes $O'Y'Z'$, $O'Z'X'$, $O'X'Y'$, $X'Y'Z'$, by their intersections with the planes OYZ, OZX, OXY, XYZ, respectively, determine four lines. These four lines lie in the plane to be defined. If, relatively to the points O, X, Y, Z the point P be (ξ, η, ζ, τ), the plane is $x\xi^{-1} + y\eta^{-1} + z\zeta^{-1} + t\tau^{-1} = 0$. This being understood, it is clear that the point S, and its polar plane σ in regard to the quadric, are in this relation with respect to the points O, X, Y, Z, determined by the planes of the three original sections α, β, γ and of the Hart section. It is easy to see that in general there are eight points whose polar planes, in regard to a given quadric, are also derivable by this construction, from four given points.

Ex. 4. There are, we have said, eight sections, ϖ, each touched by four of the common tangent planes of the sections α, β, γ. These eight sections fall into four pairs, the planes of a pair intersecting on the polar plane of O (where the planes of α, β, γ meet), being harmonic conjugates in regard to this plane and O. For the pair associated as above with the two planes ABC, $A'B'C'$ the lines OS, OS' are the same. Another pair is associated with the planes $AB'C'$, $A'BC$; a third pair with the planes $BC'A'$, $B'CA$, and a fourth pair with the planes $CA'B'$, $C'AB$.

And it may be noticed that the sections of the quadric by the planes ABC, $AB'C'$, $BC'A'$, $CA'B'$ are all touched by four planes, as follows from the fact that the lines AA', BB', CC' are concurrent. So also the sections of the quadric by the planes $A'B'C'$, $A'BC$, $B'CA$, $C'AB$ are all touched by four planes.

Ex. 5. Given two points, A, B, of a quadric, and a point O, we can define, for the moment, as *mid-axis* of AB in respect to O, the line joining O to the pole, in regard to the section OAB, of the line AB. Then, given three points, A, B, C, of the quadric, let d, e, f denote the mid-axes, in respect to O, respectively of BC, CA and AB. Let p be the line of intersection of the planes ef and BOC; similarly let q and r be the lines (fd, COA), (de, AOB), respectively. It can be shewn that the lines p, q, r lie in a plane meeting the plane ABC on the polar plane of O.

If the quadric be $(a,\ b,\ c,\ f,\ g,\ h \Xbar x,\ y,\ z)^2 = t_1^2$, where O is $(0, 0, 0, 1)$, and A, B, C are, respectively, $(1, 0, 0, a^{\frac{1}{2}})$, $(0, 1, 0, b^{\frac{1}{2}})$, $(0, 0, 1, c^{\frac{1}{2}})$, the mid-axis, d, of BC in respect to O is $yb^{\frac{1}{2}} = zc^{\frac{1}{2}}$, the plane ef is $-xa^{\frac{1}{2}} + yb^{\frac{1}{2}} + zc^{\frac{1}{2}} = 0$, and the plane pqr is

$$xa^{\frac{1}{2}} + yb^{\frac{1}{2}} + zc^{\frac{1}{2}} = 0.$$

It is easy to shew, further, that the planes (d, OA), (e, OB), (f, OC) meet in a line. This we may call the *centroid axis* of ABC in respect to O.

Consider, next, the plane joining OA to the pole of the plane BOC, and the two other similar planes. The planes are easily seen to meet in a line, which we may call the *altitude axis* of ABC in respect to O.

It can then be proved that the line OH joining O to the pole, H, of the Hart section, whose equation has been written

$$lx + my + nz - t_1 = 0,$$

lies in the plane of the centroid axis and the altitude axis.

Ex. 6. Now consider the particular case of the preceding work which arises when the point O, in which the planes of the sections α, β, γ meet, is on the quadric, so that the points A', B', C' are also at O. Then it is easily seen that also the plane σ', the polar plane of S', and one of the planes of the harmonically conjugate quadric, is the tangent plane of the original quadric at the point O. The intersections of this plane with the lines YZ, ZX, XY of the plane $t = 0$, which were previously denoted by L_1', M_1', N_1', are the harmonic conjugates of L', M', N', respectively, in regard to Y and Z, Z and X, X and Y, the points L', M', N' being points of the Hart section of the quadric, lying respectively in YZ, ZX and XY. Hence we have the following result: Consider three sections of a quadric, α, β and γ, which intersect in a point O of the quadric. Let the lines of intersection of the pairs of planes of β, γ, of γ, α and of α, β, be respectively OX, OY, OZ. In the plane YOZ take the line, l, which is the fourth harmonic, in regard to OY, OZ, of the line in which the tangent plane of the quadric at O meets the plane YOZ. Let the line l cut the section α again in L'. Construct, similarly, the points M' and N' on the sections β and γ, respectively. Let ϖ be the section of the quadric by the plane $L'M'N'$. It can be shewn that there are four plane sections of the quadric which touch the sections α, β, γ. These all touch the section ϖ.

The point of contact with ϖ, of any one of these four tangent sections, say i, can be constructed thus: Take, upon i, its three points of contact with the sections α, β, γ; take also the two intersections of i with the tangent plane of the quadric at O. Then the point of contact of i with ϖ is the apolar complement of these two intersections in respect to the three former points of i.

If we project from O on to an arbitrary plane, the section ϖ becomes the Feuerbach circle of a triangle. This touches the inscribed circle of the triangle (as also the escribed circles) at a point determined, as above, by the points of contact of the inscribed circle with the sides of the triangle and the two Absolute points on the inscribed circle. (See Vol. II, p. 114. This particular case of the general theorem of the text had been found, unknown to the writer, by F. Morley, *American Bull.* I, 1895, pp. 116-124.)

Another construction for the plane of the section ϖ may be referred to. We have mentioned above (p. 77) a construction for the plane whose equation is $x\xi^{-1} + y\eta^{-1} + z\zeta^{-1} + t\tau^{-1} = 0$, from the point (ξ, η, ζ, τ); let this plane be called the *polar* of (ξ, η, ζ, τ) in regard to the tetrad consisting of the intersections of the planes $x = 0$, $y = 0$, $z = 0$, $t = 0$. It can be shewn that if three lines be drawn through a given point, O, of a quadric, there exist three unique points, say P, Q, R, one upon each of the lines drawn through O, of which every two are conjugate in regard to the quadric. Let the three lines be the intersections, in pairs, of the planes of the sections α, β, γ. Let S be the pole of the plane PQR in regard to the quadric. Then the plane ϖ is the polar of O in regard to the tetrad P, Q, R, S.

If the point O, relatively to the tetrad P, Q, R, S, be of co-ordinates $(1, 1, 1, 1)$, the equation of the quadric referred to this tetrad may be supposed to be $ax^2 + by^2 + cz^2 - (a + b + c)t^2 = 0$. If now, to refer the quadric to the four points consisting of O and the three points in which OP, OQ, OR are met by the polar of O in regard to P, Q, R, S, we put $X = x - t$, $Y = y - t$, $Z = z - t$, $T = x + y + z + t$, the equation is found to take the form

$$aX^2 + bY^2 + cZ^2 - (b + c)\,YZ - (c + a)\,ZX - (a + b)\,XY$$
$$+ (aX + bY + cZ)\,T = 0,$$

from which it can be verified that $T = 0$ is the plane of the section ϖ arising from the sections $X = 0$, $Y = 0$, $Z = 0$. The harmonically conjugate quadric, in regard to the last tetrad, is

$$(aX + bY + cZ)\,(X + Y + Z - T) = 0,$$

consisting of the tangent plane at O and the plane $t = 0$, or *PQR*.

Ex. 7. If the sections α, β, γ be in the planes $x = 0$, $y = 0$, $z = 0$, meeting on the quadric, the plane of the Hart section ϖ being $t = 0$, the equation of the quadric is of the form

$$ax^2 + by^2 + cz^2 - (b + c)\, yz - (c + a)\, zx - (a + b)\, xy + 2\, (x + y + z)\, t = 0,$$

the harmonically conjugate quadric being the aggregate of the planes $x + y + z = 0$, $ax + by + cz - 2t = 0$.

The four planes touching the sections by $x = 0$, $y = 0$, $z = 0$, $t = 0$ are of the form

$$x\,(l - m)\,(l - n) + y\,(m - n)\,(m - l) + z\,(n - l)\,(n - m) + t = 0,$$

where $\qquad -2l^2 = b + c, \quad -2m^2 = c + a, \quad -2n^2 = a + b.$

Further, if we introduce a_1, and the coordinates X, Y, Z, by the definitions

$$a_1^{-1} + a^{-1} + b^{-1} + c^{-1} = 0, \qquad ax + a_1 X = 0,$$
$$y = Y + b^{-1} ax, \qquad z = Z + c^{-1} ax,$$

the quadric has the equation

$$a_1 X^2 + bY^2 + cZ^2 - (b + c)\, YZ - (c + a_1)\, ZX - (a_1 + b)\, XY$$
$$+ 2\, (X + Y + Z)\, t = 0,$$

which is of the same form as the original. For the original equation the points A, L on OX, YZ, of our general theory, are $(a^{-1}, 0, 0, -\frac{1}{2})$ and $(0, b^{-1}, c^{-1}, 0)$, and the lines AL, BM, CN meet in the point $(a^{-1}, b^{-1}, c^{-1}, -\frac{1}{2})$, or S, whose polar in regard to the quadric is the plane $ax + by + cz - 2t = 0$. But the transformation here given shews that the Hart section, in the plane $t = 0$, belongs equally to three other sets of three sections of the quadric, one of these sets being those in the planes $X = 0$, $Y = 0$, $Z = 0$.

Ex. 8. The reader is probably aware that in the nomenclature usual in metrical geometry, the *radius* of the Feuerbach circle of a triad of points in a plane is one-half that of the circle containing these points. This result has also been generalised to the *radius* of the Hart circle arising for a triad of points of a sphere joined by arcs of great circles (Salmon, *Geometry of three dimensions*, 1882, p. 232). We now prove a theorem for sections of a quadric which includes these results.

For this purpose we first define the *radius* of any section of a quadric, in regard to any point not lying on the quadric. It is easy to prove for a *conic*, that if two lines intersect in a point O, not on the conic, and H be one of the points in which one of these lines meets the conic, and the tangent of the conic at H meet the other line in P, then the angle between the lines, measured in regard to the conic, say ρ, is related to the interval OP, also measured in regard to the conic, say δ, by the equation $\sin\rho\sin\delta = \pm 1$, or its equivalent, $\tan\rho = \pm i\sec\delta$ (cf. Vol. II, p. 210, Ex. 6). Now, let O be any point not on the quadric, let ϖ be any section of the quadric lying in a plane whose pole is P, and let H be any point of the section ϖ. By considering the conic in which the plane POH meets the conic, we see that if ρ is the angle between the lines OP, OH, measured in regard to this conic, and δ is the interval OP, measured in regard to the quadric, then $\tan\rho = \pm i\sec\delta$; thus ρ is independent of the position of H on the section ϖ. It is this ρ which is here called the radius of the section ϖ, in regard to the point O. Evidently the relation between O and P is symmetrical.

Now let ϖ and μ be any two sections of the quadric, whose radii, as thus defined, are, respectively, ρ and R; let the plane, joining the line of intersection, l, of the planes ϖ and μ, to the point O, be called ω; and the plane joining l to the vertex of either one of the cones containing the sections ϖ, μ be called σ. If the equations of these planes be written respectively $(\varpi) = 0$, $(\mu) = 0$, $(\omega) = 0$, $(\sigma) = 0$, it is easy to shew that, with a proper constant, c, we have

$$(\sigma) = (\mu) \pm c\,(\varpi), \quad (\omega) = (\mu)\tan R + c\,(\varpi)\tan\rho.$$

This can be applied to the case when, as before, ϖ is the Hart section of three sections, α, β, γ, of the quadric, whose planes meet in a point O not lying on the quadric. For μ we take the section by the plane ABC. The vertex of one of the cones containing the sections μ, ϖ is the point S, whose polar plane in regard to the quadric is the plane σ; it has been seen (Ex. 3) that the vertex of the other cone containing the sections μ, ϖ is on the plane σ (and on OS'). Further, we have seen that (μ), (ϖ), (σ), (ω) are, respectively,

$$(l - a^{\frac{1}{2}})x + \ldots - t = 0, \quad t = 0, \quad (l - a^{\frac{1}{2}})x + \ldots + t = 0, \quad (l - a^{\frac{1}{2}})x + \ldots = 0;$$

whence $(\sigma) = (\mu) + 2\,(\varpi)$, $(\omega) = (\mu) + (\varpi)$. Comparing this with the equations above we have $c = \pm 2$, $c\tan\rho = \tan R$. Hence we have the result $\tan R = \pm 2\tan\rho$, which we desired to obtain.

Another proof may be obtained by shewing that $\Delta\tan^2 R = U$, $\Delta\tan^2\rho = V$, where U, V are given respectively (cf. p. 73) by

$$U = Aa + Bb + Cc + 2F\,(bc)^{\frac{1}{2}} + 2G\,(ca)^{\frac{1}{2}} + 2H\,(ab)^{\frac{1}{2}} - \Delta,$$
$$V = Al^2 + Bm^2 + Cn^2 + 2Fmn + 2Gnl + 2Hlm - \Delta,$$

where $A = bc - f^2$, $F = gh - af$, etc., as before, and Δ is the discriminant $abc + 2fgh - af^2 - bg^2 - ch^2$. In virtue of $2mn = f + (bc)^{\frac{1}{2}}$, etc., it can be shewn that

$$4V = U = 2\,[(bc)^{\frac{1}{2}} - f]\,[(ca)^{\frac{1}{2}} - g]\,[(ab)^{\frac{1}{2}} - h].$$

Ex. 9. By projection of sections of a quadric, prove that, if three circles in a plane intersect in pairs respectively in A, A'; B, B' and C, C', then the circles ABC, $A'B'C'$, $AB'C'$, $A'BC$, which have a common orthogonal circle and are, in pairs, inverses of one another, are all touched by four other circles which have a common orthogonal circle; also that the circles ABC, $AB'C'$, $BC'A'$, $CA'B'$ are all touched by four other circles. (A. Larmor, *Proc. Lond. Math. Soc.*, xxiii, 1892, p. 149.)

In what has preceded we have shewn that eight sections of the quadric exist which touch three given plane sections of general position, α, β, γ. We have then shewn that there are eight ways in which we can associate with these sections α, β, γ a section, ϖ, such that four of the tangent sections of α, β, γ also touch ϖ—account being now taken only of the cases in which the planes of $\alpha, \beta, \gamma, \varpi$ do not meet in a point. We have shewn that a sufficient condition for this is that another quadric, which we have called the harmonically conjugate quadric, should break up into two planes.

We consider now the converse question *whether this sufficient condition is also necessary*, by examining the most general case in which the sections of the original quadric by the planes $x = 0$, $y = 0$, $z = 0$, $t = 0$ are all touched by four other planes.

The quadric being, in the first instance, referred to the planes of three sections α, β, γ and the polar plane of their point of intersection, let its equation be $(a_1, b_1, c_1, f_1, g_1, h_1 \backslash x_1, y_1, z_1)^2 = t_1^2$; let a fourth plane be $t = 0$, where $t = t_1 - lx_1 - my_1 - nz_1$; we shall suppose that no one of l, m, n is zero, and put $x = lx_1, y = my_1, z = nz_1$. The equation of the quadric then takes a form, $\phi(x, y, z, t) = 0$, in which $\phi = (a, b, c, d, f, g, h, u, v, w \backslash x, y, z, t)^2$, where $u = v = w = d = 1$. The condition that two plane sections of the quadric should touch one another is that, of the cones enveloping the quadric along these sections, the vertex of either should lie on the other. But the cone enveloping the quadric along the section by $t = 0$ has the equation $D\phi(x, y, z, t) = \Delta t^2$, where Δ is the symmetrical discriminantal determinant of ϕ, of four rows and columns, and D is the minor of d therein; and the enveloping cones for the sections by $x = 0$, $y = 0$, $z = 0$ have similar equations. The condition that the four sections of the quadric by the planes $x = 0$, $y = 0$, $z = 0$, $t = 0$ should have a common tangent plane is that the four enveloping cones should have a point in common; this point, which will be the pole of the common tangent plane, will thus necessarily be one of the eight points

$$\pm xA^{-\frac{1}{2}} = \pm yB^{-\frac{1}{2}} = \pm zC^{-\frac{1}{2}} = tD^{-\frac{1}{2}},$$

and the condition is that, with properly chosen signs, we should have

$$\phi(\pm A^{\frac{1}{2}}, \pm B^{\frac{1}{2}}, \pm C^{\frac{1}{2}}, D^{\frac{1}{2}}) = \Delta.$$

If the four sections by $x = 0$, $y = 0$, $z = 0$, $t = 0$ are to have four common tangent planes, this last equation must be satisfied for each of four sets of signs of the radicals. These four conditions must be satisfied in virtue of proper choice of the three quantities, l, m, n, without limitation of the values of the original coefficients a_1, b_1, c_1, f_1, g_1, h_1. We suppose that the sections of the quadric by $x = 0$, $y = 0$, $z = 0$, $t = 0$ do not degenerate, in either case, into a pair of lines; and so we exclude the possibilities

$$A = 0,\ B = 0,\ C = 0,\ D = 0.$$

By taking a particular case we can infer that if the equation $\phi(\pm A^{\frac{1}{2}}, \dots) = \Delta$ is satisfied by four sets of signs of $A^{\frac{1}{2}}$, $B^{\frac{1}{2}}$, $C^{\frac{1}{2}}$, $D^{\frac{1}{2}}$, then three of these sets must be derivable from one of them by change in the signs of two of the three $A^{\frac{1}{2}}$, $B^{\frac{1}{2}}$, $C^{\frac{1}{2}}$. We take $f_1 = 0$, $g_1 = 0$, $h_1 = 0$, leading to $f = 1$, $g = 1$, $h = 1$. Then it is easy to verify that

$$\Delta = (a + 1) A + (b + 1) B + (c + 1) C - 2D,$$

and, if we put ξ, η, ζ, τ respectively for $A^{\frac{1}{2}}$, $B^{\frac{1}{2}}$, $C^{\frac{1}{2}}$, and $-D^{\frac{1}{2}}$, the equation $\phi(\pm A^{\frac{1}{2}}, \dots) = \Delta$ becomes

$$(\xi + \eta + \zeta - \tau)^2 = 2(\xi^2 + \eta^2 + \zeta^2 - \tau^2).$$

We shew that if this equation, and three others chosen from

$$(\pm \xi \pm \eta \pm \zeta - \tau)^2 = 2(\xi^2 + \eta^2 + \zeta^2 - \tau^2),$$

are all satisfied by the same set of values for ξ, η, ζ, τ, then the three others are those in which the linear functions on the left side are $\xi - \eta - \zeta - \tau$, $-\xi + \eta - \zeta - \tau$, $-\xi - \eta + \zeta - \tau$. The equation $(\pm \xi \pm \eta \pm \zeta - \tau)^2 = 2(\xi^2 + \eta^2 + \zeta^2 - \tau^2)$ represents, if ξ, η, ζ, τ be regarded as coordinates, the enveloping cone drawn to the quadric $\xi^2 + \eta^2 + \zeta^2 - \tau^2 = 0$ from the point $(\pm 1, \pm 1, \pm 1, 1)$; there are eight such cones; it is easy to find all the intersections of every three of these cones; then it appears that, excluding intersections on $\xi = 0$, $\eta = 0$, $\zeta = 0$, $\tau = 0$, there are only two sets of four cones which have four common points, the vertices of one of these sets being the points $(1, -1, -1, 1)$, $(-1, 1, -1, 1)$, $(-1, -1, 1, 1)$, $(1, 1, 1, 1)$, the vertices of the other set being $(-1, 1, 1, 1)$, $(1, -1, 1, 1)$, $(1, 1, -1, 1)$, $(1, 1, 1, -1)$. This proves the statement made.

This result being assumed, the equation $\phi(\pm A^{\frac{1}{2}}, \dots) = \Delta$, which, writing $(BC)^{\frac{1}{2}}$ for $B^{\frac{1}{2}} . C^{\frac{1}{2}}$, etc., is

$$aA + bB + cC + dD + 2[f(BC)^{\frac{1}{2}} + u(AD)^{\frac{1}{2}}]$$
$$+ 2[g(CA)^{\frac{1}{2}} + v(BD)^{\frac{1}{2}}] + 2[h(AB)^{\frac{1}{2}} + w(CD)^{\frac{1}{2}}] - \Delta = 0$$

leads, by combination of the forms it takes when the signs of two of $A^{\frac{1}{2}}$, $B^{\frac{1}{2}}$, $C^{\frac{1}{2}}$ are changed, to the four equations

$$aA + bB + cC + dD - \Delta = 0, \qquad f(BC)^{\frac{1}{2}} + u(AD)^{\frac{1}{2}} = 0,$$
$$g(CA)^{\frac{1}{2}} + v(BD)^{\frac{1}{2}} = 0, \qquad h(AB)^{\frac{1}{2}} + w(CD)^{\frac{1}{2}} = 0.$$

The last three of these equations require that A, B, C, D should be in the ratios of fvw, gwu, huv, fgh, to one another, that is the ratios of the values which A, B, C, D would have if, in the determinant Δ, we had $a = 0$, $b = 0$, $c = 0$, $d = 0$. Conversely, if proper signs be given to $A^{\frac{1}{2}}$, $B^{\frac{1}{2}}$, $C^{\frac{1}{2}}$, $D^{\frac{1}{2}}$, the last three of these equations are a consequence of the equations $A = \sigma fvw$, $B = \sigma gwu$, $C = \sigma huv$, $D = \sigma fgh$. We require to shew that if, beside these, we also have $aA + bB + cC + dD - \Delta = 0$, then the harmonically conjugate quadric, whose equation is obtained from that of the original by the change of the signs of f, g, h, u, v, w, breaks up into two planes. It will be found that this involves $\sigma = 4$.

To carry through the algebra, suppose $u = v = w = d = 1$, which is quite general; and, for brevity, put

$$\xi = 1 - a, \ \eta = 1 - b, \ \zeta = 1 - c, \ p = 1 - f^2, \ q = 1 - g^2, \ r = 1 - h^2,$$
and also

$$\alpha = f + g + h, \quad \beta = gh + hf + fg, \quad \gamma = fgh, \quad \epsilon = \alpha - \gamma,$$
$$\lambda = \epsilon(f + f^{-1}), \quad \mu = \epsilon(g + g^{-1}), \quad \nu = \epsilon(h + h^{-1}).$$

Then it is found that $A = \eta\zeta - (1 - f)^2$, etc., and the equation $aA + bB + cC + dD - \Delta = 0$ is the same as

$$-3\xi\eta\zeta + 2\Sigma\eta\zeta - \Sigma p\xi - 2\Sigma f^2 + 2\Sigma gh = 0.$$

Hence, the four equations consisting of this and the three

$$A/fvw = \ldots = D/fgh,$$

are found, introducing a quantity θ, to be the same as the aggregate

$$\theta = \eta\zeta\epsilon f^{-1} - \lambda = \zeta\xi\epsilon g^{-1} - \mu = \xi\eta\epsilon h^{-1} - \nu = \xi\eta\zeta + \Sigma p\xi - 4,$$
$$2\epsilon\Sigma p\xi = \theta(\alpha - 3\gamma) + 6\epsilon - 2\epsilon\beta.$$

These lead to

$$(\theta + \lambda)(\theta + \mu)(\theta + \nu) = \tfrac{1}{4}\epsilon\gamma^{-1}[\theta(\alpha + \gamma) + 2\epsilon + 2\epsilon\beta]^2;$$

it is necessary to solve this equation for θ, and then, putting

$$\rho = \tfrac{1}{2}\epsilon^{-1}[\theta(\alpha + \gamma) + 2\epsilon + 2\epsilon\beta],$$

together with

$$\xi = \rho\epsilon/f(\theta + \lambda), \quad \eta = \rho\epsilon/g(\theta + \mu), \quad \zeta = \rho\epsilon/h(\theta + \nu),$$

to test whether the equation

$$\Sigma p\xi = \theta - \rho + 4$$

is satisfied. If this proves to be so, the original equations can be satisfied.

In fact, the three roots of the cubic equation are $\theta = 2\epsilon$, $\theta = -2\epsilon$, $\theta = \epsilon\gamma^{-1}(\tfrac{1}{4}\epsilon^2 - \beta)$; these lead, respectively, to the three values for ξ,

$$\xi = (1+g)(1+h)(1+f)^{-1}, \quad \xi = (1-g)(1-h)(1-f)^{-1},$$
$$\xi = 4\rho\gamma f^{-1}(\epsilon - 2f)^{-2},$$

with corresponding values of η and ζ. The first and second of these roots lead to verification of the equation $\Sigma p\xi = \theta - \rho + 4$, but the second leads to $A = 0$, $B = 0$, $C = 0$, which we exclude. The third root leads to verification of the required equation only if f, g, h be such as to satisfy

$$\Sigma gh(1-f^2)(\epsilon - 2g)^2(\epsilon - 2h)^2$$
$$= 16\left[\gamma + \tfrac{1}{8}\epsilon^2(\alpha + \gamma) - \tfrac{1}{2}\epsilon\beta\right]\left[3\gamma + \tfrac{1}{8}\epsilon^2(\alpha - 3\gamma) - \tfrac{1}{2}\epsilon\beta\right];$$

this however is not generally true, as we may see, for example, by taking $f = 1$, $g = 2$, $h = 3$, which lead to a negative value for the left side, and a positive value for the right side.

There remains thus only the possibility of the first root, $\theta = 2\epsilon$. We compute then the harmonically conjugate quadric, and find that it consists of the two planes expressed, if $m = (1+f)(1+g)(1+h)$, by

$$m\left[x(1+f)^{-1} + y(1+g)^{-1} + z(1+h)^{-1}\right]^2 = (x+y+z-t)^2.$$

Further, retaining $u = v = w = d = 1$, the four planes touching the sections of the quadric by $x = 0$, $y = 0$, $z = 0$, $t = 0$, are the polars of the points $(f^{\frac{1}{2}}, g^{\frac{1}{2}}, h^{\frac{1}{2}}, -f^{\frac{1}{2}}g^{\frac{1}{2}}h^{\frac{1}{2}})$, in which all possible signs are taken for $f^{\frac{1}{2}}, g^{\frac{1}{2}}, h^{\frac{1}{2}}$. This completes the proof that the conditions found sufficient in the early part of this chapter are also necessary.

Ex. 1. In terms of four quantities a, b, c, d let $f = bc - ad$, $g = ca - bd$, $h = ab - cd$, $p = bc + ad$, $q = ca + bd$, $r = ab + cd$; also

$$P = -a + b + c + d, \qquad Q = a - b + c + d,$$
$$R = a + b - c + d, \qquad S = a + b + c - d.$$

Prove that, if the quadric have the equation

$$p^{-1}qr\,x^2 + q^{-1}rp\,y^2 + r^{-1}pq\,z^2 + 2fyz + 2gzx + 2hxy = t_1{}^2,$$

where $t_1 = 0$ is the polar plane of $(0, 0, 0, 1)$, then one Hart section for the three sections $x = 0$, $y = 0$, $z = 0$, or α, β, γ, is given by $ax + by + cz - t_1 = 0$; also, that the four planes touching α, β, γ and this Hart section are the polars of the points $(-1, 1, 1, P)$, $(1, -1, 1, Q)$, $(1, 1, -1, R)$, $(1, 1, 1, S)$. Prove also that the change of a, b, c, d, respectively, into $-d$, c, b, $-a$, leaves the equation of the quadric unaltered. This change consists of the interchange of b and c, taken with the interchange, and simultaneous change of sign, of a and d. Similarly, the equation of the quadric is unaltered by the change of a, b, c, d respectively into c, $-d$, a, $-b$; or into b, a, $-d$, $-c$. The equation is also unaltered by the change of sign of t_1. Hence obtain all the eight Hart sections, and, in each case, the four planes touching this and the sections α, β, γ. The form of the equation of the quadric here taken was suggested by Dr G. T. Bennett.

Ex. 2. If the quadric be $fyz + gzx + hxy + t(ux + vy + wz) = 0$, the sections by $x = 0$, $y = 0$, $z = 0$, $t = 0$ are touched by the plane $l^{-1}x + m^{-1}y + n^{-1}z + p^{-1}t = 0$, where

$$l = \epsilon_1 (fu)^{\frac{1}{2}}, \quad m = \epsilon_2 (gu)^{\frac{1}{2}}, \quad n = \epsilon_3 (hu)^{\frac{1}{2}},$$

in which $\epsilon_1 = \pm 1$, $\epsilon_2 = \pm 1$, $\epsilon_3 = \epsilon_1 \epsilon_2$, and $p = lmn/uvw$, provided that $l^{\frac{1}{2}} + m^{\frac{1}{2}} + n^{\frac{1}{2}} = 0$. (Cf. Vol. III, p. 145.)

Ex. 3. For the quadric $(a, b, c, d, f, g, h, u, v, w \mathop{\rangle} x, y, z, t)^2 = 0$, the pole of a plane touching the three sections by $x = 0, y = 0, z = 0$, is of coordinates

$$A^{\frac{1}{2}}, B^{\frac{1}{2}}, C^{\frac{1}{2}}, \quad d^{-1}[M^{\frac{1}{2}} - uA^{\frac{1}{2}} - vB^{\frac{1}{2}} - wC^{\frac{1}{2}}],$$

where M is such that

$$d^{-1}[(uA^{\frac{1}{2}} + vB^{\frac{1}{2}} + wC^{\frac{1}{2}})^2 - M] = (a, b, c, f, g, h \mathop{\rangle} A^{\frac{1}{2}}, B^{\frac{1}{2}}, C^{\frac{1}{2}})^2 - \Delta.$$

Ex. 4. As the condition that two plane sections of a quadric should touch one another is symmetrical, there are symmetrical relations between the planes of four sections which are all touched by four other planes, and these latter planes. It is of interest then to construct the relations of the two tetrads of planes. The following result, in this regard, is due to Study, *Math. Annal.*, XLIX, 1897, pp. 497 ff.

As before, let the planes joining the points O, X, Y, Z meet the quadric in sections all touched by four other planes; let the points of the quadric on the lines OX, OY, OZ, YZ, ZX, XY be, respectively, A, A'; B, B'; C, C'; L, L'; M, M'; N, N', the notation being such that the lines AL, BM, CN meet in a point, S, and $A'L'$, $B'M'$, $C'N'$ in a point, S'. Let the planes $BMNC$, $CNLA$, $ALMB$, $B'M'N'C'$, $C'N'L'A'$, $A'L'M'B'$ be denoted, respectively, by θ, ϕ, ψ, θ', ϕ', ψ'; let the points in which the line (ϕ, ψ') meets the quadric be denoted by P_{23}, Q_{23}, those in which the line (ϕ', ψ) meets the quadric being denoted by P_{32}, Q_{32}; a similar notation being used for the other cases, namely P_{31}, Q_{31} on the line (ψ, θ'), P_{13}, Q_{13} on (ψ', θ), P_{12}, Q_{12} on (θ, ϕ'), P_{21}, Q_{21} on (θ', ϕ). Consider now four planes touching the sections by the planes XYZ, OYZ, etc.; take the intersections in threes of these four planes, and the polar planes of these four points of intersection; let these polar planes be ξ, η, ζ, τ, so that these intersect in threes in the poles of the four tangent planes.

Then the twelve points P_{ij}, Q_{ij}, R_{ij} described above are in fact the intersections of the quadric with the lines of intersection of the pairs of the planes ξ, η, ζ, τ. With a proper choice of notation the line (η, ζ) is the same as the line (P_{23}, P_{32}), and the line (ξ, τ) the same as the line (Q_{23}, Q_{32}), and so on; the sets of six of these points lying on the planes being, respectively, given by

$$\xi\,(Q_{23}, Q_{32}, P_{31}, P_{13}, P_{12}, P_{21});\quad \eta\,(Q_{31}, Q_{13}, P_{12}, P_{21}, P_{23}, P_{32});$$
$$\zeta\,(Q_{12}, Q_{21}, P_{23}, P_{32}, P_{31}, P_{13});\quad \tau\,(Q_{23}, Q_{32}, Q_{31}, Q_{13}, Q_{12}, Q_{21}).$$

As we have not distinguished geometrically between the points P_{23}, Q_{23} of the line (ϕ, ψ'), etc., each pair of opposite lines of intersection of the planes ξ, η, ζ, τ is capable of two determinations; there are therefore eight possibilities for this set of four planes, as we have seen.

These statements can be verified easily. We can suppose the quadric to have the equation

$$x^2\,(1-a^2)+y^2\,(1-b^2)+z^2\,(1-c^2)+2yz\,(bc-1)+2zx\,(ca-1)$$
$$+\,2xy\,(ab-1)+2\,(x+y+z)\,t+t^2=0,$$

the harmonically conjugate quadric being then

$$(x+y+z-t)^2-(ax+by+cz)^2=0.$$

Then, with $f=bc-1$, $g=ca-1$, $h=ab-1$, if we put $x_1=xf^{-\frac{1}{2}}$, $y_1=yg^{-\frac{1}{2}}$, $z_1=zh^{-\frac{1}{2}}$, $t_1=t\,(fgh)^{-\frac{1}{2}}$, the poles of four planes which touch the sections by $x=0$, $y=0$, $z=0$, $t=0$, have, by what we have seen, for coordinates x_1, y_1, z_1, t_1, respectively, $(-1, 1, 1, 1)$, $(1, -1, 1, 1)$, $(1, 1, -1, 1)$, $(1, 1, 1, -1)$; thus the planes ξ, η, ζ, τ have, respectively, the equations

$$-x_1+y_1+z_1+t_1=0,\quad x_1-y_1+z_1+t_1=0,\quad x_1+y_1-z_1+t_1=0$$

and $x_1+y_1+z_1-t_1=0$. The points A, A', L, L' are given, respectively, by

$$y=0=z,\ x\,(1-a)+t=0;\quad y=0=z,\ x\,(1+a)+t=0;$$
$$x=0=t,\ y\,(1-b)=z\,(1-c);\quad x=0=t,\ y\,(1+b)=z\,(1+c),$$

and so on; and the plane $BMNC$, or θ, is given by

$$-x\,(1-a)+y\,(1-b)+z\,(1-c)+t=0,$$

and so on. Now put

$$\rho_{23}=-\,(g^{\frac{1}{2}}-h^{\frac{1}{2}})\,a^{-1}f^{-\frac{1}{2}},\quad \sigma_{23}=-\,(g^{\frac{1}{2}}+h^{\frac{1}{2}})\,a^{-1}f^{-\frac{1}{2}},$$

with similar notations; then we easily verify that the coordinates, (x_1, y_1, z_1, t_1), of the twelve points P_{ij}, etc., are given, respectively, by the scheme

$P_{23}\,(\rho_{23}, 1, 1, -\rho_{23})$, $Q_{23}\,(\sigma_{23}, 1, -1, \sigma_{23})$, $P_{32}\,(-\rho_{23}, 1, 1, \rho_{23})$, $Q_{32}\,(\sigma_{23}, -1, 1, \sigma_{23})$,
$P_{31}\,(1, \rho_{31}, 1, -\rho_{31})$, $Q_{31}\,(-1, \sigma_{31}, 1, \sigma_{31})$, $P_{13}\,(1, -\rho_{31}, 1, \rho_{31})$, $Q_{13}\,(1, \sigma_{31}, -1, \sigma_{31})$,
$P_{12}\,(1, 1, \rho_{12}, -\rho_{12})$, $Q_{12}\,(1, -1, \sigma_{12}, \sigma_{12})$, $P_{21}\,(1, 1, -\rho_{12}, \rho_{12})$, $Q_{21}\,(-1, 1, \sigma_{12}, \sigma_{12})$.

From these the statements made easily follow. It can be verified also that the pole of the plane θ lies on the line $\phi'\psi'$, etc.

Ex. 5. With the equation of the quadric taken in Ex. 4, prove that the harmonically conjugate quadric of the given one, in regard to the four planes touching the sections by $x = 0$, $y = 0$, $z = 0$, $t = 0$, consists of the two planes which are the polars, in regard to the original quadric, of the points whose coordinates (x, y, z, t) are

$$f[a(m+\sigma)+1], \quad g[b(m+\sigma)+1], \quad h[c(m+\sigma)+1], \quad -fgh,$$

where $m = \frac{1}{2}(abc - a - b - c)$, $\sigma^2 = m^2 - 1$. The discriminant, Δ, of the equation of the original quadric is $16\sigma^2$.

Ex. 6. As before, let the sections of a quadric by four planes OYZ, OZX, OXY, XYZ be touched by four planes. Let the intervals OX, OY, OZ, measured in regard to the quadric, be denoted by $i\alpha$, $i\beta$, $i\gamma$, and the similar intervals YZ, ZX, XY be denoted by $i\alpha'$, $i\beta'$, $i\gamma'$ (cf. Vol. ii, pp. 179, 205). Similarly, let the angular intervals of the pairs of planes meeting, respectively, in OX, OY, OZ, be denoted by (A), (B), (C), and those between the plane XYZ and the planes OYZ, OZX, OXY, respectively, by (A'), (B'), (C'). Taking ϵ such that

$$2\tanh\epsilon = \tanh\alpha\tanh\beta\tanh\gamma - \tanh\alpha - \tanh\beta - \tanh\gamma,$$

prove that, with suitable regard to the ambiguities involved in the definitions of these intervals,

$$\alpha' = i\pi + \beta - \gamma, \quad \beta' = i\pi + \gamma - \alpha, \quad \gamma' = i\pi + \alpha - \beta,$$

$$\tan(A) = \sinh\alpha/\cosh(\epsilon+\alpha), \tan(A') = \sinh(\beta-\gamma)/\cosh(\epsilon+\beta+\gamma),$$

and so on. These lead to

$$(A') = (B) - (C), \quad (B') = (C) - (A), \quad (C') = (A) - (B).$$

Prove that if three circles in a plane, α, β, γ, intersect in pairs at angles (A), (B), (C), and ϖ be the circle which intersects them at angles $(B) - (C)$, $(C) - (A)$, $(A) - (B)$, respectively, then a circle can be drawn to touch α, β, γ, ϖ. (Lachlan's *Modern Geometry*, 1893, p. 250.)

CHAPTER III

THE PLANE QUARTIC CURVE WITH TWO DOUBLE POINTS

In this chapter we obtain some properties of a plane curve by projection of a curve which lies in space of three dimensions. The plane curve is one which meets an arbitrary line in four points, and has two double points, or points where the curve crosses itself. The curve in space is the curve of intersection of two quadric surfaces. The matter is dealt with in more detail than is required by its difficulty, because the theory is a model for the subsequent theory of the Cyclide, a quartic surface in three dimensions, regarded as the projection of the intersection of two quadrics of fourfold space. (Chap. vi, below.)

The generation of the curve. The four principal circles. Consider two quadrics which have a common self-polar tetrad. They intersect in a quartic curve, through which there pass four quadric cones. Let V be the vertex of one of these cones, and Ω be one of the quadrics; the curve may be defined by the intersection of the cone and the quadric. Taking upon Ω an arbitrary point, U, we project the curve from U upon an arbitrary plane, ϖ. As each of the generators at U, of the quadric Ω, meets the cone of vertex V in two points, we obtain in ϖ a quartic curve having a double point at each of the points, say I and J, in which these generators meet ϖ. Conversely, any plane quartic curve having double points at the intersections of the line $z = 0$ with the two lines $x \pm iy = 0$ has an equation of the form

$$(x^2 + y^2)^2 + (x^2 + y^2) z (lx + my + nz)$$
$$+ z^2 (ax^2 + by^2 + cz^2 + 2fyz + 2gzx + 2hxy) = 0,$$

or say $(x^2 + y^2)^2 + (x^2 + y^2) zP + z^2 S = 0$, as we may assume. This curve is the projection, from the point $(0, 0, 0, 1)$, of the intersection of the two quadrics $zt = x^2 + y^2$, $t^2 + tP + S = 0$.

Now consider a section of the quadric Ω by the polar plane of V, in regard to Ω. We call this a *principal* section, and its projection upon ϖ a *principal circle*; the centre of this circle is the projection of V (above, p. 7). There are four quadric cones through the curve on Ω; there are thus four principal circles in the plane ϖ; the polar plane of V, in regard to Ω, passes through the vertices of the other three cones, the four vertices forming a self-polar tetrad. Thus (above, p. 8) every principal circle cuts the other three principal circles at right angles. If V_1, V_2, V_3 be the vertices of the

other three cones, the line VV_1 is the polar line of V_2V_3 in regard
to Ω, and the planes UVV_1, UV_2V_3 are conjugate, and are thus
harmonic in regard to the generators at U. Thus the centres of the
four principal circles, in the plane ϖ, are such that the join of any
two of them is at right angles, in regard to I, J, to the join of the
other two; or, any one of these centres is the orthocentre of the
other three.

The curve as an envelope of circles. The director conic.
Next, consider a section of Ω by a tangent plane of the cone of
vertex V. It projects into a circle cutting at right angles the
principal circle corresponding to V. The locus of the poles, in
regard to Ω, of the tangent planes of the cone (V), is a conic, say
v, lying on the polar plane of V, this being the polar reciprocal of
the cone (V), in regard to Ω. This conic, v, projects into a conic
in the plane ϖ, say σ. The circles, obtained by projecting the
sections of Ω by the tangent planes of the cone (V), thus have their
centres on a conic, σ, beside cutting at right angles the corresponding principal circle. We may call the conic σ a *director* conic.
Every one of the circles, arising from the tangent planes of (V),
has in fact two points of contact with the quartic curve of the plane
ϖ. For, if H' be an intersection, of the quartic curve on Ω, with
the section of Ω by a tangent plane of (V), then H' is on (V), as
being on the quartic curve, and is therefore at a point where the
tangent plane touches (V); the line of intersection of the tangent
plane of Ω at H' with the tangent plane of (V) at this point, is
thus both the tangent line of the quartic curve and of the section
of Ω by the tangent plane of (V). These two curves thus touch at
this point, and the curves in ϖ obtained by the projection of these
equally touch. The line VH' meets the quadric Ω in a further
point, which also projects into a point of contact of the plane
quartic curve with the same circle. The total intersections of the
circle with the plane quartic curve count therefore for eight, being
two at each point of contact, and two at each of the double
points I, J.

    Ex. 1. The facts will perhaps be clearer if we consider the
equations of a conic and a circle, in a plane, say, respectively,
$a^{-2}x^2 + b^{-2}y^2 = z^2$, and $(x - hz)^2 + (y - kz)^2 = r^2z^2$. A circle, with
centre at the point $(a\cos\theta, b\sin\theta, 1)$, of the former, which cuts
the latter at right angles, is given by the equation (cf. Vol. II,
pp. 110, 111)

$$(x - a\cos\theta \cdot z)^2 + (y - b\sin\theta \cdot z)^2$$
$$= [(h - a\cos\theta)^2 + (k - b\sin\theta)^2 - r^2]\, z^2,$$

or $P\cos\theta + Q\sin\theta = R$, where $P = 2az\,(x - hz)$, $Q = 2bz\,(y - kz)$,
$R = x^2 + y^2 - z^2\,(h^2 + k^2 - r^2)$.

This circle then touches the curve expressed by $P^2 + Q^2 = R^2$, as is easy to verify by combining the equations of the circle and curve, namely at the points where the circle is met by the line given by $P \sin \theta - Q \cos \theta = 0$. The equation of the curve is

$$[x^2 + y^2 - z^2(h^2 + k^2 - r^2)]^2 - 4z^2[a^2(x - hz)^2 + b^2(y - kz)^2] = 0,$$

and the points of contact are given by

$$a(x - hz) = \lambda z \cos \theta, \quad b(y - kz) = \lambda z \sin \theta,$$

where λ is either root of the equation

$$\lambda^2(a^{-2}\cos^2\theta + b^{-2}\sin^2\theta) + 2\lambda(a^{-1}h\cos\theta + b^{-1}k\sin\theta - 1) + r^2 = 0.$$

Cassini's oval is obtained when $h = k = 0$, $r^2 = a^2 + b^2$.

Ex. 2. Prove that the four points of contact, with the quartic curve in ϖ, of any two of the tangent circles which have their centres on the same director conic, lie upon a circle, likewise cutting the associated principal circle at right angles.

The four director conics are confocal. The quartic curve in the plane ϖ is thus the envelope of the circles described, with centres on the conic σ, to cut the corresponding principal circle at right angles. (More generally, it appeared, Vol. III, p. 200, that a general quartic curve in a plane may be regarded as the envelope of a system of conics, in various ways. In that general case, each of the conics had four points of contact with the quartic curve; in the present case, the circles each touch the quartic curve in two points, beside passing through the double points.) It is clear that there are three other ways in which the quartic curve in the plane ϖ may be regarded as the envelope of a system of circles, each system consisting of circles cutting a principal circle at right angles, with their centres on a certain conic. Let these other conics, the projections from U of the conics which are the polar reciprocals, in regard to Ω, of the four quadric cones containing the quartic curve on Ω, be called $\sigma_1, \sigma_2, \sigma_3$. It can be shewn that the four conics $\sigma, \sigma_1, \sigma_2, \sigma_3$ have four common tangents, passing in pairs through the points I and J; or, as we may say, that these four conics are confocals. To prove this, it is only necessary to shew that there are four planes through U, two through each of the generators of Ω at U, which touch the four conics in the principal planes, of which $\sigma, \sigma_1, \sigma_2, \sigma_3$ are the projections; or, reciprocating in regard to Ω, that there are two points, on each of the generators of Ω at U, which lie on all the four cones such as (V). The two points in which a generator at U meets the cone (V), lying as they do on the quartic curve upon Ω, lie, however, equally on the other three cones which contain this curve. The four points in question are thus the intersections of the generators of Ω at U with the cone (V); and the four common tangents of the conics $\sigma, \sigma_1, \sigma_2, \sigma_3$ are the intersections with ϖ of the tangent planes of Ω at these

four points, these tangent planes being the polars of their points
of contact. Hence it is clear also that the common tangent lines of
the conics σ, σ_1, σ_2, σ_3 are, in pairs, the tangent lines, of the
quartic curve in ϖ, at the points I and J. The *foci* of the confocals
σ, σ_1, σ_2, σ_3 are the intersections, other than at I and J, of their four
common tangent lines. They are, then, the intersections, with ϖ,
of the lines of intersection, other than the generators of Ω at U,
of pairs of the tangent planes of Ω, at the four points where the
cone (V) is met by the generators of Ω at U. These lines of intersection are polar lines, in regard to Ω, of joins of pairs of these
four points.

The four director conics σ, σ_1, σ_2, σ_3 are the projections, from U,
of four conics, v, v_1, v_2, v_3, which are the polar reciprocals, in
regard to Ω, of the four cones (V), (V_1), (V_2), (V_3), with the property
that a common point of any two of these cones lies on the others.
The conics v, v_1, v_2, v_3 are thus such that a common tangent plane
of any two of them is equally a tangent plane of the other two,
and they form such a set as those called the focal conics of a set of
confocal quadrics (Vol. III, p. 94). It is a familiar fact that of the
common tangent planes of such a system of conics there are four
which pass through any point (such as U).

The *centre* of the conic σ, namely the pole of the line IJ in
regard thereto, is the projection from U, of the pole, in regard to
the conic v, (in the principal plane; of which σ is the projection),
of the line in which this plane is met by the tangent plane of Ω at
U. As this principal plane is the polar plane of V, this line of
meeting is the polar line of UV in regard to Ω; and the conic v is
the polar reciprocal of the cone (V). The point in the plane of v,
which projects into the centre of the conic σ, is thus the pole, in
regard to Ω, of the polar plane of U taken in regard to the cone
(V). If the polar plane of U, in regard to the cone (V), meet the
tangent plane of Ω, at U, in the line l, the pole, in regard to Ω, of
this polar plane, lies on the line, λ, which is the polar line of l in
regard to Ω, and this passes through U. The centre of the conic σ
is the intersection of λ with the plane ϖ. In fact the polar planes
of a point, in regard to all quadrics having a quartic curve in
common, intersect in a line. Thus the line l, and also the line λ,
equally arise from any of the cones (V_1), (V_2), (V_3), or the conics
σ, σ_1, σ_2, σ_3 have, as we know, a common centre. It is easy to see
also that the foci lie in pairs on lines through this centre; as we
also know.

**The sixteen foci of the quartic curve. Are intersections
of a principal circle and a director conic.** In general for an
algebraic curve in a plane, considered in relation to two Absolute
points, I and J, a point which is the intersection of a tangent of

the curve drawn from I, with a tangent of the curve drawn from J, is called a *focus* of the curve. For a curve of class m, that is, of which m tangents pass through a general arbitrary point of the plane, there will then be m^2 foci; unless the curve contain one or other of the points I, J. For a quartic curve with two double points it can be shewn, and will appear below, that four tangents can be drawn to the curve from either double point, other than the two tangents of the curve at each of these points. There will therefore be sixteen foci. We prove that these are the intersections of the principal circles, each with its corresponding director conic.

We may first verify that such an intersection is a focus, namely that the lines which join such a point to I and J are both tangents of the plane quartic curve. Such a point is the projection of a point, say F, in which a principal section of the quadric Ω, say by the polar plane of V, is met by the conic, v, in this plane, which is the polar reciprocal, in regard to Ω, of the cone (V). The tangent plane of Ω at the point F, being the polar of F in regard to Ω, is a plane passing through F which touches the cone (V); namely, F is the point of contact with Ω of a common tangent plane of the cone (V), and the enveloping cone of Ω drawn from V. The two generating lines of Ω at this point F, lying in the tangent plane of Ω at this point, will both touch the cone (V), and the points where these touch this cone, as they lie both on Ω and on (V), are points of the quartic curve on Ω; moreover, as these lines touch (V), they equally touch this quartic curve, at the points where they touch (V). These generating lines thus project, from the point U of Ω, into two lines which are tangents of the plane quartic curve in ϖ. These generating lines, however, intersect the generators of Ω at U. Thus they project into lines passing, one through I and one through J. This shews that an intersection of a principal circle with a director conic is a focus of the plane quartic curve, as has been said. It may be called a point-circle of the enveloping system.

To shew that all the foci of the plane quartic curve are thus obtained, it will be sufficient to shew that there are four tangents of the curve which pass through either I or J. Now a tangent through I, of the plane curve, is the intersection with ϖ of a plane through the generator UI, of Ω, which contains also a point where it touches the quartic curve on Ω, this being the point of Ω which projects into the point of contact on the plane curve; the line UI is a chord of the quartic curve on Ω, meeting this curve in the two points where it meets the cone (V). It will be enough then to shew that, through this chord of the quartic curve on Ω, there can be drawn four planes of which each, in place of having two other points of intersection with the curve, has a point of contact with it, that is, contains a tangent line of the curve. Let such a plane through UI touch the curve in H; as the tangent line of the curve at H meets UI, and touches Ω at H, it is a generator of Ω; as this tangent line touches the cone (V) at H, a tangent plane of (V) can be drawn through it, which, as containing the generator of Ω constituted by this tangent line, is also a tangent plane of Ω. Thus the points such as H arise from common tangent planes of the cone (V) and the quadric Ω, which are four in number. They have already been considered; and we see that the foci already found are all that exist. There are (Vol. III, p. 69) four generating lines of Ω, of either system, which touch the quartic curve on Ω. A plane through one of these eight lines, which touches one of the quadric cones containing the curve, will contain another of the lines; and will touch Ω. The sixteen foci are the projections of the intersections of these eight generators. Incidentally we see that to a general cubic curve in a plane there can be drawn four tangents from any point of the curve, forming a pencil related to the four from any other point.

The equation of the general tangent plane of the cone (V), as for a conic, may be taken in the form $\theta^2 P + 2\theta Q + R = 0$, where $P = 0$, $R = 0$ are two arbitrary tangent planes, and $Q = 0$ is the plane containing the two generators along which these touch the cone. In particular, we may take, for $P = 0$, $R = 0$, two tangent planes of (V) which also touch Ω. There will then be two values of θ, beside $\theta = 0$ and $\theta = \infty$, for which the tangent plane of the cone also touches Ω. If one of these be θ_1, and the corresponding plane be $T = 0$, we have an identity of the form $\theta_1^2 P + 2\theta_1 Q + R = \rho^2 T$, where ρ is a constant. If, in this identity, we suppose the coordinates to be those of a point on the cone (V), on which $Q = (RP)^{\frac{1}{2}}$, we infer $\theta_1 P^{\frac{1}{2}} + R^{\frac{1}{2}} = \rho T^{\frac{1}{2}}$, an equation true for points of (V), in which each of $P = 0$, $R = 0$, $T = 0$ represents a plane touching both the cone (V) and the quadric Ω.

If we suppose the quadric Ω referred to four points of which one is the point from which the figure is projected on to the plane ϖ, taken as $(0, 0, 0, 1)$, another is the point $(0, 0, 1, 0)$, and the two others are the intersections of the generators at the former point with the generators at the latter, we may suppose the equation to be $x^2 + y^2 - 2zt = 0$. Now, we easily verify the identity

$$(xz' - x'z)^2 + (yz' - y'z)^2 + 2zz'(xx' + yy' - zt' - z't) = z^2(x'^2 + y'^2 - 2z't') + z'^2(x^2 + y^2 - 2zt)$$

wherefore, if $T' = 0$, $T'' = 0$ be the equations of the tangent planes at two points of the quadric, (x', y', z', t') and (x'', y'', z'', t''), we may suppose, at an intersection of the tangent planes with the quadric, that $T'/T'' = (D')^2/(D'')^2$, where

$$(D')^2 = [(xz' - x'z)^2 + (yz' - y'z)^2]/z^2 z'^2.$$

Thus (cf. Vol. II, p. 184), we can infer from what is proved above that *the distances of any point of the quartic curve in the plane ϖ, from three of its foci lying on the same principal circle, are connected by a linear equation.*

Construction of the plane quartic curve given four foci

lying on a circle. If the Absolute points, I, J, and also four points, be prescribed, the six points lying on a conic (circle), two of the plane quartic curves, with double points at I and J, can be found, having the four points as foci, to pass through a further given point. And these cut one another at right angles at this point; that is, the tangents of the curves at this point meet the line IJ in points harmonically conjugate in regard to I and J. That a finite number of such curves is to be expected appears by counting constants: That a plane curve should have a given point as double point requires three linear conditions for the coefficients in its equation; that the quartic should have a given focus, or should touch two given lines, requires two conditions, and in the present case there is a reduction because the four foci are not independent points. The number of prescribed conditions for a quartic curve through a given point is thus $6 + 7 + 1$, or fourteen; while the number of available coefficients in the equation of a plane curve of order n, or $\frac{1}{2}n(n+3)$, is also fourteen when $n = 4$.

Let P, Q, R, S denote the given foci, and O the given point, all in the plane ϖ. Take an arbitrary point U, not in this plane ϖ, and let Ω be a quadric containing the lines UI, UJ (other than the two planes IUJ, ϖ). Let the lines UP, UQ, UR, US, UO meet this quadric respectively in the points P', Q', R', S', O'. The section of Ω by the plane P', Q', R' projects from U, on to ϖ, into the circle PQR, which, by hypothesis, contains S; thus the plane $P'Q'R'$ contains S'. The plane $P'Q'R'S'$ is to be a principal plane, in our previous notation. Let its pole, in regard to Ω, be V. The tangent planes of Ω at P', Q', R', S' pass through V; it is possible to describe a quadric cone, with vertex at V, to touch these four planes, and have the line VO' for generator. In fact, two such cones can be described, and their tangent planes along VO' are harmonic conjugates in regard to the two tangent planes drawn to Ω from VO' (Vol. II, p. 25). Either of these cones meets Ω in such a quartic curve as we have considered; and the tangent line of this curve at O' lies on the tangent plane of the cone at O', which contains the line $O'V$, being the intersection of this plane with the tangent plane of Ω at O'; also, the two tangent planes of the quadric Ω, which can be drawn through $O'V$, contain the two generators of Ω at O'. There is thus, on the tangent plane of Ω at O', a harmonic pencil consisting of the two generators of Ω at O', and the two tangent lines of the two quartic curves on Ω obtained by the two cones (V); for we have seen that the tangent planes to Ω from $O'V$ are harmonic in regard to the two tangent planes of these cones along $O'V$. If now the curves on Ω are projected into two quartic curves on ϖ, this pencil becomes that formed by the tangents at O, of two plane quartics through O, and the lines OI, OJ.

We have thus constructed two quartic curves in ϖ, as desired. Conversely, consider any quartic curve, in the plane ϖ, that may be possible, with I and J as double points, having the same four given foci, and passing through O. Taking U as before, and the quadric Ω, we can shew that this curve arises by the construction we have made. Let the given points P, Q, R, S, O give, as before, the points P', Q', R', S', O', of Ω; project the quartic curve supposed, from U, on to the quadric. The curve on Ω, so obtained, will meet each of the lines UI, UJ in two points other than U; it will touch the two generators of Ω which intersect in P', and also the pairs of generators at Q', R', S'; and it will pass through O'. As this curve lies on Ω it will meet the tangent plane at P' only on the generators at P'; thus, as it touches each of these, its intersections with this plane count for four in number; and similarly at Q', R', S'. The curve is thus of the fourth order, and does not pass through U; it meets every generator of Ω in two points. Thus it lies on another quadric, beside Ω; and therefore lies on a quadric cone, say (W). A tangent line of the curve must thus lie in a tangent plane of this cone (W). In particular, the generators of Ω at P' must touch (W), and equally those at Q', R', S'. We may thus conclude that (W) may be taken to be the same as (V), and thus recover the preceding construction.

Inversion of the curve into itself. We have explained, in Chapter i (above, p. 12), the process of inversion, in regard to the quadric Ω; whereby, being given a point L, and its polar plane, λ, in regard to Ω, we pass from any point, P, to the point P', of the line LP, such that P, P' are harmonic conjugates in regard to L and the point (LP, λ). To a quadric cone there arises, as its inverse, another quadric cone, intersecting the plane λ in the same conic, with vertex the inverse of the vertex of the first cone.

Consider then the curve, on Ω, which lies on the cone (V). The curve which is the inverse of this, in regard to L, is the locus of the second intersections, with Ω, of the lines joining L to the points of the first curve; and this inverse curve lies also on a quadric cone, (V'), obtained by inversion from (V). The tangent planes, and generators, of these two cones, meet on the polar plane, λ, of L, in regard to Ω; and their vertices, V and V', are harmonic inverses in regard to L and λ. By projection from a point, U, of Ω, the curves (Ω, V) and (Ω, V') become two quartic curves in ϖ, both having I and J for double points, though in general with different tangents; and these curves are inverses of one another in regard to the point of ϖ obtained by projection of L, that is, in regard to the circle obtained by projection of the section (Ω, λ). If, in particular, L be taken at V, the two quartic curves on Ω coincide; thus the plane quartic inverts into itself in regard to any one of the principal circles.

Another particular case is when L is so taken that the line UL meets the original quartic on Ω; then the inverse quartic on Ω passes through U, and projects into a cubic curve on ϖ, having each of I and J as a simple point. The original quartic curve in ϖ can thus be inverted into such a cubic curve, namely by taking the centre of inversion on the curve itself. Conversely, the properties of the quartic curve in ϖ are obtainable by inverting a cubic curve which passes through the Absolute points I, J, with respect to any circle in its plane. And such a cubic curve is obtained by projecting a quartic curve on Ω, obtained by the intersection of this with another quadric, from a point U of this curve, the points I, J being the intersections of the generators at U with the plane, ϖ, on which we project.

Ex. 1. We may consider what are the properties of a cubic curve, in a plane ϖ, passing through two given Absolute points, I, J, of this plane, which arise as particular cases of those obtained above for a quartic curve having I, J for double points. There will be four families of circles (conics through I and J), each touching the cubic curve in two points. The circles of one family will have their centres on a parabola, and will all cut a fixed circle at right angles. The four parabolas will have a common focus, the intersection of the tangent lines of the cubic curve at I and J; they will also have a common axis, at right angles (in regard to I and J) to the tangent line of the cubic curve at its third intersection with the line IJ. The four intersections of a parabola with its associated principal circle are foci of the cubic curve.

Ex. 2. To obtain a formulation by equations, we may take as points of reference for coordinates x, y, z, t; first, the centre of projection U, on Ω, as point $(0, 0, 0, 1)$; by hypothesis this point lies on the cone (V); then, the points where the cone (V) is again met by the generators of Ω at U, as points $(1, 0, 0, 0)$ and $(0, 1, 0, 0)$; and last, as point $(0, 0, 1, 0)$, the intersection of the generators of Ω at these two points. The equation of Ω is then capable of the form $xy - zt = 0$; and the cone (V) may be regarded as the envelope, for varying θ, of a plane with the equation

$$\theta^2 (x + y + \sigma z) + 2\theta (lx + my) + l^2 x + m^2 y + \rho z + t = 0,$$

the plane $lx + my = 0$ being that which joins the point $(0, 0, 1, 0)$ to the generator of the cone (V) which passes through $(0, 0, 0, 1)$. The coefficients l^2, m^2 are chosen so that the cone contains the points $(1, 0, 0, 0)$, $(0, 1, 0, 0)$. By eliminating t, between the equation of the cone and $xy - zt = 0$, we thus find, for the equation of the cubic curve,

$$xy (x + y + zD) + z^2 (Px + Qy + Rz) = 0,$$

where $P = \rho + \sigma l^2$, $Q = \rho + \sigma m^2$, $R = \rho\sigma$, $D = \sigma + (l - m)^2$. This curve touches $x = 0$ and $y = 0$ on $z = 0$, and contains the point $x + y = 0$, $z = 0$. One family of enveloping circles is given, with varying ϕ, by

$$xy + xz(\phi + l)^2 + yz(\phi + m)^2 + z^2(\phi^2\sigma + \rho) = 0;$$

the centres of these circles lie on the parabola

$$x^{\frac{1}{2}} + y^{\frac{1}{2}} + (l - m)(-z)^{\frac{1}{2}} = 0,$$

and they cut at right angles the circle given by

$$xy + xz\sigma n + yz\sigma(1 - n) - z^2(\sigma lm + \rho) = 0,$$

where $n = l/(l - m)$.

If, instead of ρ, σ, l, m, we regard P, Q, R, D as given, we find, for σ,

$$(\sigma^2 - \sigma D + P + Q)^2 = 4(PQ - RD),$$

and the equations above given then determine ρ, l^2, m^2, and lm.

Ex. 3. The cubic curve just considered arises by taking the centre of projection, U, to lie on the quartic curve on Ω. Another particular case, for the plane curve obtained by projection from U, arises by taking U on one of the principal sections, say, on the polar plane of V, so that V is on the tangent plane at U, of the quadric. In this case the plane quartic curve in ϖ has four foci lying on a line; the conic in ϖ, which is in general the locus of centres of a family of enveloping circles, is in this case a line coinciding with the circle to which these enveloping circles are orthogonal. Taking

$$L = lx + my + nz, \quad C = l\xi + m\eta, \quad L_r = l_r x + m_r y + n_r z, \quad C_r = l_r \xi + m_r \eta,$$

for $r = 1, 2$, where ξ, η as well as l, m, n, l_r, m_r, n_r, are constants, such a quartic curve is given by

$$(C_1 xy - zL_1)^2 = (Cxy - zL)(C_2 xy - zL_2),$$

and is the envelope of circles

$$\theta^2(Cxy - zL) + 2\theta(C_1 xy - zL_1) + C_2 xy - zL_2 = 0,$$

whose centres lie on the line $2(x\xi + y\eta) - z = 0$.

Ex. 4. If the centre of projection, U, of the quartic curve on Ω, be taken at one of the four points of a principal section which, in general, project into foci of the plane quartic, namely at a point of contact with Ω of a tangent plane of the cone (V), so that the cone (V) touches the tangent plane at U, of the quadric, another particular case is obtained; then, the intersections of the cone (V), with the generators of Ω at U, coincide in pairs, in two points which lie on the generator of contact of (V) with the tangent plane of Ω at U. The quartic curve in ϖ then has, at each of I and J, a double point with coincident tangents, namely a cusp; and, of one set of four foci of the curve lying on a circle in general, one focus disappears, and the other three lie on a line. The quartic curve is then the curve known as a *Cartesian*. Its equation can be found from that in Ex. 3 by supposing $l_2 = 0$, $m_2 = 0$, $n_2 = 1$, so that $C_2 = 0$ and $L_2 = z$. The *Limaçon*, and, thence, the *Cardioid*, are special cases.

Angles of intersection of enveloping circles of the plane quartic which belong to different families. Returning now to the general case, we prove that, if we take two families of enveloping circles of the quartic curve in ϖ, the angles which a varying circle of one of these families makes with two fixed circles of the other family, have a constant sum or difference (Jessop, *Quarterly Journal of Math.*, XXIII, 1889, p. 375). The angles are measured in regard to the points I, J (as in Vol. II, p. 167).

The angle, between two circles obtained as projection of two plane sections of the quadric Ω, depending on the relation of the tangents of these sections, at one of their points of intersection, to the generators of Ω at this point, is equal to the angle between the planes of the sections, measured in regard to the quadric (Vol. II, p. 193); it is therefore equal to the interval between the poles of the planes of the section, measured by the quadric. The result to be obtained is therefore this: If we take upon the conic v, which is the polar reciprocal of the cone (V), a variable point, P, and take upon the conic v_1, which is the polar reciprocal of the cone (V_1), two fixed points, A, B, the sum, or difference, of the intervals PA, PB, measured in regard to the quadric Ω, is independent of P. As the conics v, v_1, v_2, v_3 are such a system as the focal conics of a system of confocal quadrics, this result has already been proved, Vol. III, p. 96. We return to this point of view below.

The proposition is however a descriptive one. Suppose that we have four planes, α, β, γ, δ, which we take in order; take the four lines of intersection of consecutive pairs of these, say (α, β), (β, γ), (γ, δ), (δ, α). Consider the intersections of these four lines with a given quadric, say, respectively, P, P'; Q, Q'; R, R'; S, S'. It may happen that four of these points, one on each line, lie on a plane; say P, Q, R, S lie on a plane, θ. It is then also the case that the other four points, P', Q', R', S', lie on a plane, say ϕ; for, since the eight points lie on the three quadrics consisting of, (a) the given quadric, (b) the degenerate quadric consisting of the plane-pair (α, γ), (c) the degenerate quadric consisting of the plane-pair (β, δ), and these three quadrics are evidently such that the third does not contain the complete intersection of the first two, it follows (Vol. III, pp. 148, 154) that the quadric consisting of the plane θ and the plane $P'Q'R'$, which contains seven of the eight points, likewise contains the eighth point S', which we suppose not to lie on the plane θ.

This being understood, the descriptive theorem above referred to is that, if (V_1) be a quadric cone passing through the curve of intersection of a quadric Ω and a quadric cone (V), and α, γ be any two tangent planes of the cone (V), while β, δ are any two tangent planes of the cone (V_1), and we consider, as here, the four lines of intersection of the pair of planes α, γ, each with both the planes β, δ, then the eight points, in which these meet the quadric Ω, lie, in two sets of four, upon two planes, say θ and ϕ. The lines of contact with the cone (V), of the two planes α, γ, lie on a plane, say ϵ; likewise the lines of contact of β, δ with (V_1) lie on a plane, say ζ; the planes θ, ϕ intersect in the line of intersection of ϵ and ζ, and are harmonic in regard to these. In order to make this proposition clear we may give at once a very simple proof with the symbols: Denoting by $\alpha = 0$ the equation of the plane α, and so on, the equation of the cone (V) may be supposed to be $\alpha\gamma - \epsilon^2 = 0$; the equation of the cone (V_1) may similarly be supposed to be $\beta\delta - \zeta^2 = 0$. Wherefore the equation of the quadric Ω, which passes through the curve of intersection of these two cones, may be supposed to be $\alpha\gamma - \epsilon^2 - m^2(\beta\delta - \zeta^2) = 0$, where m is a proper constant; this is the same as $\alpha\gamma - m^2\beta\delta = (\epsilon - m\zeta)(\epsilon + m\zeta)$. This equation shews that a point of Ω which is on the line (α, β) lies on one of the two planes $\epsilon - m\zeta = 0$, $\epsilon + m\zeta = 0$, the same being true of points of Ω on any of the lines (β, γ), (γ, δ), (δ, α). This proves the statement made.

Now consider the relation of this descriptive proposition with the metrical theorem above given for the angles of intersection of enveloping circles, of different families, of the plane quartic curve. Let α, β, γ, δ be four planes such that the four lines (α, β), (β, γ), (γ, δ), (δ, α) meet the quadric Ω in, respectively, P, P'; Q, Q'; R, R'; S, S', with the property that P, Q, R, S lie on a plane, θ, and P', Q', R', S' lie on a plane, ϕ; let the angle between the planes α, β, measured in regard to the quadric, that is in regard to the tangent planes of the quadric drawn from the line (α, β), be denoted by $[\alpha, \beta]$; as has been remarked, this is the same as the angle between the tangent lines at a point of intersection of the circles into which the sections by α, β project, measured in regard to I and J. We can then shew that $[\alpha, \beta] + [\gamma, \delta]$ is equal to $[\beta, \gamma] + [\delta, \alpha]$. It is to be understood that the statement is cast in this form in deference to traditional metrical geometry; the symbol $[\alpha, \beta]$ is ambiguous in sign, and by additive multiples of π, unless more particularly defined; the geometrical theorem, which is quite free from ambiguity, is one in regard to the symbols, once before called cross-ratios, such as are defined in Vol. II, p. 166, arising from the pairs of planes. See Ex. 4 below. This being borne in mind, the proof of the equation follows at once from the remark that, if λ, μ, ν be three planes meeting in a point, O, of the quadric,

Ω, the three angles $[\mu, \nu]$, $[\nu, \lambda]$, $[\lambda, \mu]$ are together equal to π, or one of these angles, properly taken, is equal to the sum of the other two ; this is only the statement that if the lines of intersection of the three planes, with the tangent plane at O, be represented in terms of the generators, i, j, at this point, by $i + lj$, $i + mj$, $i + nj$, then $(m/n)\,(n/l)\,(l/m) = 1$. Using this remark, and considering the two points, P, R, where the plane θ meets the lines (α, β) and (γ, δ), we have

$$\pi = [\theta, \alpha] + [\alpha, \beta] + [\beta, \theta], \quad \pi = [\theta, \gamma] + [\gamma, \delta] + [\delta, \theta],$$

and hence

$$[\alpha, \beta] + [\gamma, \delta] = 2\pi - [\theta, \alpha] - [\theta, \beta] - [\theta, \gamma] - [\theta, \delta],$$

where we have replaced $[\beta, \theta]$ by $[\theta, \beta]$, and $[\delta, \theta]$ by $[\theta, \delta]$. The same symmetrical form arises for $[\beta, \gamma] + [\alpha, \delta]$.

Ex. 1. For circles in a plane there follows, from what has been said, that, if four circles, $\alpha, \beta, \gamma, \delta$, be such that α, β meet in P, P' ; while β, γ meet in Q, Q' ; γ, δ meet in R, R' and δ, α meet in S, S', and, if P, Q, R, S lie on a circle, then also P', Q', R', S' lie on a circle (Vol. ii, p. 72).

Ex. 2. Also that, in this case, the two opposite angles $[\alpha, \beta]$, $[\gamma, \delta]$ have the same sum as the two opposite angles $[\beta, \gamma]$, $[\alpha, \delta]$. Conversely, this theorem of angles is sufficient to ensure that the points P, Q, R, S lie on a circle.

This simple theorem of metrical geometry may well be regarded as fundamental. In Todhunter-Leathem, *Spherical Trigonometry*, 1901, p. 132, there is quoted from Lexell, *Acta Petropolitana*, 1782, the theorem "If, on a sphere, the corners of a spherical quadrilateral lie on a small circle, the sums of its pairs of opposite angles are equal." See also M^{c}F. Orr, *Trans. Camb. Phil. Soc.*, xvi, 1897, p. 95.

Ex. 3. Also it is true that, if the sum of the angles of intersection of the pairs of three circles be π, the circles meet in a point.

The condition that the sum of three angles, α, β, γ, should be π, is the vanishing of the determinantal discriminant of the form

$$x^2 + y^2 + z^2 + 2yz \cos \alpha + 2zx \cos \beta + 2xy \cos \gamma.$$

It is easy to prove that the angle, in regard to the quadric $x^2 + y^2 + z^2 + t^2 = 0$, between two planes, of equations

$$a_r x + b_r y + c_r z + d_r t = 0 \ (r = 1, 2),$$

has, for cosine, $\sigma_{12}/\sigma_1\sigma_2$, where $\sigma_{12} = a_1 a_2 + \ldots + d_1 d_2$ and

$$\sigma_r{}^2 = a_r{}^2 + \ldots + d_r{}^2.$$

Using this result for every two of three planes, the signs attached to $\sigma_1, \sigma_2, \sigma_3$ being the same throughout, it is easy to deduce that the point of intersection of the three planes lies on the quadric.

Ex. 4. Supposing that the planes above denoted by α, γ, β, δ, θ, ϕ have, respectively, the equations $x = 0$, $y = 0$, $z = 0$, $t = 0$, $\xi = 0$, $\eta = 0$, where there will be two identities, say

$$x + y + z + t + \xi + \eta = 0, \quad bx + ay + dz + ct + g\xi + f\eta = 0,$$

the equation of the quadric, Ω, will be of the form

$$Pxy + Qzt + R\xi\eta = 0.$$

Let the roots of the quadratic equation in λ,

$$P^{-1}(\lambda - a)(\lambda - b) + Q^{-1}(\lambda - c)(\lambda - d) + R^{-1}(\lambda - f)(\lambda - g) = 0,$$

be denoted by λ_1, λ_2, and define twelve quantities a_1, b_1, ..., g_2 by the equations

$$Pa_r = a - \lambda_r, \quad Pb_r = b - \lambda_r, \quad Qc_r = c - \lambda_r, \quad Qd_r = d - \lambda_r,$$
$$Rf_r = f - \lambda_r, \quad Rg_r = g - \lambda_r,$$

for $r = 1, 2$. Speaking of $x = 0$, $y = 0$ as opposite planes, and of $z = 0$, $t = 0$ as opposite planes, and of $\xi = 0$, $\eta = 0$ as opposite planes, there will, among the twenty points of intersection of three of these six planes, be eight points through which no two opposite planes pass; these are the points previously denoted by P, Q, ..., R', S'. The equations of the generators of the quadric at any one of these eight points can be written down at once; for instance, at the point $x = 0$, $z = 0$, $\xi = 0$, these generators are given by

$$xa_1^{-1} = zc_1^{-1} = \xi f_1^{-1}, \quad xa_2^{-1} = zc_2^{-1} = \xi f_2^{-1};$$

and at the point $x = 0$, $z = 0$, $\eta = 0$ by similar equations, obtained by writing ηg_1^{-1} and ηg_2^{-1} respectively for ξf_1^{-1} and ξf_2^{-1}; and so for all.

Further, considering first, for example, the two planes $x = 0$, $z = 0$, the tangent planes of the quadric through their line of intersection are given, by what we have just said, by $xa_1^{-1} = zc_1^{-1}$, $xa_2^{-1} = zc_2^{-1}$. Thus the angle, ω, between the planes $x = 0$, $z = 0$, measured in regard to the quadric, is given by $e^{2i\omega} = a_1 a_2^{-1}/c_1 c_2^{-1}$. In general, if we define angles, α, β, γ, δ, θ, ϕ, by means of

$$e^{2i\alpha} = a_1 a_2^{-1}, \quad e^{2i\beta} = b_1 b_2^{-1}, \quad e^{2i\gamma} = c_1 c_2^{-1}, \quad e^{2i\delta} = d_1 d_2^{-1},$$
$$e^{2i\theta} = f_1 f_2^{-1}, \quad e^{2i\phi} = g_1 g_2^{-1},$$

then any two of the six planes, $x = 0$, $y = 0$, ..., $\eta = 0$, other than two opposite planes, make with one another an angle, relatively to the quadric, given by the difference of the two corresponding angles chosen from α, ..., ϕ. This is the general form of the theorem, for the angles between enveloping circles of different families of the plane quartic curve, from which we started.

Ex. 5. In Ex. 4, denoting $P^{-1} + Q^{-1} + R^{-1}$ by A,

also $\qquad\qquad P^{-1}ab + Q^{-1}cd + R^{-1}fg$ by C,

and $\qquad P^{-1}(a + b) + Q^{-1}(c + d) + R^{-1}(f + g)$ by B,

and $\qquad\qquad (B^2 - 4AC)A^{-1}$ by H,

prove that the condition that a plane, whose equation is written $lx + my + nz + pt + u\xi + v\eta = 0$, should touch the quadric Ω is

$$H\,(P^{-1}lm + Q^{-1}np + R^{-1}uv) + U_1 U_2 = 0,$$

where $\quad U_r = a_r l + b_r m + c_r n + d_r p + f_r u + g_r v \quad (r = 1, 2).$

Ex. 6. The preceding results in regard to the angles between planes lead to results as to the intervals between points, as has been indicated. We indicate some of these, referring, for a less summary account, to a note, *Camb. Phil. Proc.*, xx, 1920, pp. 122–130.

Let two quadrics, V, W, intersect in plane curves; they thus touch one another in two points. It is possible then to find another quadric, U, touching V at all the points of a plane section, and also touching W at all the points of a plane section; the planes of contact necessarily pass through the points of contact of V and W, and are harmonic in regard to the planes in which V and W intersect. Two such quadrics, U, having, we may say, *ring* contact with V and W, can be drawn through an arbitrary point.

Prove that, if AT be any tangent to the quadric V, drawn from any point A, and touching V in T, the interval AT, measured in regard to the quadric U, depends only on A. Deduce that, if A, B, C, D be four points such that the four joins AB, BC, CD, DA all touch V, the sum of the intervals AB, BC, CD, DA, measured in regard to the quadric U, and suitably interpreted, is zero.

Next, let K be any quadric, other than U, which likewise has ring contact with both V and W. Then prove that, if from any point, P, of the quadric K, tangents PX, PY be drawn respectively to V and W, touching these in X and Y, the sum, or difference, of the intervals PX, PY, measured in regard to U, is independent of the position of P upon K.

This result includes many others (cf. Vol. ii, p. 211, Ex. 8; Vol. iii, p. 96, Ex. 2).

Ex. 7. If $u^2 = b - c$, $v^2 = a - c$, $w^2 = a - b$, the four lines, joining each of the points of coordinates ($w \cosh \alpha$, 0, $u \sinh \alpha$, 1), ($w \cosh \gamma$, 0, $u \sinh \gamma$, 1), to both the points ($v \cos \beta$, $u \sin \beta$, 0, 1), ($v \cos \delta$, $u \sin \delta$, 0, 1), touch the two enveloping cones, of the quadric $x^2 (a - \rho)^{-1} + y^2 (b - \rho)^{-1} + z^2 (c - \rho)^{-1} = t^2$, whose vertices have the tangential equation $P^2 (b - \rho)^{-1} + Q^2 (c - \rho)^{-1} = 0$, where

$$P \sinh \tfrac{1}{2}(\alpha - \gamma) = lw \cosh \tfrac{1}{2}(\alpha + \gamma) + nu \sinh \tfrac{1}{2}(\alpha + \gamma) - p \cosh \tfrac{1}{2}(\alpha - \gamma),$$

$$Q \sin \tfrac{1}{2}(\beta - \delta) = lv \cos \tfrac{1}{2}(\beta + \delta) + mu \sin \tfrac{1}{2}(\beta + \delta) - p \cos \tfrac{1}{2}(\beta - \delta).$$

Ex. 8. From the section of a quadric with a cone of order n, having an r-ple generator, obtain a plane curve of order $2n - r$, with one r-ple point, and two $(n - r)$-ple points, I, J. This curve has $2(n - r)(n + r - 1)$ foci, lying on a circle, and is its own inverse in regard to this. (In particular, the invariant of a plane cubic curve is thus clear.)

CHAPTER IV

A PARTICULAR FIGURE IN SPACE OF FOUR DIMENSIONS

WE consider now a figure, in space of four dimensions, which is remarkable in that the numbers of the elements which it contains, and the mutual relations of these elements, when properly interpreted, are the same as arise for the figure of the twenty-seven lines of a cubic surface. The figure contains twenty-seven points, corresponding to the twenty-seven lines of the cubic surface. In space of four dimensions, four independent points determine a space of three dimensions; for such a space we shall, in this chapter, use the word *solid*, so that the figure will contain a certain number of points, lines, planes and solids. The twenty-seven points of the figure lie, in sixes, in seventy-two solids, corresponding to the rows of the thirty-six double-sixes of lines of the cubic surface; every one of these rows is in fact determined by four non-intersecting lines lying therein (Vol. III, p. 166). The figure contains two hundred and sixteen lines, each joining two points of the figure; these correspond to two lines of the cubic surface which do not intersect one another (*ibid.*). In a similar way, three lines of the cubic surface of which no two intersect correspond to a plane of the figure; there are 720 such planes. With the notation employed in Vol. II, p. 212, the elements being points, lines, planes and solids, the structure of the figure is represented by the symbol

$$27\,(\,.\,,\,16, 80, 16)\,216\,(2, .\,,\,10, 5)\,720\,(3, 3, .\,,\,2)\,72\,(6, 15, 20, .\,)\,;$$

this means, for example, that sixteen lines, eighty planes and sixteen solids pass through every one of the twenty-seven points; and that every one of the two hundred and sixteen lines contains two points, and lies in ten planes and in five solids; and so on. It should be said that the existence of the figure does not depend on the Propositions of Incidence alone, but requires also the truth of Pappus' Theorem.

In order to simplify the description, the points are represented by the same letters as are used for the lines of a cubic surface, a_1, a_2, ... a_6, b_1, b_2, ... b_6, c_{12}, c_{23}, ... The solids are denoted by capitals, O, P_{123}, P'_{123}, O', P_{12}, etc., where, in P_{123}, P'_{123}, the order of the three suffixes is indifferent, but P_{12}, with two suffixes, is to be distinguished from P_{21}. Each of the numbers occurring in the suffixes is one of the six, 1, 2, ... 6. The relation of the points and the seventy-two solids is, in fact, that represented by the symbols

O $(a_1, a_2, \ldots a_6)$, (one case); P_{123} $(a_1, a_2, a_3, c_{56}, c_{64}, c_{45})$, (twenty cases);

O' $(b_1, b_2, \ldots b_6)$, (one case); P'_{123} $(b_1, b_2, b_3, c_{56}, c_{64}, c_{45})$, (twenty cases);

$$P_{12} (a_1, b_1, c_{23}, c_{24}, c_{25}, c_{26}), \text{ (fifteen cases)};$$

$$P_{21} (a_2, b_2, c_{13}, c_{14}, c_{15}, c_{16}), \text{ (fifteen cases)};$$

these evidently correspond to the rows of the double-sixes of lines for a cubic surface (Vol. III, p. 161). It will be unnecessary to have a notation for all the lines of the figure ; but the six lines containing the pairs of points (a_1, b_1), (a_2, b_2), ... (a_6, b_6) will be denoted, respectively, by $n_1, n_2, \ldots n_6$. A set of five points which do not lie in a solid will be described as a *simplex* ; it will be found that there are two hundred and sixteen simplexes in the figure. And, for clearness, it may be stated that a simplex corresponds to five non-intersecting lines of the cubic surface which have two transversals ; five non-intersecting lines of the cubic surface which have only one common transversal are represented in the figure by five points belonging to the same solid.

The case here considered, for its interest in connexion with the lines of a cubic surface, may serve as an introduction to a wide literature. In a remarkable paper, "Circles, Spheres and Linear Complexes," *Trans. Camb. Phil. Soc.*, XVI, 1898, pp. 181–188, Mr J. H. Grace, having in mind chains of theorems given by Clifford (*Math. Papers*, 1882, pp. 51, 52), considers theorems in regard to spheres passing in sixes through points. Dr W. Burnside (*Proc. Camb. Phil. Soc.*, XV, 1909, pp. 71–75) remarks that, if the space of the spheres be inverted from a point not lying therein, theorems are thence deducible for points lying on a sphere in Euclidean space of four dimensions; he remarks, however, that the condition of lying on a sphere, or quadric, is not necessary for the figure; and proves the theorems to which we presently proceed. Professor P. H. Schoute, "On the relation between the vertices of a definite six-dimensional polytope and the lines of a cubic surface," *R. Ac. of Sc. of Amsterdam*, 24 Sep. 1910, pp. 375–383, gives, after Mr E. L. Elte, the co-ordinates of the twenty-seven vertices of a regular figure lying on a sphere in Euclidean space of six dimensions. Such a figure is capable of a group of rotations. For relations with the theory of groups see Burnside, "Groups of rational linear substitutions of finite order, etc.," *Proc. Lond. Math. Soc.*, X, 1911, pp. 300–308 (and, *ibid.*, XI, 1912, pp. 295–299). For the group of the lines of a cubic surface, see also Burnside, *Theory of Groups* (Cambridge, 1911, Second Ed.), pp. 485–488 ; and *Proc. Roy. Soc.*, LXXVII (1906), p. 182. Also Jordan, *Traité des Substitutions* (1870), pp. 316–319; Burkhardt, *Math. Annal.*, XXXVIII (1891), p. 185, and XLI (1893), p. 320, where the group is represented by linear equations, derived from the theory of hyperelliptic functions. Also L. E. Dickson, *Linear Groups* (1901), Chap. XIV.

We proceed now to shew how the figure can be constructed. We shall require two lemmas which, for clearness, may be given first :

(I) In four dimensions, let $a_1, a_2, \ldots a_6$ be six given points lying in a solid, which we denote by O. Through every three of these points let another definite solid pass; this other solid, through a_1, a_2, a_3, for example, will be denoted by P_{123}. If we consider four of the six given points there will then be such a solid through every three of these points, beside O; consider, for instance, the solids, P_{234}, P_{314}, P_{124}, P_{123}, passing through the threes of the four points a_1, a_2, a_3, a_4; these determine, by their intersection, a fifth point, not lying in O. This we denote by c_{56}, or c_{65}, the suffixes 5, 6 being those not arising in the construction. And so in general. The five points, a_1, a_2, a_3, a_4, c_{56}, constitute a simplex; the solids of this simplex, each containing four of these points, are, respectively, P_{234}, P_{314}, P_{124}, P_{123} and O. In all, there are twenty solids such as P_{123}, and fifteen points such as c_{56}; these points, which do not lie in O, belong to simplexes whose other vertices are in O. The solid P_{123} will be used, not only in the construction of c_{56}, but also, when we consider the points a_1, a_2, a_3, a_5, in the construction of c_{64}; and, similarly, in the construction of c_{45}. Thus P_{123} contains the six points a_1, a_2, a_3, c_{56}, c_{64}, c_{45}. If we now consider five of the six given points, say a_1, a_2, a_3, a_4, a_5, then, of the five simplexes arising by considering the fours of these five points, there are four simplexes each having a_1 as one vertex, the completing vertices being, respectively, c_{62}, c_{63}, c_{64}, c_{65}. These completing vertices themselves determine a solid. It can be shewn that this solid contains the point a_1. Denoting this solid by P_{16} (not the same as P_{61}), and using a similar notation in general, it can be shewn that the five solids P_{12}, P_{13}, P_{14}, P_{15}, P_{16} have a line in common, which therefore passes through a_1. This line we denote by n_1. These five solids are those which arise by considering the fives from a_1, a_2, ..., a_6 which, in turn, do not include a_2, a_3, a_4, a_5, a_6; just as P_{16} arose from the five other than a_6.

To prove this result, notice that every one of the simplexes used, in obtaining the points c_{rs} here involved, has one vertex at a_1, and has four solids, including O, passing through a_1. Take an arbitrary solid not passing through a_1, say U, and let ϖ be the plane in which this meets the solid O. The five lines joining a_1, respectively, to a_2, a_3, ... a_6, which are lines of O, meet the solid U in points, lying in the plane ϖ; let these points be called, respectively, B_2, ... B_6. The solid P_{123}, containing a_1, a_2, a_3, meets the solid U in a plane, say ϖ_{23}, which passes through B_2 and B_3. Thus, in the solid U, we have a plane, ϖ, and, therein, five points B_2, ... B_6; and through the join of every two of these points is drawn a plane, lying in U, that through B_2 and B_3 being ϖ_{23}. The point of intersection of the planes ϖ_{23}, ϖ_{34}, ϖ_{42}, which arise from the points B_2, B_3, B_4, may appropriately be named C_{56}; for it lies in the solids P_{123}, P_{134}, P_{142}, which contain a_1 and c_{56}, and is thus the point where the line of intersection of these solids, joining a_1 to c_{56}, meets the solid U. By considering every three of the four points B_2, B_3, B_4, B_5, and the three planes, ϖ_{rs}, through the joins of these three points, we obtain four points, such as C_{56}, namely $C_{62}, C_{63}, C_{64}, C_{65}$. It has been shewn above (p. 18) that these four points lie in a plane, which will then be in the solid U. We denote this plane by Π_6. These four points are on the lines joining a_1 to $c_{62}, c_{63}, c_{64}, c_{65}$, respectively; thus, the solid containing $c_{62}, \dots c_{65}$ also contains a_1, as we desired to prove, this being the solid containing a_1 and the plane Π_6. We denote this solid by P_{16}. It has, however, also been shewn above (p. 29) that the five planes, $\Pi_2, \Pi_3, \dots \Pi_6$, obtained by considering every four of the five points $B_2, B_3, \dots B_6$, meet in a point, in the solid U. Wherefore, the five solids $P_{12}, P_{13}, \dots P_{16}$ meet in a line, passing through a_1, as we also desired to prove. This is the line we denoted by n_1.

(II) The second lemma we require deals with the relations of the solids $P_{12}, P_{13}, \dots P_{16}$, which intersect in a line, n_1, passing through the point a_1, and the solids $P_{21}, P_{23}, \dots P_{26}$, which, similarly, intersect in a line, n_2, passing through the point a_2. We build up the result we require from the beginning. Suppose we have two lines, n, n', in space of four dimensions, and three solids passing through each of these; namely, the solids P, Q, R passing through the line n, and the solids P', Q', R' passing through the line n'. The solids P, P' meet in a plane, say α; similarly, Q, Q' meet in a plane, β, and R, R' in a plane, γ. Let A be the point of intersection of the planes β, γ, or the point common to the solids Q, Q', R, R'; similarly, let B be (γ, α), or (R, R', P, P'), and C be (α, β), or (P, P', Q, Q'). Through the plane ABC can be drawn a range (∞^1, a pencil) of solids; if H be the intersection with the line n of one of these solids, and H' the intersection of the same solid with n', the range of various positions of H on n is clearly related to that of the resulting positions of H' on n'. In particular, let the solid P, through n, meet the line n' in U', and the solid P', through n', meet the line n in U, so that U, U' are the intersections of the plane α respectively with n and n'; similarly, let the plane β meet n and n' in V and V', respectively, and the plane γ meet n and n' in W and W', respectively. The line UU', in the plane α, meets the line BC, which lies in this plane; thus U and U' are in a particular solid containing the plane ABC, that, namely, containing the plane of the lines UU', BC, and the point A. So V, V' are in a solid containing the plane ABC, as are W and W'. Wherefore, for any position of H, and its corresponding point H', the range U, V, W, H is related to the range U', V', W', H'.

Now let S be a fourth solid through the line n, meeting the line n' in T', and S' a fourth solid through the line n', meeting the line n in T, and let δ be the plane (S, S'); also let A' be the point

(α, δ), or (P, P', S, S'); and, similarly, B' be the point (β, δ), or (Q, Q', S, S'), and C' the point (γ, δ), or (R, R', S, S'). As, then, we considered the points A, B, C, arising by taking the solids P, Q, R with the solids P', Q', R', so we can consider the points A, B', C', arising by taking the solids S, Q, R with the solids S', Q', R'; thus we can infer that if a solid through A, B', C' meet the line n in H, and the line n' in H', the range V, W, T, H is related to the range V', W', T', H'. For these points H, H' to be the same as those previously so denoted, since we had U, V, W, H related to U', V', W', H', it is necessary, as a condition for the solids S, S', that U, V, W, T should be related to U', V', W', T'. Conversely, suppose this is so; let H be taken arbitrarily on n; then a point, H', can be taken on n' so that the range U', V', W', T', H' is related to U, V, W, T, H; and these points H, H' will be such that the five points A, B, C, H, H' lie in a solid, and also the five points A, B', C', H, H' lie in a solid. Therefore, similarly, the five points B, C', A', H, H' lie in a solid, and so do C, A', B', H, H'. Thus we have proved that, if solids P, Q, R, S, through the line n, respectively, meet the line n' in U', V', W', T', and solids P', Q', R', S', through the line n', meet n in a related range U, V, W, T; if, further, $\alpha, \beta, \gamma, \delta$ be the planes $(P, P'), (Q, Q'), (R, R'), (S, S')$, respectively, and, respectively, A, B, C, A', B', C' be the points $(\beta, \gamma), (\gamma, \alpha), (\alpha, \beta), (\alpha, \delta), (\beta, \delta), (\gamma, \delta)$; and if H, H' be any two points, respectively, on n, n', such that the range U, V, W, T, H is related to the range U', V', W', T', H'; then each of the four sets of five points $(A, B, C, H, H'), (A, B', C', H, H'), (B, C', A', H, H')$, (C, A', B', H, H') lies in a solid.

We now make application of the results of these lemmas. Take, for the lines n, n' of lemma II, the lines previously denoted, in lemma I, by n_1 and n_2. Take, for the solids P, Q, R, S, respectively, those denoted by $P_{13}, P_{14}, P_{15}, P_{16}$; and, for the solids P', Q', R', S', respectively, those denoted by $P_{23}, P_{24}, P_{25}, P_{26}$. Then, for example, the solids P_{13}, P_{23}, respectively, contain the points $(a_1, c_{32}, c_{34}, c_{35}, c_{36})$ and $(a_2, c_{31}, c_{34}, c_{35}, c_{36})$. Thus we have the respective intersections

$$(P_{14}, P_{15}, P_{24}, P_{25}) = c_{45}, \ (P_{15}, P_{13}, P_{25}, P_{23}) = c_{53}, \ (P_{13}, P_{14}, P_{23}, P_{24}) = c_{34},$$
$$(P_{13}, P_{16}, P_{23}, P_{26}) = c_{36}, \ (P_{14}, P_{16}, P_{24}, P_{26}) = c_{46}, \ (P_{15}, P_{16}, P_{25}, P_{26}) = c_{56}.$$

Therefore the points which, in lemma II, were called A, B, C and A', B', C', are, respectively, c_{45}, c_{53}, c_{34} and c_{36}, c_{46}, c_{56}. These points A, B, C, however, lie in the solid P_{126}, which meets the lines n_1 and n_2, respectively, in the points a_1 and a_2; and these points A, B', C', or c_{45}, c_{46}, c_{56}, lie in the solid P_{123}, which meets the lines n_1 and n_2, respectively, in the same points a_1 and a_2. Thus, by lemma II, we see that the solids $P_{13}, P_{14}, P_{15}, P_{16}$ meet the line n_2 in a range which is related to that in which the line n_1 is met by the solids

P_{23}, P_{24}, P_{25}, P_{26}. And in these related ranges a_2 and a_1 are corresponding points.

But we can also infer that, if any other solid be drawn through A, B, C, or c_{45}, c_{53}, c_{34}, meeting n_1, n_2, respectively, in the points b_1 and b_2, then b_2, b_1 will also be corresponding points of these two related ranges, respectively. Let such a solid be denoted by P'_{126}. Further, we can infer that the five points b_1, b_2, A, B', C', that is, b_1, b_2, c_{45}, c_{46}, c_{56}, lie on a solid, say P'_{123}; also, that the five points b_1, b_2, B, C', A', or b_1, b_2, c_{53}, c_{56}, c_{36}, lie on a solid, say P'_{124}; and that the five points b_1, b_2, C, A', B', or b_1, b_2, c_{34}, c_{36}, c_{46}, lie on a solid, say P'_{125}. Fully to justify the notation, however, it must be proved that the order of the suffixes here used is indifferent. Consider, for example, P'_{126}, containing b_1, b_2, c_{34}, c_{45}, c_{53}. Let the point in which this solid meets the line n_6, whose definition is analogous to that of n_1 and n_2, be denoted by b_6. Taking together the lines n_1, n_6, there are, through the former, the solids P_{12}, P_{13}, P_{14}, P_{15}; and, through the latter, the solids P_{62}, P_{63}, P_{64}, P_{65}; also we have the intersections represented by

$$(P_{13}, P_{14}, P_{63}, P_{64}) = c_{34}, \quad (P_{14}, P_{15}, P_{64}, P_{65}) = c_{45}, \quad (P_{13}, P_{15}, P_{63}, P_{65}) = c_{53}.$$

Thus, as before, we can define b_6 directly from b_1, and then obtain b_2 by a solid which, with parity of notation, would be denoted by P'_{162}. This is then the same as P'_{126}. And so on.

We have thus justified the propriety of all the symbols obtainable, by change of the suffixes, from

$$P_{123}\,(a_1, a_2, a_3, c_{56}, c_{64}, c_{45}), \quad P_{12}\,(a_1, b_1, c_{23}, c_{24}, c_{25}, c_{26}),$$
$$P'_{123}\,(b_1, b_2, b_3, c_{56}, c_{64}, c_{45}).$$

Finally, we prove that the six points b_1, b_2, ... b_6 lie in a solid. For this, consider, for instance, the five points b_1, c_{62}, c_{63}, c_{64}, c_{65}, which lie in the solid P_{16}. With every four of these we can form a simplex, whose completing vertices lie, by lemma I, in a solid which contains the point b_1. The simplexes in question, with the solids containing the sets of four vertices of these, are, in fact,

b_1, c_{63}, c_{64}, c_{65}	b_1, c_{62}, c_{64}, c_{65}
P_{26}, P'_{123}, P'_{124}, P'_{125}, P_{16};	P_{36}, P'_{123}, P'_{134}, P'_{135}, P_{16};
b_1, c_{62}, c_{63}, c_{65}	b_1, c_{62}, c_{63}, c_{64}
P_{46}, P'_{124}, P'_{134}, P'_{145}, P_{16};	P_{56}, P'_{125}, P'_{135}, P'_{145}, P_{16},

where the completing vertices, not written, are, respectively, b_2, b_3, b_4, b_5. Thus it appears that the five points b_1, b_2, b_3, b_4, b_5 lie in a solid. By a similar argument it appears that b_1, b_2, b_3, b_4 and b_6 lie in a solid. This is then the same as before. We can then complete the symbols used, by writing, also, $O'\,(b_1, b_2, ... b_6)$.

The statement of the structure of the figure, which has been given, can now be justified. The 27 points of the figure are those which have been named. The 216 lines of the figure consist of the joins of the pairs of the six points lying in each of the 72 solids which have been obtained; this would give 72.15 lines; but each line lies in five of the solids, so that there are 72.3, or 216 lines. In fact, the notation has involved that the line n_1, joining the points a_1 and b_1, lies in the five solids $P_{12}, \ldots P_{16}$. That any other line also lies in five of the solids may be verified; for instance, the line joining the points b_1 and b_2 lies in the solids $O', P'_{123}, P'_{124}, \ldots P'_{126}$; or, again, the line joining the points a_1 and c_{56} lies in the solids $P_{123}, P_{124}, P_{134}, P_{15}, P_{16}$. The 720 planes of the figure are, similarly, those defined by three points lying in the same solid of the figure. But it will be seen, on examination, that every such plane lies in two of the solids, so that there are $72.20 \div 2$, or 720 such planes. For example, the plane defined by a_1, a_2, a_3 lies in the solids P_{123} and P'_{123}; or again, the plane defined by a_1, c_{64}, c_{65} lies in P_{123} and also in P_{16}. The other numbers of the structure symbol can also be justified. For instance, that a line lies in ten planes: the line joining a_1 and b_1 lies in the ten planes containing a_1, b_1, c_{rs}, where r, s are any two of the numbers 2, 3, ... 6. Or, again, the line joining a_1 and a_2 is in the planes joining these two points to $a_3, a_4, a_5, a_6, c_{ij}$, where i, j are any two of the numbers 3, 4, 5, 6.

That a point lies in sixteen solids arises from $72.6 \div 27 = 16$; that a point lies in 80 planes similarly arises from $720.3 \div 27 = 80$; and that there are sixteen lines through a point can be similarly shewn, the point a_1, for example, lying in the lines joining this point to $a_2, a_3, \ldots a_6, b_1, c_{rs}$, where r, s are any two of the numbers 2, ... 6.

It is also of interest to consider the possible simplexes, of which there are 216. Of these, 40 exist with any one of the points as one vertex, 10 exist with two of the points, joined by a line of the figure, as two vertices, 3 exist with three of the points, lying in a plane of the figure, as three vertices, and 1 exists with any four of the points, lying in a solid of the figure, as four vertices. In fact, taking any four of the points which lie in one of the 72 solids, there will, through every three of these points, be another solid; and the four solids so obtained give, by their intersection, the fifth vertex of the simplex. Taking three of the points of a solid, there are three other points in this solid of which each may make a fourth vertex of a simplex, taken with the first three; the three completing vertices of the three simplexes so obtained lie in the other solid which contains the first three points. Or, the number is obtained by remarking that each of the 216 simplexes has ten sets of three vertices, while there are in all 720 planes in the figure; since

$$216 \cdot 10 \div 720 = 3.$$

Again, each of the simplexes has ten edges, while there are 216 lines in the figure. Thus there are $216 \cdot 10 \div 216$, or ten, simplexes with two vertices common. Also, the 216 simplexes give $216 \cdot 5$ points, while there are 27 points in the figure; thus there are $216 \cdot 5 \div 27$, or 40, simplexes with vertex at any point of the figure. There is in fact a correspondence between the simplexes and the lines of the figure. For example, the simplex a_1, a_2, a_3, a_4, c_{56} corresponds to the line joining the points b_5, b_6, the correspondence consisting in the fact that no one of the ten lines joining b_5, b_6 to the vertices of the simplex is a line of the figure.

The figure, we see, depends on the assignment of 4 points to determine the solid O, requiring 16 constants; then on two other points therein, requiring 6 more; then on another constant to determine each of the 20 solids such as P_{123}; and then, finally, on another constant to determine b_1. In all upon $16 + 6 + 21$, or 43, constants.

Ex. 1. The points A, B, C, D lie in one plane, ϖ, in space of four dimensions, and the points A', B', C', D' in another plane, ϖ'; these planes intersect in the point O. The condition that the range (or pencil) of four solids, joining the plane ϖ' to A, B, C, D, respectively, should be related to the range of four solids joining the plane ϖ to A', B', C', D', respectively, is that the range (flat pencil) of lines joining O to A, B, C, D should be related to the range of lines joining O to A', B', C', D'.

Ex. 2. Consider four general planes, 1, 2, 3, 4, in space of four dimensions. Let the points of intersection of the plane 4 with the planes 1, 2, 3 be denoted, respectively, by A, B, C, the points of intersection of the planes $(2, 3)$, $(3, 1)$, $(1, 2)$ being, respectively, A', B', C'. Thus, in terms of the six points, the planes 1, 2, 3, 4 are, respectively, (A, B', C'), (B, C', A'), (C, A', B'), (A, B, C). It is a known fact (Vol. I, p. 92; below, pp. 118, 120) that all lines meeting the four given planes meet another plane. Prove, in fact, that, if A, B, C, A', B', C' be symbols for the six points, subject to

$$A + B + C + A' + B' + C' = 0,$$

then all lines meeting the four given planes meet also the plane which contains the four points whose symbols are $A + B' + C'$, $B + C' + A'$, $C + A' + B'$, $A + B + C$.

It can be shewn that the unique line which can be drawn, from the point $xA + yB + zC$, to meet the first three planes, contains the point

$$x^{-1}(A + B' + C') + y^{-1}(B + C' + A') + z^{-1}(C + A' + B').$$

In threefold space, given two arbitrary lines and two arbitrary planes, let the lines meet the first plane in B, C, and meet the second plane in B', C'. Let $B'C'$ meet the first plane in A, and BC meet the second plane in A'. It can be proved that the points $xA + yB + zC$, $x^{-1}A' + y^{-1}B' + z^{-1}C'$ are on a transversal of the two given lines.

Ex. 3. The figure dual to that considered in the text is of considerable interest. This will consist of seventy-two points and twenty-seven solids. Starting with a point O, and six solids, a_1, a_2, ... a_6, passing through this, every three of these solids meet in a line. On each of these lines is taken a point, such as P_{123}; there are twenty such points. The four points P_{234}, P_{314}, P_{124}, P_{123} determine a solid, c_{56}; and so on. The four solids c_{61}, c_{62}, c_{63}, c_{64} meet in a point lying in the solid a_1, namely the point P_{16}. The five points P_{12}, P_{13}, P_{14}, P_{15}, P_{16} lie in a plane, which is in the solid a_1. This plane is n_1. If N_{12} denote the common point of the planes n_1, n_2 (cf. Ex. 1, above), the two flat pencils of lines $N_{12}(P_{12}, ..., P_{16})$, $N_{12}(P_{21}, ..., P_{26})$ are related to one another. And so on.

Ex. 4. If in Ex. 3, the point O, and the assumed twenty points such as P_{123}, lie on a quadric threefold of the fourfold space, then all the derived points, $72 - 21 = 51$, in number, lie on this quadric. By projection, of the sections of this quadric threefold by the twenty-seven solids which arise, from a point of this quadric, on to an arbitrary solid, we derive a figure consisting of 27 spheres meeting in sixes in 72 points. We may initiate the figure with six arbitrary spheres having a common point; every three of these have then another common point. This is the point of view of Mr Grace's paper referred to·(*loc. cit.*, p. 182).

CHAPTER V

A FIGURE OF FIFTEEN LINES AND POINTS, IN SPACE OF FOUR DIMENSIONS; AND ASSOCIATED LOCI

The figure of fifteen lines and points. Let a, b, c, d be four lines of general position, in space of four dimensions. Any two of these will then define a threefold space. For a threefold space we shall here use the name *solid*. There will then be six such solids, say, $[A]$, or $[a, d]$; $[B]$, or $[b, d]$; $[C]$, or $[c, d]$; $[A']$, or $[b, c]$; $[B']$, or $[c, a]$; $[C']$, or $[a, b]$. Conversely, if six such solids be given, of which no three have a plane in common, they define four lines, each the intersection of three of these solids, namely a as the intersection of $[A]$, $[B']$, $[C']$, b as the intersection of $[B]$, $[C']$, $[A']$, c from $[C]$, $[A']$, $[B']$, and d from $[A]$, $[B]$, $[C]$. The three lines b, c, d have a common transversal, which we denote by a'; this is the line common to the solids $[A']$, $[B]$, $[C]$. Similarly, let b', c', d' denote, respectively, the common transversals of c, a, d; of a, b, d; and of a, b, c. There are, then, twelve points of intersection of the eight lines now obtained, which we denote as follows:

$A = (b', c)$, $B = (c', a)$, $C = (a', b)$; $A' = (b, c')$, $B' = (c, a')$, $C' = (a, b')$,
$P = (a, d')$, $Q = (b, d')$, $R = (c, d')$; $P' = (a', d)$, $Q' = (b', d)$, $R' = (c', d)$.

The point A, as containing a point of the line c, is in the solid $[b, c]$; as containing a point of the line b', which lies wholly in the solid $[a, d]$, this point A is also in the solid $[a, d]$. Similarly, the point A' is in both the solids $[b, c]$ and $[a, d]$; and each of the points P, P' is in both these solids $[b, c]$, $[a, d]$. Denote the plane common to these solids by λ. The lines AA' and PP', both lying in this plane, have a common point; denote this point by L. By similar reasoning, the plane, say μ, common to the solids $[c, a]$ and $[b, d]$, contains both the lines BB' and QQ'; let the point of intersection of these lines be called M. Finally, the plane, say ν, common to the solids $[a, b]$ and $[c, d]$, contains the lines CC' and RR'; let N be the common point of these lines.

We can now prove that the three points L, M, N are in line, namely, in the line which is common to the three solids $[a, a']$, $[b, b']$, $[c, c']$. For, consider the point L: this point lies on the line PP', joining a point of the line a to a point of the line a'; thus L lies in the solid $[a, a']$; again, L is on the line AA', joining a point of the line b to a point of the line b', so that L is in the solid $[b, b']$; the line AA', however, equally joins a point of the line c to a

point of the line c', so that L is also in the solid $[c, c']$. Thus L is
in the three solids $[a, a']$, $[b, b']$, $[c, c']$. By similar reasoning M
and N are also in these three solids. Thus L, M, N are in a line.
This line we denote by e. See the diagram given as frontispiece to
the Volume. But further, the point L, as it lies on the line PP',
joining a point of the line d to a point of the line d', lies also in
the solid $[d, d']$; and, similarly, M and N are also in this solid.
Wherefore, the four solids $[a, a']$, $[b, b']$, $[c, c']$, $[d, d']$ meet in a
line; and, as was previously remarked (Vol. I, p. 92), the proof of
this depends on the Propositions of Incidence only. We denote
the lines $AA'L$, $BB'M$, $CC'N$, respectively, by l, m, n, and the
lines $PP'L$, $QQ'M$, $RR'N$, respectively, by p, q, r. The figure now
contains fifteen points and fifteen lines, three points lying on each
line, and three lines passing through each point. In each case,
taking any one of the lines, and the six lines, other than this,
which pass, in couples, through the three points lying on this line,
the eight remaining lines consist of four non-intersecting lines and
of the four transversals, each meeting three of these lines; the
figure is entirely symmetrical. We shall speak of the set, a, b, c,
d, e, of five lines, as being *associated*; any four of these may be
regarded as primary and the fifth determined from them, as e was
determined from a, b, c, d. But equally, e is determinable from
a', b', c', d', just as e was determined from
a, b, c, d, and the five lines a', b', c', d', e
are also associated. In all, the figure con-
tains $15 . 2 \div 5$, or six, sets of associated
lines, every line of the fifteen being common to two of these sets. As is easily
seen, these sets are given by the adjoined
scheme, which may be read either in rows
or columns.

	I	II	III	IV	V	VI
I	.	e	d	a	b	c
II	e	.	d'	a'	b'	c'
III	d	d'	.	l	m	n
IV	a	a'	l	.	r	q
V	b	b'	m	r	.	p
VI	c	c'	n	q	p	.

This scheme suggests another notation for the lines, which will
be found to be very useful. Namely, the line of which the symbol
here occurs in the r-th row and the s-th column may be denoted
either by rs or by sr. For instance, the line e is then 12 or 21,
and p is 56 or 65. In this notation, *two lines* of the fifteen *intersect if, and only if, their duad symbols have no number in common*,
as may be verified easily. Thus the three lines which meet in any
point have duad symbols which, together, employ all the six numbers 1, 2, ..., 6; for instance, the three lines 12, 34, 56 meet in a
point. Such a set of three duads was called by Sylvester a syn-
theme; the fifteen points of the figure correspond to the fifteen
synthemes which are possible. As another illustration of the notation, no two of the three lines 23, 31, 12 intersect, but each is met
by all the three lines 56, 64, 45; thus these six lines are generators

of a quadric surface, and lie in a solid. There are, then, *ten* such solids, each corresponding to one of the ways of dividing the six numbers into two triads; for instance, the solid we have considered corresponds to the triads 123, 456, and may be denoted, appropriately, by either of these triads; and so for the other solids. These ten solids are, in fact, as is seen at once, the ten previously denoted by $[b, c]$, ..., $[a, d]$, ..., $[e, a]$, ..., each defined by a pair of the five associated lines a, b, c, d, e; each contains, likewise, one pair from every set of five associated lines. We shall call these solids the ten *singular* solids. Each of the fifteen lines lies in four of these solids; for instance the line 12 lies in the four solids 123 (or 456), 124 (or 356), 125 (or 346), 126 (or 345). Further, through any one of the fifteen points where three of the lines intersect, there pass six of the singular solids, two of these intersecting in the plane of any two of these three lines. For instance, through the point 12.34.56 there pass the solids 134 (or 256) and 156 (or 234), beside the four which meet in the line 12; the former two solids contain both the lines 34, 56, but not the line 12.

Proof of the incidences with the help of the symbols. If desired, the incidences of the figure may be proved very briefly by use of symbols. If the symbols of the points A, B, C, A', B', C' be represented by the same letters, these symbols, since the six points are in space of four dimensions, must be connected by a syzygy, which, by proper choice of the symbols, may be supposed to be $A + B + C + A' + B' + C' = 0$. Thereafter, no further multiplication of these symbols by an algebraic symbol is legitimate, save by one the same for all (cf. Vol. i, p. 71). We suppose the six symbols not to be connected by any further syzygy, the six points not being in a space of three dimensions. The symbols of the points P, Q, R, which lie, respectively, on the lines BC', CA', AB', are then each expressible linearly by the symbols of the two points on whose join it lies, in such forms as $P = mB + nC'$. The points P, Q, R are, however, in line, and their symbols are connected by a syzygy; by the expressions for P, Q, R, this becomes a syzygy for the six symbols A, B, ..., C', which must then agree with the fundamental syzygy for these. Hence, by absorption of proper algebraic multipliers in the symbols P, Q, R, we have, as is easily seen, the results

$$P + B + C' = 0, \quad Q + C + A' = 0, \quad R + A + B' = 0.$$

By a similar argument, the symbols for the points P', Q', R' may be taken so that

$$P' + B' + C = 0, \quad Q' + C' + A = 0, \quad R' + A' + B = 0.$$

These, however, lead to $P + P' = A + A'$, expressing that there is a point of the line PP' which is also a point of the line AA'. The symbol, L, of this point, may then be taken so that $L + A + A' = 0$. The points of intersection of BB' and QQ', and of CC' and RR', are similarly established, with symbols M and N such that $M + B + B' = 0$ and $N + C + C' = 0$. These equations, however, in virtue of the fundamental syzygy, lead to $L + M + N = 0$; this proves that the points L, M, N lie on a line, e.

If, instead of the six points A, B, ..., C', we introduce other six points, F, G, ..., H', with respective symbols

$$F = \tfrac{1}{2}(B + C - A), \quad G = \tfrac{1}{2}(C + A - B), \quad H = \tfrac{1}{2}(A + B - C),$$

$$F' = \tfrac{1}{2}(B' + C' - A'), \quad G' = \tfrac{1}{2}(C' + A' - B'), \quad H' = \tfrac{1}{2}(A' + B' - C'),$$

so that $F + G + H + F' + G' + H' = 0$, then it is seen at once that the symbols of the original fifteen points of the figure are the fifteen sums of twos of the six symbols F, G, ..., H'. This is equivalent to saying that, if, in one of the fifteen possible ways, the six points F, G, ..., H' be divided into three pairs, and the joins of the points of the pairs be taken, then the common transversal of the three joins is one of the fifteen lines of the figure, and the points where the transversal meets the joins are three of the original fifteen points of the figure. Or again, any one of the original fifteen points, say $F + F'$, is the intersection, of the join of the two points F and F', with the solid defined by the four remaining points, G, H, G', H'. The three lines of the figure through $F + F'$ are the transversals, drawn in this solid (G, H, G', H'), to meet one of the three pairs of opposite joins of these four points, G, H, G', H'. This shews that the six points F, G, ... are entirely symmetrical in regard to the figure of fifteen points and lines, and, though they do not belong to it, may be used to construct the figure. Essentially, only one such figure exists (Vol. I, p. 152).

The figure of six general points in space of four dimensions was investigated by Mr H. W. Richmond, *Quart. J. of Math.* xxxi, 1900, pp. 125—160; *Math. Annal.* liii, 1900, pp. 161—176. To the present writer the figure of fifteen points and lines was independently suggested, as arising from four lines, by the problem of a double-six of lines (*Roy. Soc. Proc.* lxxxiv, 1911, p. 599; *Proc. Camb. Phil. Soc.* xx, 1920, p. 133). The locus of the third order, associated with the figure, to which reference will be made below, was considered by Segre (*Atti...Torino*, xxii, 1887, p. 547, and *Memorie...Torino*, xxxix, 1889, pp. 3—48). It has been remarked above (p. 36) that spheres and circles in threefold space may be regarded as derived from points and lines in fourfold space. Thus the theory of the figure of fifteen lines, in fourfold space, may be stated in terms of circles in threefold space. From this point of view an ample introduction, which includes the two number notation for the lines, and the symmetrical equation for the cubic locus studied by Segre, is given by Stéphanos, *Compt. Rend.* xciii, 1881, p. 634 (and p. 578). For him, five associated lines have their (ten) coordinates linearly connected (p. 120, below). It is shewn below (Chap. vii) that the original fifteen lines of the figure may be regarded as arising by transformation from the fifteen joining lines of six points in space of *five* dimensions.

Ex. 1. Any two of five associated lines determine a solid; this meets each of the three other lines in a point. Prove that these three points are in line.

Ex. 2. Any six points, in space of four dimensions, taken in order, form a skew hexagon (say $AB'CA'BC'$); of this hexagon there will be two sets each of three alternate sides (BC', CA', AB' and $B'C$, $C'A$, $A'B$); and there will be three diagonals (AA', BB', CC'). Prove that the two transversals, each of one set of alternate sides, form, with the three diagonals, a set of five associated lines.

Ex. 3. In space of four dimensions, the three lines which are the transversals, of one set of alternate sides, of the other set of alternate sides, and of the diagonals, of a skew hexagon, lie in a solid. If we take the two hexagons $AB'CA'BC'$ and $AB'BA'CC'$, of which the latter is obtained from the former by the interchange of B and C, the three transversals so obtained, from one of these hexagons, all meet the three transversals so obtained from the other hexagon, and the two solids are the same. (Cf. Ex. 13, p. 144, below.)

Ex. 4. Referring to the notation used above, let λ, μ, ν denote, as before, respectively, the planes $LAA'PP'$, $MBB'QQ'$, $NCC'RR'$. Then the point of intersection of the planes μ, ν is the point of symbol $F - F'$ (or $A' + B + C$); and so on. Thus the plane λ contains the points $G - G'$ and $H - H'$. Further, the solid (λ, e), defined by the plane λ and the line e meeting this plane, is the solid defined by the four points G, H, G', H'; and so on. Now, in space of four dimensions, given a point, O, and a solid, Π, we can define a transformation, a *harmonic inversion*, from a point, P, to a point P', by taking the point, M, where OP meets Π, and then taking the point, P', on the line OP, which is the harmonic conjugate of P in regard to O and M. Shew that, if this transformation be employed, with O as the point (μ, ν), or $F - F'$, and Π as the solid (λ, e), then the points F, F' are interchanged, but each of the points G, H, G', H' is unaltered. The figure of fifteen points and lines is thus unaltered, in its aggregate, by this harmonic inversion. In fact, the four lines BQ, $B'Q'$, CR', $C'R$ meet in the point (μ, ν); and the four points which are the harmonic conjugates of the point (μ, ν) in regard, respectively, to the pairs B, Q; B', Q'; C, R'; C', R, all lie in the solid (λ, e).

The plane λ is the intersection of the solids $[b, c]$, $[a, d]$; as has been remarked. Cf. Ex. 21, p. 148, below.

Ex. 5. In space of four dimensions, the dual of a line is a plane. Consider, briefly, the dual of what precedes, arising when we begin with four arbitrary planes, of general position, say α, β, γ, δ. Every two of these planes have a point in common; let the six points so obtained be denoted as follows:

$$A = (\alpha, \delta), \quad B = (\beta, \delta), \quad C = (\gamma, \delta),$$
$$A' = (\beta, \gamma), \quad B' = (\gamma, \alpha), \quad C' = (\alpha, \beta);$$

being in space of four dimensions, these six points will not be independent. Conversely, if the six points be supposed to be given, arbitrarily, the four planes are determined, each by three of these, as follows:

$$\alpha = (A, B', C'), \quad \beta = (B, C', A'), \quad \gamma = (C, A', B'), \quad \delta = (A, B, C).$$

The dual of the meeting of two lines in a point is the lying of two planes in a solid, or the meeting of these two planes in a line (instead of a point merely). Thus any three of the planes are met, each in a line, by another plane. In fact, each of the four planes

$$\alpha' = (A', B, C), \quad \beta' = (B', C, A), \quad \gamma' = (C', A, B), \quad \delta' = (A', B', C')$$

meets three of the given planes in a line. For instance, the plane δ' meets α in the line $B'C'$, meets β in $C'A'$, and meets γ in $A'B'$; and so on. As the dual of what is proved above, the four points (α, α'), (β, β'), (γ, γ'), (δ, δ') lie in a plane; we may also prove this anew, by a method not the dual of that used above. Recall first, in space of three dimensions, that, if three lines be given, of which no two intersect, and, from a variable point, P, of one of these lines, the transversal be drawn to the other two lines, and thereon the point, Q, be taken, which is harmonically separated from P by the two lines, then the locus of Q is a fourth line. In space of four dimensions, if two planes, ξ, η, be given, intersecting in a point O, and H be any point upon a line, l, which does not meet ξ nor η, the two solids (ξ, H), (η, H) meet in a plane, passing through O; this plane, as lying in the solid (ξ, H), meets ξ in a line; and, likewise, meets η in a line. Thus lines can be drawn, through the point H, meeting the planes ξ, η; let such a line meet ξ and η, respectively, in X and Y, and let K be the harmonic conjugate of H in regard to X and Y. Then the locus of K, as H varies on l, is a plane passing through O. This we see by projecting the figure, from O, on to any threefold space. With this in mind, consider, in the figure described above, the three lines AA', BB', CC', and the common transversal of these, say l, meeting AA', BB', CC', respectively, in L, M, N. Let L', on AA', be the harmonic conjugate of L in regard to A and A'; similarly, let M', N', respectively on BB' and CC', be the harmonic conjugates of M and N. Then, it follows from what has been said that the plane $L'M'N'$ contains the point common to the two planes ABC and $A'B'C'$; and, for the same reason, that this plane contains the point common to the two planes $A'BC$ and $AB'C'$; and so on. Thus the four points (α, α'), (β, β'), (γ, γ'), (δ, δ') all lie in the plane $L'M'N'$, which is thus the plane *associated with* the planes α, β, γ, δ; or associated with the planes α', β', γ', δ'.

If we use the symbols $A, B, \ldots$ for the points $A, B, \ldots$, choosing them so that the necessary syzygy for these six symbols is $A + B + C + A' + B' + C' = 0$, we see, for instance, that the symbol of the point (δ, δ') is $A + B + C$ (or its equivalent, $A' + B' + C'$). Thus the theorem we have proved is that the four points whose symbols are

$$A + B' + C', \quad B + C' + A', \quad C + A' + B', \quad A + B + C$$

lie in one plane; as follows also because these four symbols have a vanishing sum. The second and third of these symbols have, for sum, $A - A'$, which is equally the sum of the first and fourth symbols; the symbols of L, M, N are, respectively, $A + A'$, $B + B'$, $C + C'$. Thus the symbols of L', M', N' are, respectively, $A - A'$, $B - B'$, $C - C'$; and each of these points is the intersection of two of the six joins of the four points (α, α'), (β, β'), (γ, γ'), (δ, δ').

Ex. 6. We have considered a process of harmonic inversion, in space of four dimensions, in which a given point and a given solid were fundamental; we have also considered a process of harmonic inversion in which two given planes were fundamental. Suppose now that we are given a fixed line, l, and a fixed plane, ϖ, not intersecting the line l. The plane joining l to any point, P, meets the plane ϖ in a point. Thus a line, say p, can be drawn from any point P to meet the given line l and also the given plane ϖ; thereon the point, Q, can be taken which is the harmonic conjugate of P in regard to the points where p meets l and ϖ. This defines a transformation from P to Q (or conversely).

If X, Y, Z be points of the plane ϖ, and T, U be points of the line l, it can easily be shewn that, with coordinates relative to these five points, the transformation is expressed by the equations

$$x' = x, \; y' = y, \; z' = z, \; t' = -t, \; u' = -u.$$

This transformation is evidently the combination of two such as are considered in Ex. 4, each defined with reference to a fixed point and a fixed solid.

In Ex. 5, the planes δ, δ' correspond to one another by such a transformation as that considered here, in which the fixed elements are the line l and the plane $L'M'N'$ (as do the planes α, α', etc.). Similarly, in the original figure of the text, the lines d, d' correspond to one another by such a transformation, of which the fixed elements are the line e, and the plane which meets in a line the three planes λ, μ, ν (i.e. the planes $LAA'PP'$, etc.; see Ex. 4); and the lines a, a' correspond in the same transformation, as do b, b' and c, c'. The plane meeting, in a line, each of λ, μ, ν, is the plane defined by the three points $A - A'$, $B - B'$, $C - C'$ (or by the three points $F - F'$, $G - G'$, $H - H'$). Cf. Ex. 21, p. 148, below.

Ex. 7. The figure of fifteen points and fifteen lines can be separated (in fact in ten ways) into two parts, of which one part consists of two triads of points together with the nine lines joining the points of one triad to the points of the other triad, and the other part consists of two triads of lines together with nine points in which the lines of one triad intersect the lines of the other triad. Of the lines and points named, no one is common to the two parts, but the two parts together exhaust the whole figure. The nine lines of the first part can be arranged in three sets, of three each, such that the transversal line of a set is one line of a triad of lines of the second part, and the set forms, with the other two lines of this triad, an associated system of five lines. The other triad of the second part is similarly obtainable from another arrangement of the nine lines of the first part.

Ex. 8. An illustration of the transformation considered in Ex. 6 is as follows:—Let A, A', B, B', C, C' be six points of a rational quartic curve in space of four dimensions; let L, M, N be the points where the chords AA', BB', CC', of this curve, are met by their common transversal, l; let L', M', N' be, respectively, the harmonic conjugates of L, M, N in regard to A, A'; to B, B'; and to C, C'; and denote the plane $L'M'N'$ by ϖ. By applying the transformation of Ex. 6, with l and ϖ as fundamental elements, the given quartic curve gives rise to another quartic curve having, likewise, AA', BB', CC' as chords. Prove that any plane, containing one of these chords, which meets one of the two curves, also meets the other curve (the second point of meeting not corresponding to the first, however, by the transformation in question). (*Proc. Camb. Phil. Soc.* xxi, 1923, p. 684.)

The associated line of a set of five is met by all planes meeting the first four lines. We return now to our original point of view, in which we suppose four non-intersecting lines of general position to be given, say a, b, c, d, and prove that all planes which meet a, b, c, d also meet the associated line, e. It will appear that there are ∞^2 such planes. It will also appear that, of such planes, there are two which pass through an arbitrary point, O, of general position. For some positions of O these two planes coincide; the locus of such points O is of great importance. For some positions of O the number of these planes is infinite; for instance when O is on the associated line e. The proof does not follow from the Propositions of Incidence alone, but requires Pappus' theorem.

It will add to clearness to interpolate here some remarks in regard to a construct occurring in space of four dimensions, which we shall call a *quadric point-cone*; and in regard to an associated conception, that of a *quadric line-cone*. In space of three dimensions the dual of a conic, considered as the envelope of its tangent lines, which lie in a plane, is the quadric cone, or conical sheet, considered as consisting of its generating lines, which pass through its vertex. In space of four dimensions, lying in a particular solid in this space, we may have an ordinary quadric surface. This may be considered to consist of its two sets of generating lines, of which any line of either set meets all the lines of the other set. The dual conception, in the space of four dimensions, will be that of two sets of planes, all passing through a point, say V, of which any plane of either set meets every plane of the other set in a line, passing through V, while two planes of the same set meet only in V. For the original figure, of a quadric surface lying in a solid of the fourfold space, there pass, through every general line in this solid, two planes each containing two lines of the quadric surface; this line also contains two points at each of which two of these four generators meet. Thus, in the dual figure, an arbitrary plane through V contains two lines, passing through V, each of which is the intersection of a plane of one set with a plane of the other set; and there are two solids containing this plane, each of which contains two of the four planes so arising. It is the aggregate of points lying in these two sets of planes which we call a quadric *point-cone*, the point V being the *vertex*. In the original case, of the quadric surface, two generating lines which intersect determine a point of the surface, and a tangent plane; so, in the case of the point-cone, two generating planes which meet in a line lie in a solid, called the *tangent solid* of the point-cone, the line of intersection of the planes being the *line of contact* of this tangent solid with the point-cone. There are ∞^2 tangent solids, each passing, with its line of contact, through the vertex. It is clear that the generating planes of the point-cone meet an arbitrary solid in the generating lines of a quadric surface lying in that solid. Conversely, a point-cone is generated by the planes which join the generating lines of a quadric surface to a point, not lying in the solid in which the quadric surface lies.

We may equally consider the dual, in space of four dimensions, of the tangent lines of a conic. This will consist of ∞^1 planes passing through a line, say l, two of these planes lying in an arbitrary solid which contains the line l. It is the aggregate of the points of these planes which we call a quadric *line-cone*, l being the *axis*. A solid through the axis may contain only one generating plane of the quadric line-cone; the solid is then said to touch the line-cone, at every point of this plane; of such *tangent-solids* there are ∞^1. The lines of section of the generating planes of the line-cone, with an arbitrary solid, are the generating lines of a quadric conical sheet, lying in this solid, whose vertex is on the axis of the line-cone. Conversely, a line-cone consists of the planes joining a line to the points of a conic whose plane does not meet the line.

The general homogeneous quadratic function, of the five coordinates which are appropriate for space of four dimensions, can be written, in infinitely many ways, as a sum of squares of at most five independent linear functions of the coordinates (cf. Vol. III, p. 15). When this sum consists of only four squares it represents a point-cone, when equated to zero; the line-cone is obtained when the number of these squares is three; if the number is two the line-cone degenerates into two solids meeting in a plane.

Any solid, passing through the vertex of a quadric point-cone, meets this in a conical sheet. The solid meets the two generating planes of the point-cone which lie in any tangent solid, each in a line, these two lines being generating lines of the conical sheet. If the solid contain the line of intersection of the two generating planes, then its plane of intersection with the tangent solid, which touches the point-cone at the points of this line, is the tangent plane of the conical sheet at the points of this line. This is clear by remarking that the tangent solid of the point-cone is that containing the vertex and a tangent plane of the quadric surface in which the point-cone meets an arbitrary solid. Similarly if a solid be drawn through the axis of a line-cone, this meets the line-cone in two planes which coincide when the solid is a tangent solid. Further, the tangent solids of a point-cone which pass through any point, not lying on the cone, are the tangent solids of a line-cone, whose axis joins the point to the vertex of the point-cone.

Resuming our discussion, let, as before, a, b, c, d be four non-intersecting lines, in space of four dimensions, of which the transversals of threes are a', b', c', d'; and let e be the associated line, common to the four solids $[a, a']$, $[b, b']$, $[c, c']$, $[d, d']$. Let E be any point of the line e; then, as d' is the transversal of a, b, c, the plane Ed' meets these lines; and this plane meets the line d, because e is in the solid $[d, d']$. Thus, by similar reasoning, any one of the four planes Ea', Eb', Ec', Ed' meets, in a line through E, every one of the four planes Ea, Eb, Ec, Ed. Therefore, these planes meet an arbitrary solid in two sets of four lines, belonging, respectively, to the two sets of generators of a quadric surface. Thence, through E can be drawn an infinity of planes each meeting all the lines a, b, c, d, these being the planes, of one set, of a quadric point-cone, of vertex E; as well as an infinity of planes each meeting all the lines a', b', c', d'. Conversely, as a line meeting three generators of the same system of a quadric surface, equally meets all the generators of that system, it follows that a plane which meets e and meets any three of the lines a, b, c, d, likewise meets the fourth. We have seen, however, that the five lines a, b, c, d, e are symmetrical; thus any plane meeting a, b, c, d also meets e. The argument, as assuming the last quoted property of a quadric surface, depends on Pappus' theorem.

Next, let O be any point of general position. If any point, P_4, be taken on the line d, a plane can be drawn through O and P_4 to meet the lines a, b, this being the plane through OP_4 which contains the transversal of the three lines OP_4, a, b. If this transversal meet a, b in P_1 and P_2, respectively, this plane can also be described as the plane through O and P_1 which meets b and d; for the line P_2P_4 is the transversal of the three lines OP_1, b, d. Consider the line in which the solid $[a, b]$ is met by the plane Od; as P_4 varies on d, the lines OP_4 and P_1P_2 meet in a point of this line. Thus the ranges (P_1), (P_4) are related. We can similarly construct a plane through O and P_1 which shall meet the lines c and d; if this plane meet d in P_4', the ranges (P_1), (P_4') are related, as P_1 varies on a. Thus the ranges (P_4), (P_4'), on d, are related. There are, therefore, in general, two positions of P_1, on a, for which P_4 and P_4' coincide. When this is so, there is a plane through O and P_1 which meets b, c and d. There are then, in general, two planes through O which meet a, b, c, d. But these may coincide for particular positions of O. Or, if three planes can be drawn through O to meet a, b, c, d, then an infinite number of such planes is possible, and a plane drawn through O to meet three of a, b, c, d necessarily meets the fourth. We have seen that this is the case when O is on the line e; thus, also, an infinite number of planes meeting a, b, c, d, e passes through any point of any one of these five lines.

The two planes through a general position of O which meet a, b, c, d, also meet e, as we have proved; let them meet e in U and T. Then, as we have seen, considering the point-cone of vertex U, containing the planes through U which meet a, b, c, d, and those which meet a', b', c', d', there will be, through OU, a plane meeting a', b', c', d'. A plane meeting a', b', c', d' can, similarly, be drawn through OT. These are then the two planes which, by a similar argument, can be drawn through O to meet a', b', c', d'.

The six systems of planes. There are six systems of planes, those of any system being the planes which meet the five lines of one of the six associated systems of lines. A plane of a system is determined, as the argument shews, by its intersections with two of the five associated lines; thus the aggregate of the planes of each system is ∞^2. Through a general point, O, can be drawn two planes of each system; taking two of the systems, if α, β be the planes of one of these through this point, and α', β' be the planes of the other system, there are two points, T, U, upon the line common to the two associated systems of lines, from which the planes are defined, such that OT is the line of intersection of the plane α with one of α', β', say with β', while, similarly, the planes β, α' meet in the line OU. We shall denote the systems of planes by the numbers 1, 2, ..., 6, these being the same as those marking the systems of associated lines, in the scheme given above (p. 114).

It follows that, if O be such that the planes α, β, of one system, drawn through O, coincide with one another, then the planes, α', β', of any other system, drawn through O, likewise coincide. In this case the planes of the various systems which pass through O are six in number, one of each system. One of these planes, say that meeting the lines a, b, c, d, e, will be met in five lines, passing through O, by the other five planes, these lines passing to the points where a, b, c, d, e meet this plane. But, two planes meeting in a line define a solid; and another plane, which meets each of these in a line, lies in this solid. Thus, for such a position of O as now contemplated, the six planes through O lie in a solid. It will be found immediately that there are ∞^3 such positions of O, of which four lie on an arbitrary line; we shall be much concerned with the locus so determined.

The various systems of planes are also of importance in connexion with the singular solids which have been described. Each of these solids can be generated by planes of any one of the six systems, these forming an axial pencil of planes whose axis is one of the six lines lying in the solid. More precisely, taking the singular solid previously denoted by 123, or 456, this is generated by planes of the system 1, all passing through the line 23; also by planes of system 2, all passing through the line 31; and so on; and, finally, by planes of the system 6, all passing through the line 45. For instance, if a plane in the solid be drawn through the line 23, this plane, beside meeting the other five lines which lie in this solid, meets the remaining lines, of the original fifteen lines, which meet the line 23; the plane thus meets the lines 12, 13, 56, 64, 45 and 14, 15, 16; as it meets, therefore, the lines 12, 13, 14, 15, 16, it is a plane of system 1. This argument is of general character. It follows also, from this, that the planes joining any point, O, of the singular solid, to the six lines lying in the solid, are of the six systems, respectively; it will be found that every point of the singular solid is such that the two planes of any system, drawn through this point, coincide with one another.

The planes of the figure deduced with the help of the symbols. The results in regard to the planes can also be obtained with great simplicity by use of the symbols. Thereby, too, we can define with greater precision the locus of points, O, through which only one plane of each system can be drawn.

A plane meeting the lines a, b, c, not containing any one of these lines, can evidently be defined by three points, one on each of the lines, with symbols, respectively, of the forms

$$yB + zC', \quad zC + xA', \quad xA + \eta B'.$$

For this plane to meet the line d, there must be a syzygy

$$\lambda\,(yB + zC') + \mu\,(zC + xA') + \nu\,(xA + \eta B') + \rho\,(B' + C) + \sigma\,(C' + A) = 0,$$

connecting these three points with the two points, P', Q', of the line d. As this must be equivalent to the fundamental relation between A, B, ..., C', it is easily seen that $\eta = y$. Thus, any general plane meeting a, b, c, d must contain three points, respectively on a, b, c, of symbols $yB + zC'$, $zC + xA'$, $xA + yB'$. Conversely, the plane of these points contains the point whose symbol is

$$x\,(yB + zC') + y\,(zC + xA') + z\,(xA + yB'),$$

or, save for sign, $yzP' + zxQ' + xyR'$; and this is a point of the line d. This plane also contains the point whose symbol is

$$(yB + zC') + (zC + xA') + (xA + yB'),$$

or, save for sign, $xL + yM + zN$; and this is a point of the line e. In virtue of $L + M + N = 0$, this point depends only on the ratio of any two differences of x, y, z.

Now let O be any general point. We can suppose it to have a symbol

$$O = \xi A + \eta B + \zeta C + \xi' A' + \eta' B' + \zeta' C',$$

where, in virtue of the fundamental syzygy, only the ratios of the differences of ξ, η, ζ, ξ', η', ζ' are determinate. That this point should lie on the plane containing the three points $yB + zC'$, $zC + xA'$, $xA + yB'$ requires an identity, in regard to A, B, C, A', B', C', of the form

$$O = \lambda (yB + zC') + \mu (zC + xA') + \nu (xA + yB') + \rho S,$$

where S denotes the sum of A, B, ..., C'. Using this to express ξ, η, ..., ζ', we can deduce the three equations

$$x^{-1} (\xi' - \xi) + y^{-1} (\eta' - \eta) + z^{-1} (\zeta' - \zeta) = 0,$$

$$x (\eta' - \zeta) + y (\zeta' - \xi) + z (\xi' - \eta) = 0,$$

$$(\eta - \zeta') (\eta' - \zeta) + \sigma^{-1} (\zeta - \xi') (\zeta' - \xi) - (1 + \sigma)^{-1} (\xi - \eta') (\xi' - \eta) = 0,$$

where σ denotes $(z - x)/(y - z)$. The first two of these equations determine two sets of ratios of x, y, z when ξ, η, ..., ζ' are given; that is, they determine the two planes meeting the lines a, b, c, d, e which pass through O. These planes meet the line e in the two points $\sigma_1 L - M$, $\sigma_2 L - M$, where σ_1, σ_2 are the two roots of the third equation. We easily find, also, that

$$(\eta - \zeta')(\eta' - \zeta) = (\zeta - \xi') (\zeta' - \xi)/\sigma_1 \sigma_2 = (\xi - \eta')(\xi' - \eta)/(1 + \sigma_1)(1 + \sigma_2);$$

when the two points of the line e are given, these are two conditions for O which will be of interest below.

When the point O lies on the line e, so that $\xi' - \xi = 0 = \eta' - \eta = \zeta' - \zeta$, the first equation is satisfied identically; the second equation has then the solution $x = \xi + \theta$, $y = \eta + \theta$, $z = \zeta + \theta$, where θ is arbitrary. Thus an infinite number of planes meeting a, b, c, d, e can be drawn through a point of e, as was seen. This is also so when O is on any one of a, b, c, d.

The equation of an associated quartic locus. If we put p, q, r, p', q', r', respectively, for $\eta - \zeta'$, $\zeta - \xi'$, $\xi - \eta'$, $\eta' - \zeta$, $\zeta' - \xi$, $\xi' - \eta$, it is easy to see that the third (quadratic) equation, above, gives equal roots for σ provided

$$(pp')^{\frac{1}{2}} + (qq')^{\frac{1}{2}} + (rr')^{\frac{1}{2}} = 0 ;$$

this is then the condition for points O from which the two planes of any system coincide with one another; and may be regarded as the equation of the locus of such points. The equation, when rationalised, is evidently of the fourth order, and there are four such points upon an arbitrary line.

We may with advantage reach this result in another way, defining O by means of one of the planes through it which meet the lines a, b, c, d, e, and the position of O in this plane. This is done by writing the symbol of the point O in the form

$$\lambda\,(yB + zC') + \mu\,(zC + xA') + \nu\,(xA + yB').$$

The other plane, through the point O, which meets a, b, c, d, e, can then be shewn to be that containing the points $y_1B + z_1C'$, $z_1C + x_1A'$, $x_1A + y_1B'$, where

$$x_1 = (\mu - \nu)(\mu z - \nu y)^{-1},\; y_1 = (\nu - \lambda)(\nu x - \lambda z)^{-1},\; z_1 = (\lambda - \mu)(\lambda y - \mu x)^{-1}.$$

The conditions that this other plane should coincide with the first plane, which are $x_1/x = y_1/y = z_1/z$, reduce to one condition, and lead, effectively, to

$$\lambda = x\,(t - x)^{-1},\; \mu = y\,(t - y)^{-1},\; \nu = z\,(t - z)^{-1},$$

where t is arbitrary. These values express that the point O lies on a conic in the plane of the points $yB + zC'$, $zC + xA'$, $xA + yB'$, of which the points are given by varying t. This conic is, in fact, that passing through the points in which the plane is met by the lines e, d, a, b, c; for these points are given, respectively, by the values 0, ∞, x, y, z of t, as is easily seen. If the symbol of the general point of this conic be written $\xi A + \eta B + \zeta C + \xi'A' + \eta'B' + \zeta'C'$, we see that, with $x' = t - x$, $y' = t - y$, $z' = t - z$, we have

$$\xi = 1/yz',\; \eta = 1/zx',\; \zeta = 1/xy',\; \xi' = 1/y'z,\; \eta' = 1/z'x,\; \zeta' = 1/x'y.$$

These values satisfy the irrational equation above given for the locus of the point O. Conversely, they lead to the equations

$$1 - x^{-1}t = 2qr\,(pp' - qq' - rr')^{-1},\; 1 - y^{-1}t = 2rp\,(qq' - rr' - pp')^{-1},$$
$$1 - z^{-1}t = 2pq\,(rr' - pp' - qq')^{-1}.$$

Thus the locus is of the kind called *rational*, being in (1, 1) birational correspondence with the points of a threefold space, in which x, y, z, t are coordinates. Through any point of the locus there pass six conics lying on the locus, each conic lying in a plane of one of the six systems, and meeting the five associated lines of that system; in the representation here given, one of these conics corresponds to a definite set of values for the ratios of x, y, z, the various points of the conic corresponding to different values of t. It will appear incidentally below that such parameters, replacing x, y, z, t, can be taken for each of the six conics. It is easy to shew that the four points, in which the locus is met by any line lying in the plane of one of the conics, consist of the two points in which the line meets the conic, each taken doubly. Thus all the conics, and the fifteen fundamental lines also, may be spoken of as being *double* upon the locus. We shall denote the locus by Σ.

Ex. It is easy to represent the points of the six conics that
pass through any point, O, of this locus; incidentally there arises
thus a proof that the six points, on one of these conics, constituted by O and the points where the conic is met by the
associated lines, form a range related to that of the corresponding
points on any other conic, if the points be taken in proper order.
A geometrical proof of this is given below. More precisely, taking
the scheme given above (p. 114), of the six sets of five associated
lines, let each entry be supposed, now, to mean the point where
the line, given by that entry, meets the proper plane of the six
drawn through O; and imagine the point O supplied as the diagonal
element of every row of this scheme. Then the points of the six
planes through O form related ranges, upon their respective conics,
when taken in the order given by this scheme. This can be shewn
by choosing a proper parameter for each of the six conics, and
shewing that the six points in question arise, on all the conics, for
the same values of the parameter. See, also, p. 131, below.

To put the results in concise form, let a, b, c be three parameters,
and use a', b', c', respectively, for $1-a, 1-b, 1-c$. Also, use the
symbols P, Q, R, P', Q', R', respectively, for $A/bc', B/ca', C/ab',$
$A'/b'c, B'/c'a, C'/a'b$, where A, B, C, A', B', C' are the symbols
previously used for the six fundamental points. We have seen that
any point of the locus Σ is given by a symbol $A/yz' + \ldots + C'/x'y$,
where $x' = t-x, y' = t-y, z' = t-z$; in particular, suppose that
O arises for the values $a, b, c, 1$, of x, y, z, t, respectively. If, in
the symbol $A/yz' + \ldots + C'/x'y$, say Ω, we replace x, y, z by a, b, c,
respectively, the points upon the conic, through O, which meets
the lines e, d, a, b, c, are given by varying t in the symbol

$$(t-a)^{-1} a' (Q+R') + (t-b)^{-1} b' (R+P') + (t-c)^{-1} c' (P+Q'),$$

as we have seen; the six points consisting of O and the points
upon the lines e, d, a, b, c arise for the values $t = 1, 0, \infty, a, b, c$,
respectively. If, in the general symbol Ω, we replace x, y, z, t,
respectively, by $t-a, t-b, t-c, t-1$, another conic is given by
varying t in the symbol thence arising, which is

$$(t-a)^{-1} a (Q'+R) + (t-b)^{-1} b (R'+P) + (t-c)^{-1} c (P'+Q);$$

this conic meets the line e, goes through the point O, and meets
the lines d', a', b', c'. These occur, respectively, also for the values
$t = 1, 0, \infty, a, b, c$. Again, if, in the same general symbol Ω, we
replace x, y, z, t, respectively, by $a(t-a)^{-1}, b(t-b)^{-1}, c(t-c)^{-1},$
$(t-1)^{-1}$, the symbol thence arising, namely

$$(t-a)^{-1} (P+P') + (t-b)^{-1} (Q+Q') + (t-c)^{-1} (R+R'),$$

gives another conic as t varies; this meets the lines d, d', goes
through the point O, and meets the lines l, m, n, these occurring

for the same respective values, $t = 1, 0, \infty, a, b, c$. And if, in the same general symbol, we replace x, y, z, t, respectively, by $t, b, c, 1$, the symbol thence arising,

$$P + P' + (1 - t)^{-1} a' (Q + R') + t^{-1} a (Q' + R),$$

likewise represents the points of a conic as t varies; this conic meets the lines a, a', l, goes through the point O, and meets the lines r, q, for the same respective values, $1, 0, \infty, a, b, c$ of t. Similarly if, in the same general symbol, we replace x, y, z, t, respectively, by $a, t, c, 1$, the symbol thence arising,

$$t^{-1} b (R' + P) + Q + Q' + (1 - t)^{-1} b' (R + P'),$$

represents the points of a conic, meeting the lines b, b', m, r, passing through the point O, and meeting the line p, for the same respective values of t. Finally, if, in the same general symbol, we replace x, y, z, t, respectively, by $a, b, t, 1$, there is a conic, of points given by

$$(1 - t)^{-1} c' (P + Q') + t^{-1} c (P' + Q) + R + R',$$

which meets the lines c, c', n, q, p, and goes through the point O, for the same respective values of t.

The tangent solid of the quartic locus Σ. When a locus, in space of four dimensions, is given by a single equation, in coordinates X, Y, Z, T, U, say $F(X, Y, Z, T, U) = 0$, it is easy to prove that the lines drawn through a point $(X_0, Y_0, Z_0, T_0, U_0)$, of this locus, which meet the locus in two coincident points here, generate a solid, called the *tangent solid*, whose equation is $DF_0 = 0$, where D denotes the operator $X\partial/\partial X_0 + ... + U\partial/\partial U_0$, and F_0 is F, with $X_0, ...$ substituted for $X, ...$ The lines through this point which meet the locus in three coincident points here, lie on a conical sheet, obtainable as the intersection of the tangent solid with the (∞^3) quadric given by $D^2F_0 = 0$; while the lines, through this point, meeting the locus in four coincident points here, are the six which are the intersection of this conical sheet with $D^3F_0 = 0$. When, as in what precedes, we use coordinates $\xi, \eta, ..., \zeta'$, six in number, of which only the differences are effective, the operator D, it can be seen, is replaced by $\xi\partial/\partial\xi_0 + ... + \zeta'\partial/\partial\zeta_0'$. Hence it may be directly computed, by forming the rational equation for Σ from $[(\eta - \zeta')(\eta' - \zeta)]^{\frac{1}{2}} + ... = 0$, that the tangent solid of the locus Σ, at the point O, or $A/bc' + ... + C'/a'b$, has the equation

$$bc'\xi + ca'\eta + ab'\zeta - b'c\xi' - c'a\eta' - a'b\zeta' = 0.$$

This equation, however, is easily obtained otherwise, by first proving that any line, in the plane of any one of the six conics of Σ passing through O, discussed above, meets the locus in two coincident points at each of its two intersections with the conic, and thence inferring that the tangent solid of the locus, at any point of one of these conics, contains the plane of this conic. For, the general solid, through the plane containing the three points $bB + cC'$, $cC + aA'$, $aA + bB'$, has the equation

$$b(t-c)\,\xi + c\,(t-a)\,\eta + a\,(t-b)\,\zeta - c\,(t-b)\xi' - a(t-c)\eta' - b(t-a)\zeta' = 0,$$

in which t is arbitrary, as is quite clear. The value of t for which this passes through the point O is $t = 1$; and the equation then agrees with the above. It is verified at once, moreover, that the tangent solid, as given by this equation, contains the six conics passing through the point O; it is the solid in which the six planes of these conics were proved to lie. In passing, it may be remarked that the tangent solid contains the tangent lines of these conics at the point O, these being the lines through this point which meet the locus Σ in four coincident points there; these six lines lie on a quadric cone; it will be proved that the six planes touch a quadric cone.

Ex. If we put $u = bc'$, $v = ca'$, $w = ab'$, $u' = -b'c$, $v' = -c'a$, $w' = -a'b$, the equation of the tangent solid is given by

$$u\xi + v\eta + w\zeta + u'\xi' + v'\eta' + w'\zeta' = 0,$$

with the two identities

$$u + v + w + u' + v' + w' = 0, \quad uvw + u'v'w' = 0.$$

Conversely, the point of contact depends on the values a, b, c expressed by

$$a = -(v' + w)/(u' + u), \quad b = -(w' + u)/(v' + v), \quad c = -(u' + v)/(w' + w).$$

The equations of the singular solids. The tangent solid at a general point of the locus Σ, which, as we have seen, is capable of the equation $y\,(t - z)\,\xi + \ldots - (t - x)\,y\zeta' = 0$, or

$$yz\,(\xi - \xi') + zx\,(\eta - \eta') + xy\,(\zeta - \zeta')$$
$$+ tx\,(\eta' - \zeta) + ty\,(\zeta' - \xi) + tz\,(\xi' - \eta) = 0,$$

meets the locus Σ in a quartic surface. There are, however, ten tangent solids for which this quartic surface reduces to a repeated quadric surface; these are in fact the ten singular solids discussed above. They arise for the respective values of x, y, z, t given by

$$(0, y, z, 0), \qquad (x, 0, z, 0), \qquad (x, y, 0, 0), \qquad (1, 1, 1, 0),$$
$$(0, 1, 1, 1), \qquad (1, 0, 1, 1), \qquad (1, 1, 0, 1),$$
$$(x, 0, 0, t), \qquad (0, y, 0, t), \qquad (0, 0, z, t),$$

and have the respective equations

$$\xi - \xi' = 0, \quad \eta - \eta' = 0, \quad \zeta - \zeta' = 0, \quad \xi + \eta + \zeta - \xi' - \eta' - \zeta' = 0.$$
$$\eta - \zeta' = 0, \qquad \zeta - \xi' = 0, \qquad \xi - \eta' = 0,$$
$$\eta' - \zeta = 0, \qquad \zeta' - \xi = 0, \qquad \xi' - \eta = 0.$$

For instance, the solid represented by the equation $\eta - \zeta' = 0$ meets the locus Σ in the repeated quadric surface given by

$$[(\zeta - \xi')(\zeta' - \xi) - (\xi - \eta')(\xi' - \eta)]^2 = 0,$$

as follows from the irrational equation of the locus. But the equation is capable of other irrational forms; for example, two of these are

$$[(\xi - \xi')(\eta - \zeta')]^{\frac{1}{2}} + [(\eta - \eta')(\zeta - \xi')]^{\frac{1}{2}} + [(\zeta - \zeta')(\xi - \eta')]^{\frac{1}{2}} = 0,$$

$$[(\xi + \eta + \zeta - \xi' - \eta' - \zeta')(\xi' - \xi)]^{\frac{1}{2}} + [(\zeta - \xi')(\xi' - \eta)]^{\frac{1}{2}}$$
$$+ [(\zeta' - \xi)(\xi - \eta')]^{\frac{1}{2}} = 0,$$

as may be verified by the parametric forms which have been given for the general point of the locus. These shew that the ten particular solids do each meet the locus in a repeated quadric surface.

That these particular solids agree with the singular solids as previously defined, may then be verified by shewing that these quadric surfaces contain the lines, of the original fifteen, which are characteristic of the singular solids. It appears that the solids, arranged as above (p. 129), are in fact the singular solids given by the respective notations

124, or 356,   125, or 364,   126, or 345,   123, or 456
       234, or 156   235, or 146   236, or 145
       134, or 256   135, or 246   136, or 245.

Beside the six lines, from the original fifteen, characteristic of any one of the singular solids, there are, upon the quadric surface lying in this solid, infinitely many other lines, which, therefore, also lie upon the locus Σ. The fifteen original lines are distinguished from these by the fact that they are double lines of the locus; every line drawn through a point of one of the fifteen lines has two of its four intersections with Σ coincident at this point.

The intersections of a tangent solid of Σ with the fifteen fundamental lines. We consider the tangent solid of the locus Σ at a point O, and the fifteen points in which this solid is met by the fifteen fundamental lines, denoting the intersection with the line 12 by (12), and so on. We thus have, in all, sixteen points in the tangent solid. We also have sixteen planes in the tangent solid, constituted by the six planes, of the diverse systems, which pass through O, and the ten planes of intersection of the singular solids with the tangent solid. Each of the six planes, in the tangent solid, through O, contains six of the points, namely O and five points on the associated lines which meet this plane. Each of the ten planes, in which the tangent solid is met by the singular solids, likewise contains the six points lying on the lines which characterise the singular solid. And six of the planes pass through every one of the sixteen points. This has been seen for the point O. For the point (12), for example, these six planes are the intersections of the tangent solid with the singular solids 123, 124, 125, 126, all of which contain the line 12, together with the two planes, of systems 1 and 2, which lie in the tangent solid, through O; for these two planes meet the line 12 in the same point, which must lie in the tangent solid at O. But it can be shewn further that the six points, on any one of the sixteen planes, lie on a conic, and that the six planes, through any one of the sixteen points, touch a quadric cone; and also, that the ranges of six points on the conics, and the sets of six tangent planes of the cones, are thirty-two related sets, if the elements be taken in proper order. With the indices 1, 2, ..., 6, this order can be specified by a symbolical rule which is of great simplicity.

Before indicating the proof of these statements we make two remarks: (1) If two triads of points upon a conic be considered, each with its three joining lines, the six joins touch another conic; and, among the tangents of this other conic, these joins are a set related each to its opposite point on its own triad, the points being regarded as belonging to a range of the original conic (Vol. II, p. 29, Ex. 7); (2) The line which joins any two of the sixteen points of the tangent solid, is also the intersection of two of the sixteen planes lying in this solid. For instance, consider the line joining the points (12), (13). This lies in the plane of system 1 through O, and also lies in the plane (123), in which the solid 123 (or 456) meets the tangent solid. Or, again, consider the line joining the points (12), (34). This lies in both the singular solids 125 (or 346) and 126 (or 345), and, therefore, in the planes, (125) and (126), in which these solids meet the tangent solid.

The proof of the statements made depends on the following circumstance: let the six points, in one of the sixteen planes, be divided into two triads, in any one of the ten possible ways; then, among the ten remaining points, which do not lie in this plane, there is one which is such that the six planes through this point contain the six joining lines of the points of the two triads. To prove this, take, first, the six points (23), (31), (12), (56), (64), (45), lying in the plane (123), arising from the singular solid 123 (or 456). These six points lie on a conic because the six lines in the singular solid lie on a quadric surface. The join of (31) and (12) lies on the plane of system 1 through O; and so on. Thus the six planes, of the systems 1, 2, 3, 4, 5, 6, through O, touch a quadric cone; and further, as tangent planes of this cone, they are related, in this order, respectively to the points of the conic, in the order taken. Take, next, the same six points, but arranged as the two triads (31), (12), (56) and (64), (45), (23). The six joins of these then lie on the six planes passing through the point (14). For instance, the join of (12), (56) lies on the solid 124 (or 356), which contains the line 14, as well as the lines 12 and 56 ; in the same way, the join of (56) and (31) lies on the solid 314 (or 256), which contains the line 14, as well as the lines 56 and 31 ; and likewise, the join of (31), (12) lies on the plane of system 1 through O, which also contains the point (14). The other triad may be similarly considered. Then, first, we have the result that, the six points (31), (12), (56), (64), (45), (23), in the range of points of the conic on which they lie, are related, respectively, to the six planes, through the point (14), obtained in order by, the solids 124, 314, the plane of system 1 through O, the solids 145, 146, and the plane of system 4 through O, these planes being considered in the series of tangent planes of the quadric cone which they touch. But, second, rearranging the points, and using the result found before, we see that the planes through O, of systems 1, 2, ..., 6 are, respectively, related to the six planes (4), (124), (134), (1), (145), (146), through the point (14). And we notice that the notations for these latter six planes are obtainable by a symbolical multiplication, of the symbol 14, respectively with 1, 2, ..., 6. Now consider, finally, the points in the plane of system 1 through O, taken in the two triads O, (12), (13) and (14), (15), (16). We shew that the joins of these triads lie on the six planes through the point (23). In fact, the three joins (12), (13); (13), O; O, (12) lie, respectively, on the solid 123, the plane 3 through O, and the plane 2 through O; while the three joins (15), (16); (16), (14); (14), (15) lie, respectively, on the solids 234, 235, 236. Assuming that it has been proved as above, that the six planes through (23), so arising, touch a quadric cone, and are related, respectively, as planes of this cone, to the planes through O of systems 1, 2, ..., 6 (these being the numbers obtained by symbolical multiplication, with 23, of the notations for the six planes), it follows that the points O, (12), (13), (14), (15), (16), of plane 1 through O, lie on a conic, and are related to the planes through O, respectively, in the order taken.

We can thus take the six points of any one of the sixteen planes, as also the six planes through any one of the sixteen points, in such an order that they are associated, respectively, with the numbers 1, 2, ..., 6. The association is obtained, in the former case, by symbolical multiplication of the symbol of the plane, respectively, with 1, 2, ..., 6; and, in the latter case, of the symbol of the point. Thus, for example, the number 1 is associated with the point (12) lying in the plane 2 through O, and is associated with the point (23) of the plane (123), in which the tangent solid is met by the singular solid 123 (or 456); or again, the number 1 is associated with the plane 3 through *O*, regarded as a plane through the point (13), and is associated with the plane (134), through the point (34). This association enables us to make the sixteen points (as also the sixteen planes) correspond in pairs, in any one of fifteen ways. For instance, take the two numbers 1, 2. From any point, we may pass, with the number 1, to a particular plane through this point; and then, with the number 2, we may pass from this plane to a particular one of the six points which it contains. The second point is evidently obtained from the first point by symbolical multiplication with the numbers 1 and 2, in either order; for instance, if the first point be (45), the second is (36).

What has been said suggests the arrangement of the sixteen points in such a scheme as that adjoined. This is typical of ten such schemes, of which the other nine are obtainable from this by interchange of any one of the numbers 1, 2, 3 with any one of 4, 5, 6. If, in this scheme, we take any element in the last row, or in

(14),	(15),	(16),	(23)
(24),	(25),	(26),	(31)
(34),	(35),	(36),	(12)
(56),	(64),	(45),	*O*

the last column, other than *O*, the other six elements in the same row and column, as this chosen, are the points lying in one of the six planes through *O*. But, the points, given by the elements of the scheme, which are in the same row and column as any of the other ten elements, are those in one of the ten planes arising from the singular solids. Further, considering the elements of the scheme as consisting of the four tetrads which are in the four rows, the points given by any two of these tetrads are at once seen to be Moebius tetrads, inscribed to one another (Vol. I, pp. 61, 91). For instance, taking the first and second rows, the point (14) lies in the plane of the points (25), (26), (31), since these four points are in the singular solid 256 (or 314); similarly, the point (25) lies in the same plane with (14), (16), (23), these four points being in the singular solid 146 (or 235). The same relation holds for the four tetrads given by the columns of the scheme. Moreover, there is a quadric surface in regard to which the four tetrads given by the four rows are all self-polar; and, taking the four lines joining elements in the same column, in the first two rows of the scheme, and, likewise, the four lines joining elements in the same column, in the last two rows, these eight lines have two common transversals, which are generators of this quadric. By taking the four rows in two pairs, in this way, three sets each of two generators are obtained. These are all of the same system, of generators of the quadric, and the two of any set are harmonic conjugates in regard to those of either of the other two sets (cf. Vol. III, pp. 68, 138, 143).

**The intersections of a tangent solid of Σ with the planes of
the six systems.** In space of four dimensions, a plane meets a solid
in a line. Thus the ∞^2 planes of any one of the six systems,
which have been considered, give rise to ∞^2 lines in the tangent
solid of Σ at a particular point, O. As these planes all meet three
lines, not lying in the tangent solid, whose transversal does not lie
in this solid (and this in various ways), we infer that the ∞^2 lines,
obtained in the tangent solid, belong to (at least one) tetrahedral
complex of lines in this solid (p. 32, above). Through any general
point of the tangent solid there pass, we have seen, two planes of
any one of the six systems, giving rise to two lines of the system
considered. We may expect, therefore, that the lines of this
system belong also to a linear (as well as to a tetrahedral) complex.
This we now prove, in a direct way, which also shews that the six
linear complexes so arising are mutually conjugate or apolar
(cf. p. 42, above). General results in regard to quadratic congruences of lines are given below (Chap. VII); the sixteen points
and planes we have considered, in what precedes, then arise as
singular points and planes. The proof we now give is, in part,
algebraic; another direct proof, of a synthetic kind, arises below
(p. 155), from a dual point of view.

Consider any particular line, *of the second system*, obtained by
the intersection of the tangent solid with the plane of the second
system containing the three points $\eta B' + \zeta C$, $\zeta C' + \xi A$, $\xi A' + \eta B$.
We first prove that any line of the first system which meets this
line, also meets another determinate line of the second system.
Through any point, say P, of the given line of the second system,
there pass, we have remarked, two lines of the first system; we are
to prove that both these meet another line of the second system,
whose determination is independent of the position of P upon the
first line. The plane of the second system, (ξ, η, ζ), referred to,
meets the line e, or 12, in a particular point, say T; through this
point there pass ∞^1 other planes of the second system, meeting the
tangent solid, at O, in the generating lines, of one system, of a
quadric surface, as we have seen. Through T there pass, also,
∞^1 planes of the first system, giving rise to the other system of
generators of the quadric surface spoken of; one of these passes
through P; it meets all lines of the second system which lie on
planes through T. There is through P another line of the first
system; this line, in fact, is not on a plane of the first system
passing through T, and does not lie on the quadric surface, in the
tangent solid at O, obtained by planes through T. This line meets
this quadric surface in another point, say P'. Through P' there
passes a generator of the quadric surface, lying on a plane of the
second system through T; this is of the same system as the gene-

rator, (ξ, η, ζ), on which P lies. We shew that this second generator, through P', depends only on ξ, η, ζ, and not on the position of P upon the generator (ξ, η, ζ). This second generator is then met by both the lines, obtained by planes of the first system, through all points, P, of the generator (ξ, η, ζ).

For, the point common to the plane defined by the points $\eta B' + \zeta C$, $\zeta C' + \xi A$, $\xi A' + \eta B$, and the plane, of the first system, defined by the points $yB + zC'$, $zC + xA'$, $xA + yB'$, is the point whose symbol is

$$x^{-1}\xi\,(\eta B' + \zeta C) + y^{-1}\eta\,(\zeta C' + \xi A) + z^{-1}\zeta(\xi A' + \eta B),$$

its coordinates, relative to A, B, C, A', B', C', being

$$y^{-1}\eta\xi, \quad z^{-1}\zeta\eta, \quad x^{-1}\xi\zeta, \quad z^{-1}\zeta\xi, \quad x^{-1}\xi\eta, \quad y^{-1}\eta\zeta;$$

the condition that this point lies on the tangent solid at O, whose equation is $bc'X + \ldots = b'cX' + \ldots$, where $a' = 1 - a$, etc., is

$$\xi^{-1}a'x\,(bz - cy) + \eta^{-1}b'y\,(cx - az) + \zeta^{-1}c'z\,(ay - bx) = 0.$$

This assumes that the two planes do not meet in a line; that is, do not meet the line e in the same point; for which the condition is

$$\xi\,(y - z) + \eta\,(z - x) + \zeta\,(x - y) = 0.$$

In order to prove what we have stated it is thus necessary and sufficient to prove that, if these two conditions are satisfied by one set of values of ξ, η, ζ, then they are satisfied by another set of values, which depend on ξ, η, ζ only, and are independent of x, y, z. Such another set, ξ_1, η_1, ζ_1, is, in fact, given, in terms of the former set, by

$$\xi_1\,(\eta - \zeta) = a'\,(\eta c' - \zeta b'), \quad \eta_1\,(\zeta - \xi) = b'\,(\zeta a' - \xi c'),$$
$$\zeta_1\,(\xi - \eta) = c'\,(\xi b' - \eta a').$$

For the former condition is equivalent, with proper values of λ, μ, to the three

$$\xi^{-1}a' = \lambda + \mu ax^{-1}, \quad \eta^{-1}b' = \lambda + \mu by^{-1}, \quad \zeta^{-1}c' = \lambda + \mu cz^{-1},$$

leading to $\eta^{-1}b' - \zeta^{-1}c' = \mu\,(by^{-1} - cz^{-1})$, etc. Thus, what we wish to verify as to ξ_1, η_1, ζ_1 satisfying the first condition, is clear, this being that the sum of three such quantities as

$$x\,(bz - cy)\,(\eta^{-1} - \zeta^{-1})\,(\eta^{-1}b' - \zeta^{-1}c')^{-1}$$

vanishes. The latter condition, for proper ρ, σ, is equivalent to the three equations $\xi = \rho x + \sigma$, etc.; what we wish to verify for this is thus the obvious fact that the sum of $\xi_1\,(\eta - \zeta)$, $\eta_1\,(\zeta - \xi)$ and $\zeta_1\,(\xi - \eta)$ vanishes.

It can now be further verified that, if we form the same functions of ξ_1, η_1, ζ_1 that ξ_1, η_1, ζ_1 are of ξ, η, ζ, we shall obtain the quantities ξ, η, ζ, save for a common factor. These are then obtained from ξ_1, η_1, ζ_1 just as ξ_1, η_1, ζ_1 are obtained from ξ, η, ζ.

Whence, considering, first, only those lines in the tangent solid at O, that lie upon planes of the second system which pass through a particular point of the line e, these being the generators, of one system, of a particular quadric surface, we have shewn that all lines, obtained by intersection of the tangent solid with planes of the first kind, meet this quadric surface on generators which are pairs of an involution of these generators, or, are themselves generators of this quadric, of the opposite system. From this it follows that these lines of the first kind belong to a linear complex of lines, in the tangent solid at O (Vol. III, p. 67). By parity of reasoning, considering now all the lines, in the tangent solid, obtained by planes of the second kind, these will also belong to a linear complex. But further, as we have shewn that all lines of the first linear complex, which meet any particular line of the second linear complex, likewise meet another line of the second linear complex, it follows that the pair of associated lines of the second complex are polar lines of one another in regard to the first complex. Thus either complex consists of pairs of lines which are polar lines of one another in regard to the other complex. Whence it follows that the two complexes are apolar, or conjugate (Vol. III, p. 65. Above, p. 42). A similar argument applies to any two of the six systems of lines in the tangent solid. The fact that any two of the linear complexes are conjugate involves the fact, remarked above (p. 133), that we can associate together two of the sixteen points of the tangent solid, in one of fifteen ways; the converse is also true.

Ex. 1. Taking any two points, P, Q, given, relatively to A, B, ..., C', respectively by

$$P = X_1 A + ... + Z_1' C', \text{ and } Q = X_2 A + ... + Z_2' C',$$

and putting

$$S_1 = yz(X_1 - X_1') + zx(Y_1 - Y_1') + xy(Z_1 - Z_1'),$$
$$U_1 = x(Y_1' - Z_1) + y(Z_1' - X_1) + z(X_1' - Y_1),$$

with a corresponding meaning for S_2, U_2, prove that the planes of the first kind, joining three points $yB + zC'$, $zC + xA'$, $xA + yB'$, which meet the line joining the two points P, Q, are given by the values of x, y, z which satisfy the equation $S_1 U_2 - S_2 U_1 = 0$. Regarded as representing a locus in the plane (x, y, z), this equation represents a cubic curve passing through the four points $(1, 0, 0)$, $(0, 1, 0)$, $(0, 0, 1)$, $(1, 1, 1)$. When the two points P, Q are both in the tangent solid at O, this point being given, as before, by the coordinates $(bc')^{-1}$, $(ca')^{-1}$, etc., the equations $S_1 + U_1 = 0$, $S_2 + U_2 = 0$ are both satisfied by $(x, y, z) = (a, b, c)$, and the cubic curve contains the fifth point (a, b, c). When the two points P, Q, not necessarily lying in the tangent solid at O, are in a plane of the second kind, being, respectively,

$p_1(\eta B' + \zeta C) + q_1(\zeta C' + \xi A) + r_1(\xi A' + \eta B)$, and $p_2(\eta B' + \zeta C)$ + etc.,

the cubic locus breaks into two, becoming

$$[\xi uyz + \eta vzx + \zeta wxy][x(\eta - \zeta) + y(\zeta - \xi) + z(\xi - \eta)] = 0,$$

where $u = q_1 r_2 - q_2 r_1$, $v = r_1 p_2 - r_2 p_1$, etc. If the two points satisfy both conditions, being on a line of the second kind in the tangent solid at O, the ratios of u, v, w are the same as of $a(\eta c' - \zeta b')$, $b(\zeta a' - \xi c')$, $c(\xi b' - \eta a')$. Both the ratios of ξu, ηv, ζw, and the ratios of $\eta - \zeta$, $\zeta - \xi$, $\xi - \eta$, are unaltered by replacing ξ, η, ζ by ξ_1, η_1, ζ_1, where $\xi_1(\eta - \zeta) = a'(\eta c' - \zeta b')$, etc., as above.

Ex. 2. If ξ, η, ζ, ξ', η', ζ' be the coordinates of any point relatively to A, B, C, ..., then independent coordinates, for a point in the tangent solid at O, may be taken to be X, Y, Z, T, where

$$T = (c - a)(ab'\zeta - a'b\zeta') - (a - b)(ca'\eta - c'an'),$$

and $X = \xi - \xi'$, $Y = \eta - \eta'$, $Z = \zeta - \zeta'$. The tetrad of reference for these coordinates consists of the four points where the tangent solid is met by the four lines 34, 35, 36, 12 (or l, m, n, e). If then the line coordinates of any line in the tangent solid be defined by writing its equations in the forms

$$l'T + mZ - nY = 0, \quad l'X + m'Y + n'Z = 0,$$

and we put

$$P = x(b - c) + y(c - a) + z(a - b),$$
$$Q = yza(b - c) + zxb(c - a) + xyc(a - b),$$

it can be shewn that the plane of the first system, containing the points $yB + zC'$, $zC + xA'$, $xA + yB'$, meets the tangent solid at O in the line whose coordinates, l, m, n, l', m', n', are, respectively,

$$a'(b - c)xQ, \quad b'(c - a)yQ, \quad c'(a - b)zQ, \quad yzP, \quad zxP, \quad xyP.$$

These coordinates satisfy the equation

$$l/a' + m/b' + n/c' - l'a(b - c) - m'b(c - a) - n'c(a - b) = 0,$$

expressing that the line belongs to a linear complex, beside the equations $ll' + mm' + nn' = 0$, $ll'/a' + mm'/b' + nn'/c' = 0$, corresponding to a tetrahedral complex.

Similarly, a plane of the second kind, joining the points $y_1 B' + z_1 C$, $z_1 C' + x_1 A$, $x_1 A' + y_1 B$, if

$$P_1 = x_1(b - c) + \text{etc.}, \quad \text{and} \quad Q_1 = y_1 z_1 a'(b - c) + \text{etc.},$$

gives a line whose coordinates are

$$a(b - c)x_1 Q_1, \quad b(c - a)y_1 Q_1, \quad c(a - b)z_1 Q_1, \quad y_1 z_1 P_1, \quad z_1 x_1 P_1, \quad x_1 y_1 P_1.$$

These satisfy the equation

$$l/a + m/b + n/c - l'a'(b-c) - m'b'(c-a) - n'c'(a-b) = 0,$$

beside $ll' + $ etc. $= 0$ and $ll'/a + $ etc. $= 0$. Evidently the two linear complexes thus obtained are conjugate to one another.

Ex. 3. With X, Y, Z, T as in Ex. 2, the points of the quadric surface, in the tangent solid at O, obtained by planes of the second kind drawn through the point $\xi(A + A') + \eta(B + B') + \zeta(C + C')$, are expressible, in terms of the ratios of three parameters, x, y, z, in the forms

$$X = (y-z)(Q\nu - R\mu), \quad Y = (z-x)(R\lambda - P\nu), \quad Z = (x-y)(P\mu - Q\lambda),$$

$$T = \lambda(b-c)QR + \mu(c-a)RP + \nu(a-b)PQ,$$

where

$$P = bc'y - b'cz, \quad Q = ca'z - c'ax, \quad R = ab'x - a'by,$$

$$\lambda = \eta - \zeta, \quad \mu = \zeta - \xi, \quad \nu = \xi - \eta.$$

The plane sections of this quadric surface correspond to conics, in a plane (x, y, z), which pass through the two points $(1, 1, 1)$, $(a^{-1}a', b^{-1}b', c^{-1}c')$. The generators of the quadric surface, on the planes of the second kind, correspond to the lines in this plane which pass through the second of these points. The planes of the first and second kinds, defined, respectively, by

$$(yB + zC', \quad zC + xA', \quad xA + yB') \text{ and } (\eta B' + \zeta C, \quad \zeta C' + \xi A, \quad \xi A' + \eta B),$$

meet on the tangent solid at O if

$$x^{-1}\xi = a^{-1}a', \quad y^{-1}\eta = b^{-1}b', \quad z^{-1}\zeta = c^{-1}c'.$$

Ex. 4. If, instead of X, Y, Z, T in Ex. 2, we use $X_1 = X/a'$, $Y_1 = Y/b'$, $Z_1 = Z/c'$, $T_1 = T/a'b'c'$, prove that the intersection of the locus Σ with the tangent solid at O has the equation $P^{\frac{1}{2}} + Q^{\frac{1}{2}} + R^{\frac{1}{2}} = 0$ where $\quad P = a(b-c)X_1[-a'T_1/a + (c-a)Z_1 - (a-b)Y_1]$,

with similar values for Q and R. This equation represents a quartic surface. It has sixteen double points, namely at O, and at the points where the tangent solid is met by the fifteen fundamental lines. These double points lie, in sets of six, in sixteen planes, each of which touches the surface at all the points of a conic. The lines in which the tangent solid is met by the planes of any one of the six systems, belonging both to a quadratic (tetrahedral) complex, and to a linear complex, are said to form a quadratic *congruence*. As any line in one of these planes meets the locus Σ in only two points, each a pair of comcident points, the lines of the quadratic congruence are double tangents of the quartic surface. As the two planes of any system, which pass through a point of the locus Σ, coincide with one another, the points of the quartic surface are those points, in the tangent solid at O, for which the two lines of the quadratic congruence, generally possible through any point, coincide with one another. Any tangent plane of the quartic surface meets it in a plane quartic curve, to which six tangent lines can be drawn from the (double) point of contact of the plane. These are the six bitangents of the quartic surface which are possible through any point of it, arising from the six planes of the various systems through any point of Σ. See, also, p. 142, below.

The quartic surface is generally called *Kummer's quartic surface.* We return to the consideration of it below (Chap. VII), from the point of view of space of five dimensions; and it will be proved that any quadratic congruence of lines lies in forty tetrahedral complexes.

Ex. 5. A set of six linear complexes of lines, in space of three dimensions, of which every two are conjugate, can be constructed from six arbitrary points, of which no four lie in a plane. It has been said (Vol. III, p. 68, Ex. 26) that, if five points, A, B, C, D, E, be taken, in threefold space, in a particular order, there is a polar system in which each of these points has, for its polar plane, the plane through this point which contains the two contiguous points; so that the five sides of the pentagon are lines of the associated linear complex. In this polar system, the polar line of the joining line of two alternate points of this pentagon, for example of the join AD, is the transversal, drawn from the intermediate point, E, to the two sides of the pentagon, AB and CD, which are contiguous to the side, BC, opposite to E. Hence, if A, P, Q, B, C, D be any six points of the threefold space, and two polar systems be determined, respectively by the two ordered pentagons A, P, Q, B, C and A, Q, P, B, D (wherein the four points A, P, Q, B are the same, but P, Q occur in reverse orders), then these polar systems, or the associated linear complexes, are conjugate to one another. For, the line AP, which is self-polar in the first system, has, for polar in the second system, the transversal drawn from Q to PB and AD; while this transversal, drawn from Q in the plane PQB, is likewise self-polar in the first system. Namely, the linear complex, associated with the first polar system, contains two lines which are polars of one another in the second polar system. This is a sufficient condition for the two associated linear complexes to be conjugate (cf. p. 42, above, Ex. 1).

These preliminaries being clear, take, in the threefold space, six points, A, B, C, P, Q, R. We can then define six polar systems respectively by the ordered pentagons

$$B, R, Q, C, A; \quad C, P, R, A, B; \quad A, Q, P, B, C;$$
$$B, Q, R, C, P; \quad C, R, P, A, Q; \quad A, P, Q, B, R;$$

these we denote, respectively, by α, β, γ and α', β', γ', and prove that every two of them are conjugate. For this it is sufficient to prove that the four pairs, (β, γ), (β', γ'), (α, α') and (α, β'), are conjugate. But, β, γ are determined, respectively, by the two pentagons A, B, C, P, R and A, C, B, P, Q; β' and γ' are determined by the two pentagons R, P, A, Q, C and R, A, P, Q, B; α and α' are determined by the two pentagons B, R, Q, C, A and B, Q, R, C, P; while the two systems α, β' are determined, respectively, by the two pentagons R, Q, C, A, B and R, C, Q, A, P. In every case there are four common points to the two pentagons, of which two come in reverse orders. This proves the statement, by what is said above.

Recurring to the scheme given above (p. 133), it will be found that we may take, for A, B, C, a set of three points in one of the rows of this scheme, then, for P, a point of another row, in the same column as A; for Q, a point of another row, in the same column as B; and for R, the point of the remaining row which is in the same column as C.

Ex. 6. In the preceding Example, if the planes BPC, CQA, ARB meet in D, and, referred to A, B, C, D, the points P, Q, R be, respectively, $(0, q_1, r_1, k_1)$, $(p_2, 0, r_2, k_2)$, $(p_3, q_3, 0, k_3)$, and we take $\theta_1, \theta_2, \theta_3$ so that

$$q_3 r_2 \theta_1 = p_2 k_3 - p_3 k_2, \quad r_1 p_3 \theta_2 = q_3 k_1 - q_1 k_3, \quad p_2 q_1 \theta_3 = r_1 k_2 - r_2 k_1,$$

prove that the six linear complexes considered have the respective equations

$$l - \theta_1 l' = 0, \quad m - \theta_2 m' = 0, \quad n - \theta_3 n' = 0,$$

and

$$k_1 \theta_1^{-1} L + r_1 M - q_1 N = 0, \quad -r_2 L + k_2 \theta_2^{-1} M + p_2 N = 0,$$
$$q_3 L - p_3 M + k_3 \theta_3^{-1} N = 0,$$

where

$$L = l + \theta_1 l', \quad M = m + \theta_2 m', \quad N = n + \theta_3 n'.$$

It is easy to see that any two of these are conjugate.

Without loss of generality, we may replace the coordinates, x, y, z, t, of a point, respectively, by lx, my, nz, t. Thereby, for $\theta_1, \theta_2, \theta_3$, we obtain, respectively, $mn\theta_1/l, nl\theta_2/m, lm\theta_3/n$. Thus we can choose l, m, n so that $\theta_1, \theta_2, \theta_3$ have any assigned numerical values.

Ex. 7. Prove that, in the figure of fifteen lines, which we have considered, in space of four dimensions, there are sixty pairs of non-intersecting lines, and forty-five pairs of intersecting lines. Also, that there are eighty triads of non-intersecting lines, namely sixty triads with symbols such as 12, 13, 14, and twenty triads with symbols such as 23, 31, 12. These eighty triads correspond to the eighty *tetrads* of points represented by such schemes as that given above (p. 133).

Ex. 8. Prove that if, in the tangent solid of the locus Σ at the point O, we take coordinates given by $X = \eta - \zeta'$, $Y = \zeta - \xi'$,

$Z = \xi - \eta'$, $T = bc\,(\xi - \xi') + ca\,(\eta - \eta') + ab\,(\zeta - \zeta')$, for which the points of reference are the intersections of the tangent solid with the lines 14, 15, 16, 23 (or a, b, c, d'), then the line coordinates, of the line in which the plane $(yB + zC'$, $zC + xA'$, $xA + yB')$ meets the tangent solid, are

$$(b - c)\,UV'W', \quad (c - a)\,VW'U', \quad (a - b)\,WU'V',$$
$$- a'U'VW, \quad - b'V'WU, \quad - c'W'UV,$$

where $U = y - z$, etc., and $U' = yc - zb$, etc. These satisfy the equations

$$ll' + mm' + nn' = 0, \quad ll'/a' + mm'/b' + nn'/c' = 0,$$

and the equation

$$l + m + n - a\,(b - c)\,l'/a' - b\,(c - a)\,m'/b' - c\,(a - b)\,n'/c' = 0.$$

The corresponding results may be obtained when the coordinates

$$X = \eta' - \zeta, \quad Y = \zeta' - \xi, \quad Z = \xi' - \eta,$$

and $\qquad T = b'c'\,(\xi - \xi') + c'a'\,(\eta - \eta') + a'b'\,(\zeta - \zeta')$

are taken, for which the reference points are the intersections of the tangent solid with the lines 24, 25, 26, 31 (or a', b', c', d). Or when the coordinates

$$X = c\eta - b\zeta' + c'\eta' - b'\zeta, \quad Y = a\zeta - c\xi' + a'\zeta' - c'\xi, \text{ etc.},$$
$$T = \xi - \xi' + \eta - \eta' + \zeta - \zeta'$$

are taken, for which the reference points are the intersections of the tangent solid with the lines 56, 64, 45 (or p, q, r), and the point O. See the following Example.

Ex. 9. Take a set of four tetrads of points, in space of three dimensions, of which every two are mutually inscribed, consisting of the tetrad of reference and three tetrads $(0, c_r, -b_r, a_r)$, $(-c_r, 0, a_r, b_r)$, $(b_r, -a_r, 0, c_r)$, $(a_r, b_r, c_r, 0)$, for $r = 1, 2, 3$, where $a_r a_s + b_r b_s + c_r c_s = 0$. Such a set is given by the rows of the scheme above (p. 133). Let $Q = a_1 a_2 a_3 yz + b_1 b_2 b_3 zx + c_1 c_2 c_3 xy$, and $P_1 = a_1 x + b_1 y + c_1 z$. Consider any one of the ∞^2 lines given, for varying x, y, z, by the line coordinates

$$xQ, \quad yQ, \quad zQ, \quad a_2 a_3 yz P_1, \quad b_2 b_3 zx P_1, \quad c_2 c_3 xy P_1;$$

these are such that

$$ll'/a_2 a_3 = mm'/b_2 b_3 = nn'/c_2 c_3,$$

and $\qquad (l - l')\,a_1 + (m - m')\,b_1 + (n - n')\,c_1 = 0.$

The line can be joined to the points of any one of the four tetrads, so forming four axial pencils each of four planes. Prove that these four axial pencils are related to one another.

If x, y, z be regarded as coordinates in a plane, the line coordinates, equated to zero, represent cubic curves with five points in common. This is a general result (cf. Chap. VII, below).

Ex. 10. Consider any solid, Π, passing through the line e, in the figure of fifteen lines, in space of four dimensions, considered above. The eight lines, of the fifteen fundamental lines, which do not meet the line e, namely $a, b, c, d, a', b', c', d'$, will meet this solid in points which we may denote by $(a), (b), \ldots$. The plane of the three points $(a), (b), (c)$, in this solid, will meet the line e; this is a plane of the first system, and meets the line d, in (d). Similarly, the points $(a'), (b'), (c'), (d')$ are in a plane of the second system. Again, the points $(b), (c), (d'), (a')$, lying, respectively, on the lines 15, 16, 23, 24, are in the singular solid 234 (or 156), and are in the plane in which this solid meets Π. Similarly for $(c), (a), (d'), (b')$ and $(a), (b), (d'), (c')$. Thus the two tetrads of points, $(a), (b), (c), (d')$ and $(a'), (b'), (c'), (d)$, form a pair of mutually inscribed Moebius tetrads.

Using, as before, coordinates $\xi, \eta, \ldots, \zeta'$ referred to $A, B, \ldots, C'$, put
$$X = \eta - \zeta', \quad Y = \zeta - \xi', \quad Z = \xi - \eta',$$
$$P = b\xi + c\eta + a\zeta - c\xi' - a\eta' - b\zeta',$$
$$T = bc\,(\xi - \xi') + ca\,(\eta - \eta') + ab\,(\zeta - \zeta').$$
The equation $tP - T = 0$, for varying t, is the general solid containing the points $bB + cC'$, $cC + aA'$, $aA + bB'$; the solids $X = 0$, $Y = 0$, $Z = 0$ contain, respectively, the lines b, c, d', a'; c, a, d', b'; a, b, d', c'. It is easy to verify such identities as
$$a\,[P + (c - a)\,Z - (a - b)\,Y] = T - (a - b)\,(a - c)\,(\xi - \xi');$$
thus the locus Σ, in the space of four dimensions, has the equation (see the irrational form, p. 130; also pp. 138, 158)
$$[a\,(b - c)\,X\,[P - a^{-1}T + (c - a)\,Z - (a - b)\,Y]]^{\frac{1}{2}} + \text{etc.} = 0.$$

If, herein, we replace T by P, the equation represents the Kummer surface, in which the locus Σ is met by the tangent solid at the point (a, b, c), the equation of this tangent solid being $T - P = 0$. The reference points for the coordinates X, Y, Z, P are $(a), (b), (c), (d')$. Any solid containing the line e may be represented by $T = 0$. The surface in which the locus Σ is met by this solid, Π, in terms of the coordinates X, Y, Z, P, is therefore given by putting, in this equation for Σ, zero in place of T. The surface then arising is one studied by Plücker (*Geometrie des Raumes*, 1868, pp. 163 ff.), under the name of the *Meridian Surface*. It is the locus of the conics touched by the lines of a quadratic complex which meet an arbitrary line; and is the envelope of the cones generated by lines of this complex which meet this line. It will be seen (Chap. VII, below) that the lines of a quadratic complex which belong to a linear complex (or, in particular, meet a line), belong to a tetrahedral complex (to forty such). Thus Plücker's meridian surface may be regarded as generated from the tetrahedral complex $ll'/A = mm'/B = nn'/C$, where $A + B + C = 0$, by lines meeting the *particular* line (l, m, n, l', m', n'); its equation is then of the form

$$\{Ax (l't + mz - ny)\}^{\frac{1}{2}} + \text{etc.} = 0,$$

as we easily verify, and of many other forms. The surface is of the fourth order, and rational (Vol. III, p. 102, Ex. 9), as is, in fact, every quartic surface in threefold space which has a double line. If, in this equation, we take p, q, r so that $l' = qr$, $m' = rp$, $n' = pq$; put $x = pX$, $y = qY$, $z = rZ$, $t = P$; and also take a, b, c so that

$$a (Bnq - Cmr) = Amn, \quad b = a - n/r, \quad c = a + m/q,$$

it is easily seen that this form reduces to that obtained above.

The surface has the eight points (a), (b), ... for double points; these lie in pairs, (a) and (a'), (b) and (b'), etc., upon four lines, lying on the surface, which meet the line e; and there is a corresponding dual theorem. The complete investigation need not be given here. (Cf., for example, the writer's *Multiply Periodic Functions*, 1907, p. 159.)

Determine the equations of the congruences of lines in which the solid Π is met by the planes of the systems 3, 4, 5, 6. The meridian surface is evidently the locus of points from which the two lines, of any one of these congruences, are coincident; namely, is the *focal* surface of the congruence. (Cf. Chap. VII, below.)

Ex. 11. Six arbitrary lines of general position, in space of four dimensions, are met by five planes. These form an associated system of planes (as defined above, p. 118).

For, taking four of the lines, a, b, c, d, as in the preceding work, consider the planes joining three points, $yB + zC'$, $zC + xA'$, $xA + yB'$, (which therefore meet a, b, c, d), which are also such as to meet two other arbitrary lines. The condition for such a line to meet a further arbitrary line, we have seen (Ex. 1, p. 136, above), is that (x, y, z), regarded as coordinates in a plane, should be the coordinates of a point of a cubic curve passing through four given points. Two plane cubic curves have nine common points. There are, then, five other sets of values of (x, y, z), and, therefore, five planes, which meet two arbitrary lines in addition to meeting the four independent lines a, b, c, d. And, as any cubic curve, which passes through eight of the nine intersections, passes through the ninth (Vol. III, p. 156, Ex. 6), four of these planes determine the fifth. (Cf. Segre, *Rend. d. Palermo*, II, 1888, pp. 45—52.)

As is indicated in Ex. 1, p. 136 above, the theorem has exceptions when the six given lines are not of general positions.

Ex. 12. Given four arbitrary planes, in space of four dimensions, these are met, by a solid, in four lines. These four lines will lie on a quadric surface, in the solid, if this solid is any solid passing through a certain fifth plane, *associated* with the given four. Let the four given planes be α, β, γ, δ; and the four planes each meeting three of these in a line be α', β', γ', δ', the common associated plane of these two sets being ϵ. Then the solid, drawn through the plane ϵ, meets α', ..., δ' in four generators of the same quadric surface, of the opposite system. These meet the four generators in the planes α, ..., δ, respectively, at the points (α, α'), ..., (δ, δ'), which are on the plane ϵ, on the conic in which the plane ϵ meets the quadric surface. Thus the first four generators, as lines of their system of generators, are *related with* the second four generators, as lines of their system.

Dually, if a, b, c, d be four general lines, and a', b', c', d' the transversals of threes of these, and O be a point such that the planes Oa, Ob, Oc, Od belong to a quadric point-cone (and, therefore, on a certain line, e), of which the planes Oa', Ob', Oc', Od' are planes of the other system, then the first four planes meet any plane, through O, which meets a, b, c, d, in a flat pencil of lines which is related to the pencil in which the second four planes meet any plane, through O, which meets a', b', c', d'.

Ex. 13. Let three points, A, B, C, in space of four dimensions, be joined, by lines, each to one of three other points, A', B', C'. In this way there arise six triads of joining lines, according to the order in which the points A', B', C' are taken. An arbitrary solid, by its intersections with the lines of a triad, determines a plane. It may be shewn that the six planes so arising meet in a point, and touch a quadric cone.

The triads of joining lines may also be described as the two sets of alternate sides, and the set of diagonals, of the hexagon $AB'CA'BC'$, together with the three other similar triads from the hexagon $AB'BA'CC'$ (which is obtained from the former hexagon by the interchange of B and C).

Let the given points be of symbols A, B, C, etc.; and let ξ, η, ζ, etc. be coordinates relative to these points; the equation of the solid may be written

$$a^{-1}\xi + b^{-1}\eta + c^{-1}\zeta = a_1^{-1}\xi' + b_1^{-1}\eta' + c_1^{-1}\zeta'.$$

The points in which this solid meets the joins of A, B, C, respectively, to B', C', A', for example, are $aA + b_1B'$, $bB + c_1C'$, $cC + a_1A'$. The plane of these points passes through the point, say O, whose symbol is $aA + bB + cC + a_1A' + b_1B' + c_1C'$; as, similarly, do all the six planes in question. Next, consider the points, taken in order, whose symbols are

$$bB + a_1A', \ aA + b_1B', \ cC + a_1A', \ bB + b_1B', \ aA + a_1A', \ cC + b_1B'.$$

These form a hexagon of which the diagonals meet in the point

$a_1 A' - b_1 B'$; thus any side of the hexagon is met, not only by the two contiguous sides, but also by the opposite side. Wherefore, the six sides of this hexagon are generators of a quadric surface. The solid in which this surface lies may be defined by three alternate vertices of the hexagon, together with the point of intersection of the diagonals, and this space is identical with the solid by which the six planes are constructed. The planes joining O to the sides of this hexagon are, in fact, these six planes, which, therefore, touch a quadric cone, as stated.

If the six given points be in a fivefold space, and the six planes be similarly defined by a fourfold, with equation

$$a^{-1}\xi + \dots = a_1^{-1}\xi' + \dots,$$

these six planes lie upon the quadric $a^{-2}\xi^2 + \dots = a_1^{-2}\xi'^2 + \dots$, and pass through the point where this is touched by the given fourfold.

Ex. 14. The theorem, for fourfold space, proved in the preceding Example, includes the result, given above, that the six planes, of the various systems, which can be drawn through a point of the quartic locus Σ, touch a quadric cone; as is easy to see. Now let an arbitrary plane, ϖ, be drawn in one of the singular solids, say 123 (or 456). Through the two points where this plane meets a pair of lines, of one of the two triads of non-intersecting lines lying in this singular solid, say the two lines 12, 13, there can be drawn a definite plane of one of the six systems, in this case of system 1. As only a single tangent solid of the locus Σ contains the plane ϖ, because this lies in the singular solid, we infer that the six planes obtained as described, meet in a point and touch a quadric cone.

Ex. 15. We may relate the points of three planes, in space of four dimensions, by taking four arbitrary points in one of these planes, and making correspond to these, respectively, four points in each of the other two planes (Vol. i, p. 148). Then we may consider the aggregate of planes containing three corresponding points, one in each plane. A particular case is that when the three sets of four corresponding points are on four lines, each meeting the three planes. On the first plane, α, let the four points be U, V, W, K; on the second plane, α', let the points corresponding thereto be U', V', W', K'; on the third plane, α'', let the corresponding points be U'', V'', W'', K''—so that U, U', U'' are in line, as also V, V', V''; W, W', W'' and K, K', K''. Each three of these four lines has a transversal. It can be shewn, if H, H', H'' be any three corresponding points respectively lying in the related planes $\alpha, \alpha', \alpha''$, that the plane $HH'H''$ meets the four transversals so arising.

We may identify the four lines, $UU'U''$, etc., which meet the three given planes, respectively, with the lines a, b, c, d of the figure discussed above, in the text; then the points U, V, W, K, and the general point, H, of the plane α, may be taken to have symbols, respectively,

$$yB+zC', \quad zC+xA', \quad xA+yB', \quad yz(B'+C)+zx(C'+A)+xy(A'+B)$$
and $$\lambda x(yB+zC')+\mu y(zC+xA')+\nu z(xA+yB');$$

the points U, V, W, K arise from the last point, H, for the particular values $(\lambda, \mu, \nu)=(1, 0, 0);\ (0, 1, 0);\ (0, 0, 1)$ and $(1, 1, 1)$. The point, H', in the plane α', may then be taken to be, similarly, of symbol $\lambda x'(y'B+z'C')+$ etc., where λ, μ, ν are the same as for H; and the point, H'', in the plane α'', may be represented by $\lambda x''(y''B+z''C')+$ etc. The general point of the plane $HH'H''$ has, then, a symbol

$$\xi(\nu B'+\mu C)+\eta(\lambda C'+\nu A)+\zeta(\mu A'+\lambda B),$$

where ξ, η, ζ are of the respective forms $\xi=pyz+p'y'z'+p''y''z''$, $\eta=pzx+p'z'x'+$ etc., and $\zeta=pxy+$ etc. The ratios of p, p', p'' identify any individual point of this plane; in particular we obtain the points of intersection of this plane with the four transversals, spoken of, by choosing p, p', p'' so that, for these four lines, respectively, $(\xi, \eta, \zeta)=(1, 0, 0);\ (0, 1, 0);\ (0, 0, 1);\ (1, 1, 1)$.

Ex. 16. With the notation of the text, the common point of the two planes of the first kind, containing, respectively, the points $yB+zC'$, $zC+xA'$, $xA+yB'$ and y_1B+z_1C', z_1C+x_1A', etc., is

$$\frac{y_1-z_1}{yz_1-y_1z}(yB+zC')+\frac{z_1-x_1}{zx_1-z_1x}(zC+xA')+\frac{x_1-y_1}{xy_1-x_1y}(xA+yB').$$

The point of intersection of the former plane with the plane of the second system containing the points; P, or $\eta B'+\zeta C$; Q, or $\zeta C'+\xi A$; R, or $\xi A'+\eta B$, provided these two planes do not meet in a line, is

$$(y\zeta)^{-1}A+(z\xi)^{-1}B+(x\eta)^{-1}C+(\eta z)^{-1}A'+(\zeta x)^{-1}B'+(\xi y)^{-1}C',$$

which is $x^{-1}\xi P+y^{-1}\eta Q+z^{-1}\zeta R$. The other plane of the first system through this point is that given by x', y', z', where

$$x'(\eta-\zeta)=x(y\zeta-z\eta), \quad y'(\zeta-\xi)=y(z\xi-x\zeta), \text{ etc.};$$

these are such that

$$x/x'=\theta x\eta\zeta+\phi, \quad y/y'=\theta y\zeta\xi+\phi, \quad z/z'=\theta z\xi\eta+\phi,$$

where θ, ϕ are symmetrical quantities.

Ex. 17. Three planes, α, β, γ, of the first three systems, respectively, are given by the triads of points

$$(yB+zC', \quad zC+xA', \quad xA+yB'), \quad (y_1C+z_1B', \quad z_1A+x_1C', \quad x_1B+y_1A'),$$
$$(y_2A+z_2A', \quad z_2B+x_2B', \quad x_2C+y_2C');$$

prove that the conditions for the pairs of planes (α, β), (α, γ), (β, γ) to meet in lines are, respectively,

$x_1^{-1}(y-z)+\text{etc.}=0, \quad x^{-1}(y_2-z_2)+\text{etc.}=0, \quad x_2^{-1}(y_1-z_1)+\text{etc.}=0,$

and that these equations are not consistent, in general. Cf. Ex. 23.

Ex. 18. With coordinates, $\xi, \eta, \ldots, \zeta'$, referred to the six points $A, B, \ldots, C'$, as in the text, prove that the equation

$$(\eta-\zeta')(\eta'-\zeta)+\sigma^{-1}(\zeta-\xi')(\zeta'-\xi)-(1+\sigma)^{-1}(\xi-\eta')(\xi'-\eta)=0$$

represents the quadric point-cone, drawn from the point $\sigma L-M$, of the line e, whose planes are those of the first and second kinds, meeting, respectively, the lines a, b, c, d and a', b', c', d'. Prove, also, that every tangent solid of this cone is a tangent solid of the locus Σ.

Ex. 19. Consider the rational quartic curve, in space of four dimensions, whose points have coordinates, (x, y, z, t, u), of the forms $(\theta^4, \theta^3, \theta^2, \theta, 1)$; the equations of a plane meeting this curve in three points, a so-called *trisecant* plane, are of the forms

$$Ax+By+Cz+Dt=0, \quad Ay+Bz+Ct+Du=0.$$

If L, M be any two points of the space, the trisecant planes of the curve, which pass through the point $\sigma L-M$, lie upon a quadric point-cone, whose equation is of the form $U+2\sigma V+\sigma^2 W=0$. Determine the other system of planes of this cone; and shew that the cone becomes a line-cone for three values of σ.

Prove that the trisecant planes of the quartic curve which meet a line are projected, from an arbitrary point of the curve, by the tangent solids of a quadric point-cone. Deduce, for a cubic curve in space of three dimensions, that, if three tetrads of points be taken on the curve, the twelve planes, each containing three points of one tetrad, all touch a quadric surface; further, that the two tangent planes of this quadric surface, drawn from an arbitrary chord of the cubic curve, determine a further tetrad of tangent planes of this quadric surface (of which these are two); and also (as was remarked to the writer by Mr Vaidyanathaswamy), that there are three chords of the cubic curve which are generators of the quadric surface. (Cf. Vol. III, Ex. 8, p. 135; and p. 120, above.)

Ex. 20. The tangent solid of the locus Σ, whose equation, with co-ordinates referred to the points $A, B, \ldots C'$, was $u\xi+\ldots+u'\xi'+\ldots=0$, subject to $u+v+\ldots+w'=0$, $uvw+u'v'w'=0$, may be referred to the symmetrical points $F, G, \ldots$, previously used (p. 116 above). For this, let the symbol

$$\xi A+\ldots+\zeta'C', \quad \text{or} \quad \xi(G+H)+\eta(H+F)+\ldots+\zeta'(F'+G'),$$

be written $xF+\ldots+z'H'$, so that $x=\eta+\zeta, \ldots, z'=\xi'+\eta'$. The equation of the tangent solid,

$$u\xi+\ldots=0, \quad \text{or} \quad u(y+z-x)+\ldots+u'(y'+z'-x')+\ldots=0,$$

10—2

becomes $Ux + ... + W'z' = 0$, with $u = V + W, ..., w' = U' + V'$. The conditions for the coefficients then become

$$U + V + ... + W' = 0,$$
$$(V + W)(W + U)(U + V) + (V' + W')(W' + U')(U' + V') = 0.$$

For any p, q, r, however, we have

$$(p + q + r)^3 = p^3 + q^3 + r^3 + 3(q + r)(r + p)(p + q).$$

Thus, the second condition for the tangent solid, the *tangential equation* of the locus Σ, becomes

$$U^3 + V^3 + W^3 + U'^3 + V'^3 + W'^3 = 0,$$

where $U + ... + W' = 0$.

It may be proved, further, that the ten singular solids have equations $Ux + ... + W'z' = 0$, where three of $U, V, ..., W'$ are 1, and three are -1.

Ex. 21. It has been shewn (p. 117 above) that the figure of fifteen points and fifteen lines, with the associated planes and solids, is changed into itself by fifteen involutory transformations. There exists also a quadric (∞^3), in regard to which the figure is its own polar reciprocal. In particular, the polar solid of the point $F - F'$, in regard to this quadric, is the solid defined by the points G, H, G', H'. Thus, the fifteen points such as $F - F'$ are the complete intersection of the polar solids, in regard to this quadric, of the six points $F, G, ..., H'$, any four of these solids intersecting in one of these fifteen points; also, the plane of the points $F - F'$, $G - G'$, $H - H'$ is the polar plane, in regard to this quadric, of the line, e, containing the points $F + F'$, $G + G'$, $H + H'$, the polar solids of the points of this line all containing this plane. With coordinates $x, y, ...$, referred to $F, G, ...$, as above, the equation of this quadric is

$$(x + y + z + x' + y' + z')^2 - 6(x^2 + y^2 + z^2 + x'^2 + y'^2 + z'^2) = 0;$$

or, if $3X$ denote $2x + 2x' - y - y' - z - z'$, with similar forms for $3Y$, $3Z$, the equation is also

$$(x - x')^2 + (y - y')^2 + (z - z')^2 + X^2 + Y^2 + Z^2 = 0,$$

which is explicitly a function only of the differences of the coordinates. The condition that a solid, expressed by

$$Ux + ... + W'z' = 0,$$

should touch this quadric is

$$U^2 + V^2 + W^2 + U'^2 + V'^2 + W'^2 = 0,$$

subject to $U + ... + W' = 0$. In terms of coordinates ξ, η, ..., referred to $A, B, ...$, if $p = \eta - \zeta'$, $q = \zeta - \xi'$, $r = \xi - \eta'$, $p' = \eta' - \zeta$, etc., and $3P = q + q' - r - r'$, etc., the equation is

$$(p - p')^2 + (q - q')^2 + (r - r')^2 + P^2 + Q^2 + R^2 = 0.$$

Ex. 22. The reciprocity of the preceding Example is a particular case of a theorem already many times referred to (cf. Vol. ii, p. 218; Veronese, *Math. Annal.* xix, 1882, p. 161). The twenty-one points, consisting of the six F, G, etc., and the fifteen $F - F'$, $F - G'$, etc., may be regarded as consisting of two *simplexes*, each of five points, which are in perspective with one another from a centre, (which is also one of the twenty-one points), together with the ten points, lying in a solid, in which joining lines, of corresponding pairs of points of the two simplexes, intersect one another. These twenty-one points may be obtained either, by projection of the intersections of seven fourfolds, lying in a space of five dimensions which contains the fourfold space of the given figure; or, as the intersections, with this fourfold space, of the joins of seven points of this fivefold space. Taking the latter point of view, the fourteen linearly independent (∞^4) quadrics, which can be drawn through the seven points of the fivefold space, meet the fourfold space of the given figure in fourteen (∞^3) quadrics. These are all outpolar to the quadric in regard to which the specified polarity has place, and determine this quadric. (Cf. p. 10, above, Ex. 1.)

Ex. 23. Take any point O, of general position, not lying on the locus Σ. Through this point there pass fifteen planes, each joining the point to one of the fifteen lines of the figure. There also pass through this point two planes of each of the six systems, which we may denote, respectively, by $(1, 1')$, $(2, 2')$, ..., $(6, 6')$. Each of the two planes of the first system meets one of the planes of the second system in a line, intersecting the line 12; we can choose the notation so that the planes 1, 2′ meet in a line, and also the planes 1′, 2 meet in a line. And, then, so that the planes 1, 3′ meet in a line, and also the planes 1′, 3. Then we can prove that the planes 2, 3′ meet in a line, and also 2′, 3. For, if it were the planes 2, 3 which meet in a line, the planes 1′, 2, 3 would be in a solid; and, then, the lines 23, 31, 12, each of which meets a pair of these planes, would be in this solid; this would thus be the singular solid 123 (or 456). But this solid contains the point O; and a general point O does not lie in any of the ten singular solids. Thus it is not the planes 2, 3, but the planes 2, 3′, and, therefore, also, the planes 2′, 3, which meet in a line. This argument being repeated, we can infer that, with proper notation, no two of the six planes 1, 2, ..., 6 meet in a line, and no two of the six planes 1′, 2′, ..., 6′; but the plane r (for $r = 1, 2, ..., 6$) meets in a line every one of the planes 1′, 2′, ..., 6′ other than r'. Now, let the symbols 1, 2, ... denote, not the planes through the point O, but the conics in which these planes meet the locus Σ. Then, for instance, the conics 1, 2′ have a point in common, where the line of intersection of the planes 1, 2′ meets the line 12. We have, therefore, twenty-seven loci on the locus Σ, the fifteen lines and the twelve conics; and we easily see that, upon each of these twenty-seven loci, there are five points, through each of which two other of the twenty-seven pass. For instance, on the line 12 there are, first, the three points through which pass, respectively, the three pairs of lines (34, 56), (35, 64), (36, 45), and there are the two points through which pass, respectively, the two pairs of conics (1, 2') and (1', 2). Or again, on the conic 1' there are the five points in which this is met by the lines 12, 13, ..., 16; and through these there pass, respectively, also the conics 2, 3, 4, 5, 6. The total number of such intersections of three of the twenty-seven loci is, therefore, forty-five.

Now, let Π be an arbitrary solid, not containing O. Allow the planes, through O, containing the twenty-seven loci, to meet this solid. We obtain twenty-seven lines in Π, meeting in threes in forty-five points. Transferring the notation to these lines, the lines 1, 2', in the solid Π, meet in a point, as do the lines 1', 2; and the join of these points is the line 12. Also, two lines in Π, denoted by pq and rs, meet if the symbols p, q, r, s are all different, but not otherwise. The figure in the solid Π is, in fact, the *dual* of that of the twenty-seven lines of a cubic surface, explained in Vol. III, p. 160. And, as in that case, it may be shewn that there are thirty-six ways of choosing a set of twelve of the lines, in the solid Π, to play the parts here played by 1, 1', ..., 6, 6', the other fifteen being then constructed from these. If we consider, in the solid Π, beside the lines, the planes in which this solid is met by the tangent solids, of the locus Σ, which pass through O, we shall easily see that these are an aggregate dual to that of the points of a cubic surface. Three tangent solids of the locus Σ pass through an arbitrary plane of the fourfold space; in particular, three through a plane which contains O. Thus, in Π, three planes of the aggregate pass through an arbitrary line.

We may deal with the matter algebraically. If the point O be $aA + \dots + a'A' + \dots$, so that (a, b, c, a', b', c') are its coordinates referred to $A, B, \dots, C'$, the tangent solid of the locus Σ, whose equation is $y(t-z)\xi + \dots = (t-y)z\xi' + \dots$ (p. 128, above), contains the point O provided $t = S/P$, where

$$S = yz(a'-a) + zx(b'-b) + xy(c'-c),$$
$$P = x(b'-c) + y(c'-a) + z(a'-b).$$

This relation connecting x, y, z, t may be regarded as representing a quadric surface in a space (x, y, z, t). To every point of this quadric surface corresponds a tangent solid, of the locus Σ, passing through O; and, therefore, also, a plane of the cubic aggregate in the solid Π. By means of this equation, the coefficients of $\xi, \eta, \dots,$

in the equation of the tangent solid of Σ, take the forms $y(S - zP)$, $z(S - xP)$, ... When equated to zero, these represent cubic curves, in the plane (x, y, z), with six common points. (Cf. Vol. III, p. 189.)

More geometrically, we may shew that the planes, of the aggregate in Π, are the dual of the points of a cubic surface, by shewing that they meet three fixed planes in corresponding points of three related plane systems (cf. Vol. III, p. 185; and Ex. 15, above). That this is so follows from the figure in the space of four dimensions, because a variable plane of one system meets three fixed planes of another system in such related plane systems (cf. Ex. 16, above); while a tangent solid of the locus Σ, at a point of such a plane, contains this plane.

The dual of the figure which has been considered. We have already referred briefly to the dual of the figure, here considered, which arises when we begin with four planes, instead of four lines (Ex. 5, p. 117, above). There will be fifteen planes, and six systems of lines, each line meeting five associated planes. Instead of the aggregate of the tangent solids of the quartic locus Σ, each containing planes of all the six systems, meeting in its point of contact, there will be a locus of points, which we shall denote by S; this will be expressible by an equation $\xi\eta\zeta = \xi'\eta'\zeta'$, where $\xi + \eta + \zeta = \xi' + \eta' + \zeta'$, there being three points of the locus upon an arbitrary line. The six systems of lines will lie entirely upon this cubic locus S, a line of each system passing through every ordinary point of the locus. The locus S will have ten double points, corresponding to the singular solids, such that any line through one of these points meets the locus only in one further point. All the points of any one of the fifteen planes will be points of the locus. As is indicated in Ex. 20 above, the equation of the locus S can be written in symmetrical form by taking coordinates $X, Y, \ldots$ such that $\xi = Y + Z$, $\eta = Z + X$, $\zeta = X + Y$, $\xi' = -(Y' + Z')$, $\eta' = -(Z' + X')$, $\zeta' = -(X' + Y')$. These lead to

$$X + Y + Z + X' + Y' + Z' = 0,$$

and the equation is

$$X^3 + Y^3 + Z^3 + X'^3 + Y'^3 + Z'^3 = 0.$$

A plane lying entirely on the locus is then represented by $X + X' = 0$, $Y + Y' = 0$, which involve $Z + Z' = 0$; and there are fifteen such planes, each corresponding to a mode of dividing the six coordinates into three pairs. The double points of the locus are those given by $X^2 = Y^2 = \ldots = Z'^2$, and are the points of which three of the coordinates $X, Y, \ldots$ are $+1$, and three of them -1. To state the equations of a system of five planes of which no two meet in a line, let the coordinates be denoted in order by $1, 2, \ldots, 6$; let such an equation as $X + Y = 0$ be denoted by 12. There are six

ways of forming a system of five synthemes, such as 12 . 34 . 56, so
that these five synthemes contain, in their aggregate, all the fifteen
duads such as 12 (cf. Vol. ii, p. 221). Such a system of synthemes
gives such a system of five planes. When such a system of five
planes is obtained, the solids which join an arbitrary point of one
such plane, to three others of the planes, meet in a line intersecting
four of the planes; this line, therefore, lies entirely on the locus;
and it meets the fifth plane.

But, the locus S is obtained in a natural way by considering,
in space of three dimensions, all the quadric surfaces which pass
through five arbitrary points of general position. If coordinates,
in this threefold space, be x, y, z, t; and the five given points,
which we denote by A, B, C, D, E, be, respectively, $(1,0,0,0)$,
$(0,1,0,0)$, $(0,0,1,0)$, $(0,0,0,1)$, $(1,1,1,1)$, then six degenerate
quadric surfaces through these five points are $\xi = 0$, $\eta = 0$, ... etc.,
where

$$\xi = y\,(t-z), \quad \eta = z\,(t-x), \quad \zeta = x\,(t-y),$$
$$\xi' = (t-y)\,z, \quad \eta' = (t-z)\,x, \quad \zeta' = (t-x)\,y.$$

These are subject to $\xi + \eta + \zeta = \xi' + \eta' + \zeta'$, and may be interpreted
as coordinates in space of four dimensions. They are also subject
to $\xi\eta\zeta = \xi'\eta'\zeta'$, and represent a cubic (∞^3) locus, S, in the fourfold space. Between the general points of this locus, and the
points of the threefold space, there is thus set up a correspondence,
the reverse equations being

$$t^{-1}x = (\eta' - \zeta)/(\xi - \xi'), \quad t^{-1}y = (\zeta' - \xi)/(\eta - \eta'), \quad t^{-1}z = (\xi' - \eta)/(\zeta - \zeta').$$

In general, any quadric surface through the points A, B, ..., E,
with equation of the form $A\xi + B\eta + C\zeta = A'\xi' + B'\eta' + C'\zeta'$, corresponds to the section of the locus S by a solid; this is a cubic
surface. Thus, the quartic curve of intersection, of two such
quadric surfaces, corresponds to the cubic curve in which the locus
S is met by a plane. The three points, beside the five points
A, B, ..., E, in which three such quadric surfaces meet, correspond
to the three intersections of the locus S with a line, of which two
determine the third.

Though the results obtainable can only be the dual of those
already obtained for the locus Σ, it is instructive to follow out the
correspondence. Of this we give some indications:—There are,
through the five fundamental points A, B, C, D, E, of the space of
three dimensions, ∞^2 cubic curves, one such curve passing through
a point of a plane taken to contain two of the five points. Each
of these cubic curves lies on three linearly independent quadric
surfaces, which are, then, among those through the five fundamental points; thus, each of these curves corresponds to a line
lying on the locus S, of the fourfold space; and, of lines of this

system, there is one passing through every general point of S. Again, taking an arbitrary line through one of the points $A, B, ...,$ E, there are three linearly independent quadric surfaces, containing this line, passing through the other four points; these quadric surfaces correspond to solids in the fourfold space, and they have in common only the arbitrary line first taken. Thus the lines, ∞^2 in aggregate, through one of $A, B, ..., E$, correspond to a system of lines lying on the locus S; and, of these, one passes through each general point of S. We thus obtain six systems of lines lying on S. Further, upon any quadric surface through $A, B, ..., E$, there are two cubic curves passing through these five points (Vol. III, p. 129); and there are two lines of the quadric surface through any one of these five points. Thus, two of the lines of any system, of the locus S, lie in an arbitrary solid of the fourfold space—just as there are two planes of either system, in the dual figure considered above, which pass through an arbitrary point. But there are particular solids, in the fourfold space, each containing ∞^1 of the lines of S, of each of two systems. For, the degenerate quadric surface, which consists of the plane ABC and a plane through the line DE, contains lines through D and E; and the quadrics of this description are linearly dependent from two of them. They correspond, therefore, to a set of solids, in the fourfold space, having a plane in common. Each of these solids contains ∞^1 lines, of each of two systems, on the locus S, of which any line of one system meets every line of the other; this will be so if the solid meets S in a cubic surface degenerating into a quadric surface and a plane. We thus reach ten planes, lying wholly on the locus S, each corresponding to the aggregate of a plane containing three of the points $A, B, ..., E$ taken with the line joining the other two of these five points. Again, a quadric cone with vertex, say, at A, containing the other four points B, C, D, E, contains ∞^1 lines through A, and also contains ∞^1 cubic curves passing through the five points; and any one of these lines has an intersection, not at the fundamental points, with any of these curves; these cones are linearly dependent from two of them. We thus reach five other planes, lying wholly on the locus S in the fourfold space, each corresponding to the aggregate of the joins, of one of the five fundamental points, to the other four. There are then fifteen planes on S. Moreover, these fifteen planes, each taken twice over, consist of sets of five planes all met by the lines of one system on S. For, the condition that a plane, in fourfold space, should be met by a line, is that the two should lie in a solid; such solid corresponds to a quadric surface, in the threefold space, passing through the five fundamental points; and, first, through a cubic curve which contains these points can be drawn a quadric cone with

vertex at any one of them; while, next, through any line drawn
through the fundamental point *A* can be passed such a cone, with
vertex at *A*, to contain the other four points, beside four degenerate quadrics, such as that consisting of the plane *BCD* and the
plane defined by *AE* and this line. Consider, now, the joining line
of two of the five fundamental points, in the threefold space.
Through this line, and the other three of the five points, there
pass four linearly independent quadric surfaces; the corresponding
solids in the fourfold space have a point in common. Moreover,
three of these quadric surfaces, it is easily seen, have one further
point in common, not one of the fundamental points. Thus, the
point of the fourfold space, on the locus *S*, is such that an arbitrary line, through it, meets *S* in one further point. We thus
reach ten double points of the locus *S*, each corresponding to the
join of two of the five fundamental points *A*, *B*, ..., *E*. We can
then proceed to prove that each of the fifteen planes lying on *S*
contains four of the double points; that every solid through one
of these planes contains an infinite number of lines, of each of two
systems of *S*, has already appeared. In fact, the aggregate of a
plane through three of the points *A*, *B*, ..., in the threefold space,
with the line joining the other two, contains this last line and the
joins of the three points; while the joins of one of the points
A, *B*, ..., to the others, are four in number. The corresponding
dual property, for the figure studied above, was that a line, say *e*,
lay in four singular solids, [*a, a'*], [*b, b'*], [*c, c'*], [*d, d'*], each point
of this line *e* being the vertex of a quadric point-cone whose
planes belonged to two of the six fundamental systems of planes.
These singular solids, moreover, could be, respectively, generated,
we saw, by axial pencils of planes, of axis *e*, of the four other
systems of planes. In the present figure in fourfold space, for the
locus *S*, each of the four double points, in one of the fifteen planes
of *S*, is the centre of a flat pencil of lines, lying in this plane;
these are lines of *S* of systems different from one another, and
different from the two systems remarked as lying in any solid containing the plane. Again, as we likewise see from the dual figure,
through each of the ten double points of the locus *S*, there pass
six planes lying on *S*, from among the fifteen; and each of these
planes is the seat of a flat pencil of lines, these belonging to the
six systems on *S*, respectively, the double point being the vertex of
all these pencils.

Now, let *O* be an arbitrary general point of the locus *S*; we
suppose it to correspond to a point (*a, b, c,* 1) of the threefold
space. Take, in the fourfold space, a solid, Π, upon which we
project, from *O* as centre of projection. Every line lying on *S*
gives rise, by intersection of Π with the plane joining *O* to this

line, to a line on Π. We thus have six systems of such lines. Every line through O meets S in two further points; through each of these there passes a single line of any specified one of the six systems of lines of S. Thus, of any one of the six systems of lines in Π, there are two lines which pass through an arbitrary point of Π. In fact, as we know from the dual figure, the lines, of any system, in Π, belong to a quadratic congruence, being common to a linear complex and a quadratic complex. We give now a proof, independent of the preceding (p. 135, above), that all the lines of one of these systems, in Π, which meet a definite line of another system, equally meet a second line of this system; from this it follows, as before, that every two of the linear complexes are conjugate (or apolar). For this we prove, by consideration of the figure in the threefold space, the following theorem:—Let a line, say e, of the locus S, in the fourfold space, be joined to O by a plane, meeting S, further, in a conic, ϵ. From every point of e, and from every point of ϵ, let the unique line of another system, upon S, be drawn; this other system we denote by (a). Then there is a second line, e', of S, of the same system as e, which is met by all the lines (a) drawn through the points of e; and the conic, ϵ', in which the plane Oe' further meets S, is likewise met by all the lines, of the system (a), which can be drawn from the points of the conic ϵ. To see this, let the line e, of S, be that corresponding to a line, l, drawn through the point E, in the threefold space of the points A, B, ..., E. The solids which contain the plane Oe correspond to quadric surfaces in the threefold space, these passing, not only through the five points A, B, ..., E, but also through the point $(a, b, c, 1)$, and through the line l. Any two of these quadric surfaces meet in a cubic curve, which contains the four points A, B, C, D, and the point $(a, b, c, 1)$, and has the line l for chord. This cubic curve, which we denote by θ, corresponds then to the conic ϵ. We may now suppose that the lines (a), of the solid S, which meet the line e, correspond to the lines, in the threefold space, which pass through A and lie in the plane Al (which contains E); and that the lines (a), which meet the conic ϵ, correspond to the generating lines of the quadric cone, of vertex A, which contains the curve θ. What we desire to prove is, then, that there is, upon this cone, another cubic curve, say θ',—likewise containing the points A, B, C, D and the point $(a, b, c, 1)$—such that the chord, l', of this cubic curve θ', which can be drawn through E, lies in the plane Al. For this, consider the quadric cones, with vertex at the point $(a, b, c, 1)$, which contain the points A, B, C, D. Any one of these meets the first cone in a cubic curve containing the five desired points. If the line l meet the curve θ in P and Q, a single cone, of vertex $(a, b, c, 1)$, can be drawn to contain A, B, C, D and any specified point, P', of AP; the intersection, Q', of this cone, with AQ is then definite. Thus the ranges, (P'), (Q'), upon the lines AP, AQ, are related; and the lines $P'Q'$ touch a conic in the plane APQ (or Al), of which EPQ is one tangent line; for the curve θ lies upon one of the cones of vertex $(a, b, c, 1)$. Thus, we can draw through E, in the plane Al, a single line, EP_1Q_1, which is a common chord, first, of the cone which projects the curve θ from A (of which AP, AQ are two generators), and, second, of another cone, vertex $(a, b, c, 1)$, passing through A, B, C, D. Thereby we find the line l', and the curve θ', desired.

The two further points in which the locus S is met by an arbitrary line through O evidently correspond to the two remaining intersections of three quadric surfaces drawn, in the threefold space, through the six points consisting of $A, B, \ldots, E$ and the point $(a, b, c, 1)$. If these two remaining intersections, of which one determines the other, coincide, it is easy to see that their point of coincidence is the vertex of a quadric cone containing the first six points. The locus of the vertex of such a quadric cone, containing six given points, may be proved to be a quartic surface (see the Examples below). This we shall call a *Weddle surface*. This surface then corresponds to the points of the locus S which, when joined to O, are coincident intersections of S with lines from O. Writing $\alpha = bc'$, $\beta = ca'$, $\gamma = ab'$, $\alpha' = b'c$, $\beta' = c'a$, $\gamma' = a'b$, where $a' = 1 - a$, $b' = 1 - b$, $c' = 1 - c$, we easily see, by substitution of $\xi + \lambda\alpha$, $\eta + \lambda\beta$, etc. for the coordinates in the equation of S, that the points of S concerned lie on the quadric represented by

$$\alpha\eta\zeta + \beta\zeta\xi + \gamma\xi\eta = \alpha'\eta'\zeta' + \beta'\zeta'\xi' + \gamma'\xi'\eta'.$$

This, which we may call the polar quadric of the point O, or $(\alpha, \beta, \gamma, \alpha', \beta', \gamma')$, in regard to S, has the same tangent solid at O as the locus S; and the sextic surface in which it meets S has a double point at O. By projection from O, it gives rise to a quartic surface in the solid Π. This quartic surface, which we may denote here by K, can at once be seen, from the properties of the locus S, to be Kummer's quartic surface, with sixteen double points and sixteen tangent planes each touching it at the points of a conic. The plane which joins O to any line of S contains also a conic lying on S, which meets this line in two points; these are clearly upon the intersection of S with the polar quadric of O. Wherefore, every line of S projects, from O, into a line, in the solid Π, which is a double tangent of the surface K. Further, at every general point of the surface K, there are six tangent lines of K which touch this surface again, these being projections of lines of S of the six systems. The polar quadric of O, by its definition, passes through the ten double points of S; and contains, also, the six lines of S, of the various systems, which pass through O. This shews the origin of the sixteen double points of the surface K. Again, the polar quadric of O meets each of the fifteen planes, lying upon S, in a conic; thus, by projection from O, are obtained fifteen conics lying on the surface K, along each of which the surface is touched by a plane. It may be proved that the tangent solid of S, at the point O, gives rise to another such singular tangent plane of K. Further, we saw that six of the planes lying on S pass through every double point of S; thus, for ten of the double points of K, it is true that six singular tangent planes pass through a double point. This is also true for the other six double points of K. For, any one of the six lines of S, passing through O, meets, we have seen, five of the planes (of an associated system) lying on S; and this line meets the tangent solid of S at O. Again, we saw that any one of the planes of S contains four of the double points; and that any solid through this plane also holds ∞^1 lines, of each of two systems, lying on S. Defining such a solid as that which contains O, we see that the fifteen planes of S project, from O, into planes in Π, each containing six double points of K, two of these arising from lines of S which pass through O. The remaining singular tangent plane of K, lying in the tangent solid of S at O, contains double points on the six lines of S which pass through O.

In general, as has been said, there pass, through a point of the solid Π, two lines arising as projections of lines of S of a particular system. But there are, in fact, sixteen points of Π through each of which there pass ∞^1 lines of this system, forming a flat pencil. These are the singular points of the surface K, as is clear from what has been said; and the same points equally arise, with the same characteristic property, for each of the six systems of lines. There are, similarly, sixteen planes, in the solid Π, each of which, instead of containing two lines of a particular system, contains a pencil of such lines; and each of these planes contains, also, a pencil of lines of every system. ·These six points and planes arise, in fact, from the systems of lines of the space Π. From this point of view they are reconsidered below, in Chap. vii.

We see that the whole theory is analogous to that of the projection, from a point of a cubic surface, of the sextic curve, of this surface, lying on the polar quadric surface of this point, taken in regard to the cubic surface (Vol. iii, p. 200). Indeed, the present theory includes the former, if we take a section of the present figure by an arbitrary solid passing through O (cf. Vol. iii, p. 205). Thus, also, it appears that the plane quartic curve obtained, in the former theory, by projection of the 'apparent contour' of a cubic surface, which was stated to be a general plane quartic curve, may be obtained as a plane section of the Kummer surface

K; a conclusion which was enunciated by Kummer. But, further, it also appears, comparing the former theory with the present, that six intersections of suitable pairs of bitangents of a plane quartic curve, which lie on a conic (Vol. III, p. 201), may be regarded as the polar points, of the plane of the quartic curve, in regard to six linear complexes of lines which are mutually conjugate (p. 43, above).

Ex. 1. The fifteen planes of the locus S, whose equation is $\xi\eta\zeta = \xi'\eta'\zeta'$, subject to $\xi + \eta + \zeta = \xi' + \eta' + \zeta'$, are the nine expressed by such a pair of equations as $\xi = 0$, $\xi' = 0$; and the six expressed by such a pair of equations as $\xi - \xi' = 0$, $\eta - \eta' = 0$. The ten double points are $(1, 1, 1, 1, 1, 1)$, with the nine of coordinates $(\xi, \eta, \zeta, \xi', \eta', \zeta')$ in which two of ξ, η, ζ are zero, and two of ξ', η', ζ' are zero. If $x' = 1 - x$, etc., the six lines, of the various systems, lying on S, which pass through the point $(yz', zx', xy', y'z, z'x, x'y)$, are those which join this point, respectively, to the points $(z'/z, x'/x, y'/y; y'/y, z'/z, x'/x)$, $(y, z, x; z, x, y)$, $(z', x', y'; y', z', x')$, $(yz', 0, y'; y'z, z', 0)$, $(z', zx', 0; 0, z'x, x')$, $(0, x', xy'; y', 0, x'y)$.

Ex. 2. The equation of the Weddle surface, the locus of the vertex of a quadric cone, in space of three dimensions, which contains six given points, A, B, C, D, E, F, say $(1, 0, 0, 0)$, $(0, 1, 0, 0)$, $(0, 0, 1, 0)$, $(0, 0, 0, 1)$, $(1, 1, 1, 1)$, $(a, b, c, 1)$, is obtainable from

$$\alpha\eta\zeta + \beta\zeta\xi + \gamma\xi\eta = \alpha'\eta'\zeta' + \beta'\zeta'\xi' + \gamma'\xi'\eta'$$

by putting $\alpha = b(1 - c)$, $\alpha' = (1 - b)c$, $\xi = y(t - z)$, $\xi' = (t - y)z$, etc. It is capable of many forms. For instance,

$$(t - x)(y - z)(bcxt + ayz) + \ldots = 0.$$

Or, if we put

$$X/a = \alpha\xi' - \alpha'\xi, \quad Y/b = \beta\eta' - \beta'\eta, \quad Z/c = \gamma\zeta' - \gamma'\zeta,$$
$$-T/abc = a(\xi - \xi') + b(\eta - \eta') + c(\zeta - \zeta'),$$

it is capable of the form $P^{\frac{1}{2}} + Q^{\frac{1}{2}} + R^{\frac{1}{2}} = 0$, where

$$P = a(b - c)X[-a'T/a + (c - a)Z - (a - b)Y], \text{ etc.}$$

The surface has a double point at each of the six fundamental points; and contains the cubic curve which can be drawn through these. Any chord of this curve meets the surface in two points which are harmonic conjugates in regard to the end points of the chord. The surface contains the fifteen joining lines of the six fundamental points, and also contains ten other lines, each the intersection of a plane containing three of these points with the plane containing the other three. The coordinates of a point of the cubic curve, through the six points, may be written, in terms of a parameter, θ, as $a(\theta - a)^{-1}$, $b(\theta - b)^{-1}$, $c(\theta - c)^{-1}$, $(\theta - 1)^{-1}$;

thus the respective values $\theta = 1$, 0, ∞, a, b, c give the points D, E, F, A, B, C. See, also, p. 142, above.

The surface is expressible by those multiply-periodic functions of two variables which arise from the irrationality involved in the square root of a sextic polynomial in one variable (see the writer's *Multiply-Periodic Functions*, Cambridge, 1907).

Ex. 3. In the space of four dimensions, four independent solids passing through the point O, or $(\alpha, \beta, \gamma, \alpha', \beta', \gamma')$, are $X = 0$, $Y = 0$, $Z = 0$, $T = 0$, where X, Y, Z, T have the values of Ex. 2. These may be used as coordinates in the solid, Π, upon which we project from O. Putting $p = a'/a$, $q = b'/b$, $r = c'/c$, and considering a line, of one system, of the locus S, given, for proper values of the parameters x, y, z, by the equations $\xi' = yz^{-1}\xi$, $\eta' = zx^{-1}\eta$, $\zeta' = xy^{-1}\zeta$, prove that this projects from O into a line, of the solid Π, whose line coordinates, relatively to X, Y, Z, T, are

$$(b-c)\,U'VW, \quad (c-a)\,V'WU, \quad (a-b)\,W'UV,$$
$$-UV'W', \quad -VW'U', \quad -WU'V',$$

where $U = y - z$, $U' = yr - zq$, etc. The lines of the system, obtained by varying x, y, z, thus belong to the tetrahedral complex $all' + bmm' + cnn' = 0$, and also to the linear complex

$$a'l + b'm + c'n - a(b-c)\,l' - b(c-a)\,m' - c(a-b)\,n' = 0.$$

Ex. 4. Shew that, with proper constant values of A, B, C, the relation connecting u, v, w, p, in order that the quadric surface, in threefold space, expressed by $AuX + BvY + CwZ + pT = 0$, where X, Y, Z, T are the same functions of the coordinates of the space as in Ex. 2, should be a cone, is identical with the equation, in these coordinates, of the Weddle surface, given in Ex. 2.

Ex. 5. The equation $(XX')^{\frac{1}{2}} + (YY')^{\frac{1}{2}} + (ZZ')^{\frac{1}{2}} = 0$, where

$$X + Y + Z + X' + Y' + Z' = 0,$$

is identically satisfied by writing, for X, Y, Z, X', Y', Z', respectively,

$$aa'(b-c), \quad bb'(c-a), \quad cc'(a-b), \quad a'bc(b-c), \quad b'ca(c-a), \quad c'ab(a-b),$$

where $a' = 1 - a$, $b' = 1 - b$, $c' = 1 - c$. The reverse equations are

$$2aYZ = XX' - YY' - ZZ', \quad 2bZX = YY' - ZZ' - XX',$$
$$2cXY = ZZ' - XX' - YY'.$$

The equation of the tangent solid of the locus, in space of four dimensions, represented by the equation, at the point (a, b, c), is then

$$bcX + caY + abZ + aX' + bY' + cZ' = 0.$$

If the values of the parameters a, b, c, at any point common to the locus and this tangent solid, be denoted by $t^{-1}x$, $t^{-1}y$, $t^{-1}z$, these are found to satisfy the equation

$$(t - x)(y - z)(bcxt + ayz) + \text{etc.} = 0,$$

identical in form with the equation of the Weddle surface mentioned in Ex. 2.

The locus is identical with that before called Σ, which appeared with the equation $[(\eta_1 - \zeta_1')(\eta_1' - \zeta_1)]^{\frac{1}{2}} + \text{etc.} = 0$. (Above, p. 126.) This was expressed by parameters, x_1, y_1, z_1, in the forms $\xi_1 = (y_1 z_1')^{-1}$, etc. Shew that there is complete agreement if $(1 - x_1)(1 - a) = 1$, $(1 - y_1)(1 - b) = 1$, $(1 - z_1)(1 - c) = 1$.

Ex. 6. In space of three dimensions, a general quartic surface with *fifteen* double points is represented by the first two equations of Ex. 5, taken with a single further relation

$$l'X + m'Y + n'Z + lX' + mY' + nZ' = 0.$$

By the birational transformation of Ex. 5, this is changed into the quartic surface with *five* double points, which is expressed by the equation

$$(t - x)(y - z)(lyz + l'xt) + \text{etc.} = 0.$$

In general, this surface contains the ten joins of the five fundamental points, and ten other lines, each in a plane containing three fundamental points. It becomes the Weddle surface if $ll' = mm' = nn'$.

Ex. 7. The locus S, in space of four dimensions, given by $\xi\eta\zeta = \xi'\eta'\zeta'$, $\xi + \eta + \zeta = \xi' + \eta' + \zeta'$, may be obtained by projection of the (∞^3) locus, in space of five dimensions,—the intersection of two quadric line-cones,—which is represented by the equations

$$XX'/aa' = YY'/bb' = ZZ'/cc'.$$

For, putting, herein, $X = a(1 + \lambda x)$, $X' = a'(1 + \lambda x')$, etc., the elimination of λ leads to the equation

$$(y - z')(z - x')(x - y') + (y' - z)(z' - x)(x' - y) = 0.$$

This may be described, briefly, as the condition that the pairs x, x'; y, y'; z, z' should be in involution. A tetrahedral complex of lines, in space of three dimensions, can be represented, in space of five dimensions, by the intersection of two quadric line-cones. (Cf. Segre, *Atti...Torino*, xxII, 1886–7, p. 556. See, also, Chap. vII, below.)

Ex. 8. The Weddle and Kummer surfaces may be obtained, as appears from what precedes, without space of four dimensions, by considering the four quadric surfaces passing through six points in space of three dimensions. A more general theory is that which arises by considering any four quadrics whatever. See the beautiful account given by Reye in three chapters of the *Geometrie der Lage* (for other references see also *Proc. Lond. Math. Soc.* xxi, 1923, p. 121).

CHAPTER VI

A QUARTIC SURFACE IN SPACE OF FOUR DIMENSIONS. THE CYCLIDE

A quartic surface in space of four dimensions. We have shewn in detail, in Chapter III, how the theory of a plane curve of the fourth order, which has two double points, may be deduced from that of a curve, lying in space of three dimensions, defined as the intersection of two quadric surfaces. In a similar way, we may take two quadrics in space of four dimensions, each represented by the vanishing of a quadratic function of five homogeneous variables, and consider the locus of their common points; this will be an aggregate of ∞^2 points, say a *surface*, having four points on any general plane of the fourfold space. If, in analogy with the case previously discussed, this surface be projected, from an arbitrary point of one of the quadrics, on to any threefold space, there arises a quartic surface in this space. This surface passes through a certain conic of the threefold space, which depends only on the centre of projection and the quadric upon which this is taken; this conic is a double curve for the surface obtained. As in the other case, if this definite conic be regarded as the Absolute conic of the threefold space, it is easily proved, from the four dimensional figure, that the quartic surface, in the threefold space, can be generated as the envelope of *spheres*, each touching it in two points, the generation being possible in five different modes. The spheres of one system have their centres on a quadric surface, and cut a fixed sphere at right angles. Further, the five quadric surfaces, which are the loci of the centres of the spheres of the different systems, are confocal, all touching the common tangent planes of any two of them. These common tangent planes all touch the quartic surface, at points of its double conic; this conic, being the Absolute conic, lies on all the spheres. If the projection be made from a point which is common to two quadrics of the fourfold space, the locus obtained, in the space of three dimensions, is a *cubic* surface, on which the Absolute conic is a simple curve. The theory of *inversion*, which we have given previously (above, pp. 12, 96), is likewise applicable here. As inversion in a plane, in regard to a circle, was reduced to projection in space of three dimensions, so inversion in threefold space, in regard to a sphere, is reducible to projection in space of four dimensions. The fact that we can obtain, from the same quartic surface in fourfold space, either a quartic or a cubic surface in threefold space, becomes the fact that a quartic surface, in threefold space, possessing a double conic, can be inverted into a cubic surface passing through this conic. In particular, from the theory of the lines of a cubic surface, we can thus infer that such a quartic surface contains sixteen lines; the lines of the cubic surface which consist of the line lying in the plane of the Absolute conic, and the ten others meeting this, disappear in the inversion.

The theory from this point of view is very analogous to that we have given for the case of two and three dimensions; but the surface of the fourth order in space of four dimensions is in fact simpler than the curve which is the intersection of two quadric surfaces in threefold space. The latter is a curve of which the points are expressible, in single-valued form, only by elliptic functions. But the surface in fourfold space is a rational surface, whose points are in (1, 1) birational correspondence with the points of a plane. It is thus proper to develop the theory of this surface for itself. The theory of a quartic surface, in threefold space, with a double conic, and the theory of a cubic surface, arise then as consequences. This seems, in fact, the simplest approach to the theory of the cubic surface.

The surface in question is capable of an extremely simple definition. Let a, b, c be any three non-intersecting lines of the space of four dimensions, and T, U be two points, all of general position. From one of these points, say T, there can be drawn ∞^1 planes to meet the lines a, b, c; from U can also be drawn ∞^1 planes to meet a, b, c. The surface we consider is the locus of the point common to one of the planes through T and one of the planes through U.

Preliminary algebraic consideration of the surface. We first shew, directly from this definition, with help of the symbols, that the surface is the intersection of two quadric (∞^3) loci, obtain the parametric expression, and prove that the surface contains sixteen lines. This involves some repetition of work that has preceded; moreover the surface appears later (Chap. VII), as the representation, in space of five dimensions, of a quadratic congruence of lines. But perhaps this is not a disadvantage.

Given the lines a, b, c, as stated, let the equations of the solids $[b, c]$, $[c, a]$, $[a, b]$ be written, respectively, $p = 0$, $q = 0$, $r = 0$; and, denoting the line TU by e, let the equations of the solids $[a, e]$, $[b, e]$, $[c, e]$ be, respectively, $P = 0$, $Q = 0$, $R = 0$. We can suppose that, identically, $p + q + r + P + Q + R = 0$. The lines a, b, c, e are then given, respectively, by

$$q = r = P = 0, \quad r = p = Q = 0, \quad p = q = R = 0, \quad P = Q = R = 0.$$

Thus a plane through the point U, to meet the lines a, b, c, may be supposed to meet these, respectively, in points for which p, q, r, P, Q, R are in the ratios of

$$(\alpha, 0, 0; \; 0, -y, z), \quad (0, \beta, 0; \; x, 0, -z), \quad (0, 0, \gamma', -x', y', 0);$$

and here, since the plane contains U, for which $P = Q = R = 0$, we can suppose $x' = x$, $y' = y$, and put $\gamma' = \gamma$. Then $\alpha = y - z$, $\beta = z - x$, $\gamma = x - y$, so that $\alpha + \beta + \gamma = 0$. Hence we find, for the equations of the plane, any two of the three

$$-P + x(\beta^{-1}q - \gamma^{-1}r) = 0, \quad -Q + y(\gamma^{-1}r - \alpha^{-1}p) = 0,$$
$$-R + z(\alpha^{-1}p - \beta^{-1}q) = 0,$$

the plane meeting the line e, in the point U, where p, q, r, P, Q, R are in the ratios of $(\alpha, \beta, \gamma, 0, 0, 0)$. Instead of P, Q, R we may use coordinates, p', q', r', such that

$$p' + P + q + r = q' + Q + r + p = r' + R + p + q = 0,$$

and, therefore, $p + q + r + p' + q' + r' = 0$. The equations of the plane then become two of the three

$$p' + z\beta^{-1}q - y\gamma^{-1}r = 0, \quad q' + x\gamma^{-1}r - z\alpha^{-1}p = 0,$$
$$r' + y\alpha^{-1}p - x\beta^{-1}q = 0,$$

shewing, incidentally, that the plane also meets another line, say d, given by $p' = q' = r' = 0$, $p/x\alpha = q/y3 = r/z\gamma$. The equations of the plane shew that it lies on the quadric point-cone given by

$$\alpha^{-1}pp' + \beta^{-1}qq' + \gamma^{-1}rr' = 0,$$

whose vertex, U, has, for coordinates p, q, r, p', q', r', the ratios of $\alpha, \beta, \gamma, \alpha, \beta, \gamma$, respectively.

Let the point T be that for which the coordinates are α', β', γ', α', β', γ'; a plane through this, which meets the lines a, b, c, will be given by equations arising from the former by putting x', y', z', respectively, for x, y, z, where $\alpha' = y' - z'$, $\beta' = z' - x'$, $\gamma' = x' - y'$.

Now, let μ denote any one of $\beta\gamma' - \beta'\gamma$, $\gamma\alpha' - \gamma'\alpha$, $\alpha\beta' - \alpha'\beta$, which are at once seen to be equal to one another; put

$$t = \alpha'\beta\gamma(p + p') + \beta'\gamma\alpha(q + q') + \gamma'\alpha\beta(r + r'),$$
$$u = \alpha\beta'\gamma'(p + p') + \beta\gamma'\alpha'(q + q') + \gamma\alpha'\beta'(r + r'),$$

which give rise to

$$\mu^2(p + p') + \alpha't + \alpha u = 0, \quad \mu^2(q + q') + \beta't + \beta u = 0,$$
$$\mu^2(r + r') + \gamma't + \gamma u = 0.$$

The equations $t = 0$, $u = 0$ represent solids passing, respectively, through the points U, T; in terms of these the equation of the quadric point-cone found above, with vertex at U, becomes

$$\alpha^{-1}(p - p')^2 + \beta^{-1}(q - q')^2 + \gamma^{-1}(r - r')^2 + t^2/\mu^2\alpha\beta\gamma = 0.$$

There is a corresponding point-cone, with vertex at T, whose equation is found from this by putting $\alpha', \beta', \gamma', u$ in place of α, β, γ, t, respectively. The quartic surface under consideration is the intersection of these two point-cones; its equations are then given by

$$pp'/\alpha\alpha' = qq'/\beta\beta' = rr'/\gamma\gamma',$$

and these, we see, are equivalent to

$$[\mu^4(p-p')^2 - (\alpha't + \alpha u)^2]/\alpha\alpha' = [\mu^4(q-q')^2 - (\beta't + \beta u)^2]/\beta\beta'$$
$$= [\mu^4(r-r')^2 - (\gamma't + \gamma u)^2]/\gamma\gamma'.$$

These are equivalent, also, to what are obtained by changing the sign of u throughout.

Thus the surface contains the four lines a, b, c, d, given, respectively, by $q=r=p'=0$, $r=p=q'=0$, $p=q=r'=0$, and $p'=q'=r'=0$. Likewise it contains the four lines, a', b', c', d', given, respectively, by $q'=r'=p=0$, $r'=p'=q=0$, $p'=q'=r=0$, and $p=q=r=0$; we easily see that these are the transversals of triads of a, b, c, d. But the equations shew that the surface contains the sixteen lines given by

$$\mu^2(p-p')\,\theta = \alpha't + \epsilon\alpha u, \quad \mu^2(q-q')\,\phi = \beta't + \epsilon\beta u,$$
$$\mu^2(r-r')\,\psi = \gamma't + \epsilon\gamma u,$$

wherein each of $\theta, \phi, \psi, \epsilon$ is ± 1; the value $\epsilon = 1$ gives the eight lines already named, in virtue of the identities such as

$$\mu^2(p+p') + \alpha't + \alpha u = 0.$$

The equations also shew that the pentad, of five points, given by the intersections, in fours, of the five solids whose equations are $p-p'=0$, $q-q'=0$, $r-r'=0$, $t=0$, $u=0$, is self-polar in regard to all quadrics containing the quartic surface; and that all the sixteen lines may be obtained, from any one of them, by combination of the five harmonic inversions, in which one of the solids, and the opposite vertex of the *simplex*, are taken as fundamental elements. The solids, $p-p'=0$, $q-q'=0$, $r-r'=0$, and the opposite vertices, are independent of the positions of T and U upon the line e. The four lines consisting of e and the joins of any point, O, of the line e, to these three vertices of the simplex, are conjugate, in pairs, in regard to the quadric point-cone formed by the planes drawn from O to meet the lines, a, b, c.

We have said that the quartic surface is representable rationally upon a plane. The parameters for this expression are those which determine planes lying, respectively, upon two quadric point-cones containing the surface. Taking the ratios of ξ, η, ζ, such that $\xi/p\alpha^{-1} = \eta/q\beta^{-1} = \zeta/r\gamma^{-1}$, the equations

$$pp'/\alpha\alpha' = qq'/\beta\beta' = rr'/\gamma\gamma'$$

give

$$p/\alpha\xi = \ldots = \alpha'\eta\zeta + \beta'\zeta\xi + \gamma'\xi\eta,$$

$$p'/\alpha'\eta\zeta = q'/\beta'\zeta\xi = r'/\gamma'\xi\eta = -(\alpha\xi + \beta\eta + \gamma\zeta);$$

thereby the coordinates, $p, q, \ldots, r'$, of a point of the surface, are expressed by the coordinates, ξ, η, ζ, of a point of a plane; the six curves $p = 0, \ldots, r' = 0$, in this plane, are cubics through the five points $(1, 0, 0)$, $(0, 1, 0)$, $(0, 0, 1)$, $(1, 1, 1)$, $(\alpha^{-1}\alpha', \beta^{-1}\beta', \gamma^{-1}\gamma')$. The last two points are the intersections of $\alpha\xi + \beta\eta + \gamma\zeta = 0$, $\alpha'\eta\zeta + \beta'\zeta\xi + \gamma'\xi\eta$. These cubics are subject to $p + \ldots + r' = 0$. Conversely, the theory of the quartic surface may be initiated by considering the cubic curves, of which five are linearly independent, passing through five arbitrary points of a plane. (Cf. Vol. III, p. 189.)

Ex. 1. Shew that an arbitrary plane, drawn through any line of the quartic surface, meets the surface again in one point. As such a plane meets a fixed plane, of general position, in a point, it establishes a representation of the quartic surface upon a plane.

There are five lines of the quartic surface which meet the line through which the planes are drawn. These are represented by five fundamental points of the plane. The other ten lines of the surface become the joins of these five fundamental points. Obtain, on the quartic surface, the representation of a general line of the plane; and, also, of a general cubic curve passing through the five fundamental points of the plane.

Ex. 2. Any two general quadric point-cones, in fourfold space, can be expressed by the equations

$$x^2 + y^2 + z^2 + t^2 = 0, \quad a^{-1}x^2 + b^{-1}y^2 + c^{-1}z^2 + u^2 = 0.$$

Denoting $(b - c)^{\frac{1}{2}}$, $(c - a)^{\frac{1}{2}}$, $(a - b)^{\frac{1}{2}}$, respectively, by λ, μ, ν, verify that, if we take

$$p = \lambda\mu(\lambda x + \mu y + \nu z), \quad q = c(\mu x - \lambda y + \nu t), \quad r = -b\mu x + a\lambda y - \nu(abc)^{\frac{1}{2}}u,$$

$$p' = \lambda\mu(\lambda x + \mu y - \nu z), \quad q' = c(\mu x - \lambda y - \nu t), \quad r' = -b\mu x + a\lambda y + \nu(abc)^{\frac{1}{2}}u,$$

together with $\alpha = c - a$, $\beta = -c$, $\gamma = a$, $\alpha' = c - b$, $\beta' = -c$, $\gamma' = b$, so that $\alpha + \beta + \gamma = \alpha' + \beta' + \gamma' = 0$, then $p + q + r + p' + q' + r' = 0$, and the two point-cones are expressed, respectively, by

$$pp'/\alpha\alpha' = qq'/\beta\beta' \quad \text{and} \quad pp'/\alpha\alpha' = rr'/\gamma\gamma'.$$

Ex. 3. Given, in space of n dimensions, a simplex consisting of $(n + 1)$ points, and the $(n + 1)$ *primes*, or spaces of $(n - 1)$ dimensions, which contain the sets of n of these points, consider the $(n + 1)$ harmonic inversions, in each of which one of these points and the complementary prime are the fundamental elements. Shew that an arbitrary line, subjected to all possible combinations of these inversions, gives rise to a set of 2^n lines. (Cf. Vol. III, p. 80.)

Ex. 4. We may verify directly from the equations that the quartic surface, in the fourfold space, contains no other line than the sixteen we have found. Any line can be represented by equations of the forms $\rho_1 (p - p') = l_1 t + m_1 u$, $\rho_2 (q - q') = l_2 t + m_2 u$, $\rho_3 (r - r') = l_3 t + m_3 u$; the substitution of the values of $p - p'$, $q - q'$, $r - r'$ found from these, in the equations of the surface, determines all the existing lines.

Ex. 5. Prove that the two lines obtained, respectively, from the equations

$$\theta x = \alpha' t + \epsilon \alpha u, \quad \phi y = \beta' t + \epsilon \beta u, \quad \psi z = \gamma' t + \epsilon \gamma u,$$

by taking $\epsilon = \pm 1$, lie in the plane expressed by

$$(\theta x - \alpha' t)/\alpha = (\phi y - \beta' t)/\beta = (\psi z - \gamma' t)/\gamma;$$

and that this plane meets, in a line, any one of the four planes whose equations are obtained from these by changing the sign either of one, or of all, of θ, ϕ, ψ,—where $\theta^2 = \phi^2 = \psi^2 = 1$.

Resumption of the descriptive theory of the quartic surface. The sixteen lines. As before, let a, b, c, e be four lines, of which no two intersect and no three lie in a solid; and let T, U be any two points of e. We consider the surface which is the locus of the intersection of a plane drawn from T to meet a, b, c, with a plane drawn from U to meet these lines. All such planes, we have shewn, will meet another line, d, associated with a, b, c, e. Let a', b', c', d' be the transversals of the threes of the lines a, b, c, d, the line d' meeting a, b, c, and so on. Through any point of a plane, which meets a, b, c, d, there passes also a plane which meets a', b', c', d', and meets e in the same point as does the former plane (p. 123, above). Thus the quartic surface may also be defined by planes through T meeting a, b, c, taken with planes through U meeting a', b', c'; or *vice versa*; or by planes through T and U both meeting a', b', c'.

The planes from T meeting a, b, c intersect an arbitrary solid in generating lines, of one system, of a quadric surface lying in this solid (p. 122, above); another quadric surface is obtained in this solid by planes from U meeting a, b, c. The quartic surface under discussion thus meets an arbitrary solid in a quartic curve, the intersection of two quadric surfaces; and, therefore, meets an arbitrary plane in four points. We thus speak of the surface as being a quartic, or of order four; the order of a locus of r dimensions, lying in a space of n dimensions, being the number of points of this locus which lie in a (planar) manifold of $n - r$ dimensions, of general position, when this number is independent of this manifold. A plane through T (or U), which meets a, b, c, contains, however, not four, but an infinite number of points of the quartic surface, these lying on a conic; for the planes through U (or T)

which meet a, b, c determine a quadric surface in any solid containing this plane.

Every one of the eight lines a, b, c, d, a', b', c', d' lies on the quartic surface, such a line being a common generator of the two quadric surfaces determined by T and U, as above described, in any solid drawn through this line. Any other line lying on the surface must likewise be a common generator of two such quadric surfaces; this line must, therefore, be common to a plane through T meeting a, b, c, and to a plane through U which either meets a, b, c, or meets a', b', c'; or the statement must hold with interchange of T and U. Now, a plane through T, meeting a, b, c (and, therefore, d), cannot have a line in common with a plane through U meeting a, b, c, unless this line be one of a', b', c', d'. For the two planes, having a line in common, determine a solid; and a line which meets both the planes lies in this solid, unless it is a line, meeting the line of intersection of the planes, not lying in either plane. Thus the line e lies in this solid; and, as two of the lines a, b, c, d do not lie in the same solid with e, three of these lines must meet the line of intersection of the two planes. This line is, therefore, one of a', b', c', d'. Similarly, a line lying on the surface can be one of a, b, c, d. There is next the possibility of a line, lying on the surface, which is the intersection of a plane, α, through T, meeting a, b, c, d, with a plane, β, through U, meeting a', b', c', d'. For this case, consider a solid through the former plane, α, and the quadric surface, lying therein, which is determined by the planes through U of the two systems, those meeting a, b, c, d, and those meeting a', b', c', d'. As this plane α contains a generator of this quadric surface, lying on a plane through U meeting a', b', c', d', the plane α also contains a generator lying on a plane through U meeting a, b, c, d. This second generator must then, as we have seen, be one of a', b', c', d'. By a similar argument the plane β must contain one of a, b, c, d. That is, the line in question is the intersection of a plane through T containing one of a', b', c', d' (which, therefore, meets a, b, c, d), with a plane through U containing one of a, b, c, d (which, therefore, meets a', b', c', d'). The solid determined by these two planes, as it contains T and U, contains the line e. The only solids possible, containing e and also one of a', b', c', d' and one of a, b, c, d, are the four containing, respectively, the lines (a, a', e), (b, b', e), (c, c', e) and (d, d', e). Conversely if, for example, the plane Td' meet the line d in H, and the plane Ud meet the line d' in K', it is at once seen that HK' is a line of the quartic surface; and, if TH meet d' in H', and UK' meet d in K, that $H'K$ is another such. The four lines d, d', HK', $H'K$ are the degenerate quartic curve in which the solid (d, d', e) meets the quartic surface.

Three other pairs of lines are similarly obtained by drawing the transversals, from T and U, to a, a', to b, b', and to c, c'. The only remaining possibility, for a line of the quartic surface, of being the intersection of a plane through T meeting a', b', c', d' with a plane through U meeting a, b, c, d, leads to the same four pairs of lines.

Thus, beside the original eight lines, a, b, ..., d', there are eight others, two in each of the solids (a, a'), ..., (d, d'); and these are all. Another geometrical derivation of the lines arises below (p. 170). Now denote, temporarily, the two lines so obtained which meet d, d' by t and t'; and, similarly, those meeting (a, a'), (b, b'), (c, c'), respectively, by (x, x'), (y, y'), (z, z'). As the section of the quartic surface by any solid is the curve of intersection of two quadric surfaces, there cannot be three lines of the surface lying in a plane, or meeting in a point. Thus, for instance, the line x, meeting a' (which meets d), cannot meet d; and the same line x, meeting a, cannot meet d'. The intersection of x with the solid (d, d') must then be on t or t'. The six points in which x, x', y, y', z, z' meet the solid (d, d') must, in the same way, be all on t or t'. It will appear that there are three on each. Thus we see that every one of the sixteen lines is met by five others.

The self-polar pentad for the quartic surface. The quartic surface has appeared as the intersection of a quadric point-cone of vertex T, with a quadric point-cone of vertex U, each generated by planes through its vertex meeting either a,b,c,d or a',b',c',d'. By an appeal to the equations of these cones, it thus appears that the surface lies on ∞^1 quadrics, each a locus of ∞^3 points, of which the equation of any one is of the form $(T) + \lambda(U) = 0$. As in the analogous cases in a plane, or in three dimensions, there are then, in the general case considered, three other quadric point-cones containing the quartic surface; and the vertices of the five cones form a self-polar pentad for all the quadrics containing the quartic surface. It is easy to specify the vertices of these other cones. Using a preceding notation (p. 113, above), let λ, μ, ν denote, respectively, the planes of the pairs of lines (l, p), (m, q), (n, r), of which, for example, the first contains the points A, A', P, P', L; and let X, Y, Z be, respectively, the points of intersection (μ, ν), (ν, λ), (λ, μ). These are the points of which, (cf. Ex. 4, p. 117), the symbols are $F - F'$, $G - G'$, $H - H'$, or $A' + B + C$, $B' + C + A$, $C' + A + B$ (p. 117, above). It can be shewn that the self-polar pentad is formed by these points, X, Y, Z, together with T and U. It will be noticed that X, Y, Z do not depend on the positions of T and U upon the line e. For, it was seen that the points Y, Z are in the solid $[\lambda, e]$; and that, in regard to X and this solid, the two points Q, B, that is the points (b, d') and (c', a), are harmonic conjugates; as are also the points of each of the pairs (R, C'), (R', C), (Q', B') (p. 117, above). The four pairs of points all lie on the quartic surface, by what we have proved; and a solid is determined by four points. Wherefore the solid $[\lambda, e]$, which contains the four points Y, Z, T, U, is the polar solid of the point X in regard to every quadric containing the quartic surface. Similarly for Y and Z. Again, the solid containing X, Y, Z, U is the polar of T in regard to all these quadrics. For, this follows if we shew that, for the quadric point-cone constituted by the planes drawn to meet a, b, c, d, or a', b', c', d', from any point, O, of the line e, the polar solid of the line e is that containing X, Y, Z and O; and, for this, it is sufficient to shew that the plane XYZ contains three points, each of which is a harmonic conjugate of a point of the line e, in regard to a pair of points lying on the quartic surface. In fact, the point which is the harmonic conjugate of the point L, in regard to the points A, A', both of which lie on the quartic surface, lies on the plane XYZ; this point lies, indeed, on the line YZ (p. 117, above); and there is a similar statement for M and N. This discussion shews also that U is the pole of the solid containing X, Y, Z, T. The property of X, Y, Z, T, U is thus established.

It follows hence, from the harmonic property of the polar solid of a point in regard to a quadric, that to any line of the quartic surface there corresponds another, the harmonic inverse of the former in regard to any vertex of the polar pentad and its opposite solid. These two lines intersect on this solid, so that any one of the lines meets five others. All the sixteen lines are obtainable from any one of them by combination of the five possible inversions. As a line in space of four dimensions depends on six parameters, the number of parameters necessary to specify one line and the five points of the polar pentad is $6 + 5.4$, or twenty-six; this is the same number as that, $3.6 + 2.4$, required to specify the three lines a, b, c and the two points T, U.

Again, in any one of the solids of the polar pentad, say that opposite to the vertex U, there will be eight points, each an intersection of two lines of the quartic surface. These pairs of lines lie in eight planes through U, which are planes of the quadric point-cone, of vertex U, containing the quartic surface. The plane of the two lines which intersect in one of these eight points, say E, of the polar solid of U, is easily seen to lie in the tangent solid, at E, of any quadric which contains the quartic surface. Thus, this plane is tangent, at E, to the quartic curve in which the quartic surface meets the polar solid of U. This quartic curve lies on the quadric surface, say ω, in which this polar solid is met by the quadric point-cone, of vertex U, which contains the quartic surface. It is known that there are four generators, of either system, of a quadric surface in threefold space, which touch the quartic curve in which this quadric surface is met by another quadric surface (Vol. III, p. 69). The eight points such as E are thus identified as the points in which generators, of the quadric surface ω, touch the quartic curve in which the quartic surface meets the polar solid of U. For every generator of ω determines a plane of the cone of the vertex U.

Ex. 1. It has been shewn above (p. 119) that the original figure of fifteen points and lines is transformed into itself by the harmonic inversion in which the line e and the plane XYZ are fundamental elements. Thus if, as above, a line from T meet d and d', respectively, in H and H', this line will meet the plane XYZ, and the solid (X, Y, Z, T) will contain the points H, H'; we have seen that two lines of the quartic surface intersect at both these points. The same solid contains three other pairs of intersections, on (a, a'), (b, b'), (c, c'). Similarly for the solid (X, Y, Z, U).

Ex. 2. The plane ν, in a preceding notation, contains the points C, C', R, R', each of which is an intersection of a pair of lines of the quartic surface. Prove that the solid (X, Y, T, U), that is, the solid (ν, e), also contains four intersections, each of a line, such as HK', which meets d and d', with a line meeting another pair (a, a'), or (b, b'), or (c, c').

Precisely, with the symbols, the line joining the points

$$(z-x)^{-1} B - (x-y)^{-1} C' \quad \text{and} \quad (z_1 - x_1)^{-1} B' - (x_1 - y_1)^{-1} C,$$

meets the line joining the points

$$(x_1 - y_1)^{-1} C - (y_1 - z_1)^{-1} A' \quad \text{and} \quad (x-y)^{-1} C' - (y-z)^{-1} A,$$

in a point of the solid (ν, e).

Ex. 3. The fact that the sixteen lines of the quartic surface are obtainable from any one of them by repeated harmonic inversion, in respect to the five solids named, suggests a notation for the lines; in terms of which, also, a particular property of the lines can be conveniently stated. If one of the lines be denoted by 0, and those which are derived from this by harmonic inversion in the five solids be denoted by 1, 2, 3, 4, 5, respectively, then it is easy to see that inversion of the line 1, in the second solid, gives the same line as is obtained by inversion of the line 2 in the first solid; so that this line may be denoted by 12 or 21. There will be ten such lines. The notation 123 would then denote, for instance, the line obtained, by inversion in the third solid, from the line 12; it is, however, easy to see that this is the line 45; and so on, every line for which three indices might be used being the same as that denoted by the two complementary indices. The origin of the notation shews, for instance, that the lines 12, 13 both intersect the line 1; while they do not intersect one another. But the lines 12, 34 intersect, for example; and, in general, two lines represented each by two numbers, when the four numbers involved are all different. Consider now the two sets of four lines, 1, 23, 24, 25, and 2, 13, 14, 15; it is easy to see that each set consists of the four transversals of the threes of the other set. By using different numbers, such a pair of sets of four lines can be formed in ten ways. Again, the two sets of four lines, 1, 2, 3, 45 and 23, 31, 12, 0, have the same character, each consisting of the transversals of the threes of the other set; and of such pairs of sets there are also ten. We may say, then, that, from the lines of the surface, there can be formed *twenty double-fours* of lines.

The conics lying on the quartic surface. The quartic surface meets any generating plane of the quadric point-cone, of vertex U, which contains the surface, in a conic; as has been remarked. This conic is on the quadric surface in which any solid, drawn through the plane, is met by the quadric point-cone of vertex T, which contains the quartic surface. The planes of the former cone, of vertex U, are of two systems, two planes of different systems meeting in a line. Thus the two conics, in two generating planes, of the cone, of different systems, have two points in common, where the line of intersection of their planes meets the point-cone, with vertex at T, which determines these conics. In particular, the conic, in such a plane through U, may consist of two lines of the quartic surface; then these lines will meet the conic lying in any plane of the other system through U. The sixteen lines of the surface, we have seen, consist of four pairs lying in four generating planes of the same system, of the cone of vertex U, and of four other pairs lying in four generating planes of the other system. Thus a conic of the surface, generated as here, is met by eight lines of the surface. In particular, two lines of the surface, lying in one generating plane of the cone of vertex U, are each met by four other lines. Again, through any point, say O, of the quartic surface, can be drawn two of the generating planes of the cone of vertex U, meeting in a line through U. These two planes define a solid, the tangent solid at O of the cone. This solid contains the tangent plane at O of the quartic surface, which is the intersection of the tangent solids of all the quadrics containing the surface. Thus, the tangent plane of the quartic surface at O meets, in a line, each of the two generating planes of the cone, of vertex U, which pass through O. We assume, on the basis of what has been proved, that the quadric point-cones containing the quartic surface, with vertices at X, Y, Z, T, U, play entirely symmetrical parts, in general. Then it follows that, through any point, O, of the surface, there pass ten conics lying on the surface; two of these, which intersect in another point beside O, lie in planes, of different systems, of the cone of vertex U, and there are four other such pairs. Every one of the sixteen lines of the surface meets five of the ten conics, one of each of the five pairs; and each conic is met by eight of the lines. Further, each conic is met, besides at O, by the tangent plane of the surface at O. We can fix these relations by remarking the consequences which follow when we project from O on to an arbitrary solid:—The tangent plane of the quartic surface at O will meet this solid in a line, say t. Each of the conics through O will project into a line meeting the line t; and the ten lines so found will intersect in pairs. Each of these ten lines will also be met by eight of the sixteen lines which arise by projection of the lines of the quartic surface; and will thus be met by ten lines in all. A line arising by projection of one of these sixteen lines of the quartic surface will be met by five others of these, and will also be met by five of the lines arising by projection of the ten conics. As an arbitrary plane through O meets the quartic surface in three other points, the quartic surface will project into a cubic surface; and this will contain the twenty-seven lines which have been described.

The theory of the quartic surface may be studied in further detail, for its great interest and simplicity. In particular, the systems of curves which lie upon the surface may be examined; either directly, or, as in the case of other rational surfaces, by their representation upon a plane; it appears that every curve is co-residual with an aggregate formed from *six* fundamental curves. The corresponding theory of curves upon a cubic surface in threefold space (cf. Vol. III, pp. 191 ff.) may be deduced from this theory by projection. In this place only indications can be given, of an extensive theory. The most important source of information, with full references to previous literature, is Segre, *Math. Annal.* XXIV, 1884, pp. 313–444. See also Darboux, *Sur une classe remarquable de courbes et de surfaces algébriques*, Paris, 1873, and the Bibliography, to 1872, appended thereto. For an introduction to the theory of the co-residuation of curves upon the surface, reference may be made to C. V. H. Rao, *Proc. Lond. Math. Soc.* XVII, 1919, pp. 272–305.

The cubic surface in space of three dimensions; and the theory of inversion. Regard the quartic surface which we have discussed as the intersection of a quadric point-cone (U), and a particular quadric, Ω. Denote the surface by Γ. We have just considered the projection of Γ, from a point O common to (U) and Ω, upon a solid, say Π; this gave a cubic surface in Π. Through the point O, of Ω, there passes a conical sheet of lines, lying on Ω and on the tangent solid of Ω at O (p. 37, above). Each of these lines meets the cone (U) in a further point, which is, then, also on Γ. The conical sheet of lines meets Π in a conic, lying in the plane in which Π is met by the tangent solid of Ω at O. Denote this conic by ω. The cubic surface in Π, obtained by projection of Γ, contains the conic ω. Every one of the sixteen lines of Γ meets the tangent solid of Ω, at O, in a point, also lying on Ω, as does the line; thus, the sixteen lines of the cubic surface which arise by projection of the sixteen lines of Γ intersect the conic ω. Other ten lines of the cubic surface, arising by projection of the conics of Γ which pass through O, meet the line of the cubic surface, lying in the plane of ω; this arises, as we have seen, from the tangent plane of Γ at O.

In this derivation of the cubic surface, the centre of projection, O, lies on the cone (U), as well as on Ω. We may, however, obtain, on the quadric Ω, another quartic surface, say Γ', exactly similar to Γ, but not passing through O. Take any point, H, not lying on Ω, and the polar solid of H in regard to Ω; to any point, P, of the fourfold space, make correspond the point, P', of the line HP, which is the harmonic conjugate of P, in regard to H and the polar solid of H. In particular, when P is on Ω, the point P' is the second intersection of the line HP with Ω. Then, as in the case of threefold space (p. 13, above), to the quadric point-cone (U) will correspond another quadric point-cone, say (U'), whose generating planes meet those of (U) on the polar solid of H. To the surface Γ will then correspond the intersection of Ω with this cone (U'); this will be a surface, Γ', exactly similar to Γ, but not in general containing O. If the surface Γ' be projected from O, on to the solid Π, we obtain, therein, a surface of the fourth order. This may, then, be obtained, from the cubic surface into which Γ projects, by a process of inversion in this space Π, in regard to a *sphere* lying in this solid (the Absolute conic being ω). The centre of this sphere is the projection, from O, of the point H (pp. 14, 37 ff., above). By the process of harmonic inversion, with H and its polar solid, the sixteen lines of Γ, which lie upon Ω, become lines of Γ'; and these are projected from O into lines of the quartic surface, in Π, into which Γ' projects. Thus the sixteen lines of the cubic surface, in the space Π, invert into lines. But the other eleven lines of the cubic surface, as we see in a similar way, do not invert into lines.

Ex. 1. A cubic surface, in the threefold space of coordinates x, y, z, u, which meets the plane $u = 0$ in the conic $x^2 + y^2 + z^2 = 0$, and also in the line $lx + my + nz = 0$, is expressible by an equation $(x^2 + y^2 + z^2)(lx + my + nz) - u\phi = 0$, where $\phi = 0$ is a quadric cone of vertex $(0, 0, 0, 1)$. This surface arises, by projection, from the point, in the fourfold space (x, y, z, u, t), whose coordinates are $(0, 0, 0, 0, 1)$, of the quartic surface which is the intersection of the quadric $x^2 + y^2 + z^2 + tu = 0$, and the point-cone

$$t(lx + my + nz) + \phi = 0,$$

whose vertex is $(0, 0, 0, 1, 0)$.

Ex. 2. With the notation of the text, let the conic of Γ, lying in a plane through O and T which meets the lines a, b, c, d, be called (5′); and the conic, in a plane through OT, which meets $a′, b′, c′, d′$, be called (6). Through OU there is a plane, meeting a, b, c, d, which contains a conic of Γ which we call (6′), and a plane, meeting $(a′, b′, c′, d′)$, containing a conic (5). The two sets of six elements, a, b, c, d, (5), (6), and $a′, b′, c′, d′$, (5′), (6′), project from O into a double-six of lines (cf. Vol. III, p. 160). The conics of Γ in planes from O to the vertex X, of the self-polar pentad, may be called (23) and (14); those similarly arising for Y and Z being, respectively, (31), (24) and (12), (34). The eight other lines of Γ may then be denoted by symbols $(r5)$, $(r6)$, for $r = 1, 2, 3, 4$.

Ex. 3. Prove by projection in fourfold space that, by inversion in threefold space, a line meeting the Absolute conic inverts into another line meeting this conic, and meeting the former line.

The Cyclide, or quartic surface with a double conic, in threefold space. Consider now more particularly the character of the surface, in the threefold space Π, which is obtained by projecting, from the point, O, of the quadric Ω, the intersection of Ω with the quadric point-cone (U), not passing through O. Denote the quartic surface constituting this intersection by Γ; and let the surface in Π, obtained by projection of Γ, be called the *Cyclide*. As Γ is met by an arbitrary plane, of the fourfold space, in four points, the Cyclide is met by an arbitrary line of the threefold space in four points, and is a quartic surface. The lines of the conical sheet, in which Ω is met by its tangent solid at O, each meet the cone (U) in two points. Thus the Cyclide has the conic, ω, in which this conical sheet meets Π, as a double conic. Every line lying on Γ, since it meets the tangent solid at O, and lies itself on Ω, meets the conical sheet; thus the Cyclide has sixteen lines, which all meet the conic ω.

The poles, in regard to Ω, of the tangent solids of the cone (U), are an aggregate of ∞^2 points; they lie in the solid which is the polar of the vertex, U, of the cone (U), in regard to Ω; they describe a quadric surface in this polar solid. This we denote by $Q(U)$. A tangent solid of the cone (U) touches this cone along a line; this line meets Ω in two points. The tangent solid of (U) contains the tangent plane of Γ at each of these two points. The section of Ω, by this tangent solid of (U), projects from O into a quadric surface in Π, containing the conic ω; or, as we shall say, into a *sphere*, the conic ω being regarded as Absolute conic of the space Π. The *centre* of this sphere, being the projection of the pole, in regard to Ω, of the tangent solid of (U), lies on the quadric surface in (U) which is the projection of the quadric $Q(U)$,

(p. 37, above). This sphere, then, touches the Cyclide in two points. The section of Ω by the polar solid of the vertex U, of the point-cone, also projects into a sphere, in Π. This sphere, which depends only on U and O, is cut at right angles by the sphere arising from the chosen tangent solid of (U), because this tangent solid and the polar solid of U are conjugate in regard to Ω (p. 40, above). Thus it appears that the Cyclide is touched, in two points, by an aggregate of spheres, each defined as having its centre on a certain fixed quadric surface, in the space Π, and cutting at right angles a certain fixed sphere of this space. Or, we may regard the matter in a slightly different way: Any plane, in the fourfold space, meets the quadric Ω in a conic. Thus the section of Ω, by a plane, projects from O, on to the solid Π, into a conic having two points of intersection with the Absolute conic, ω, that is, into a *circle*. When the plane, not passing through O, meets Ω in two lines, these become, on projection, two lines meeting the conic ω. If the plane be a generating plane of the cone (U), so consisting wholly of points of this cone, this plane meets the quartic surface Γ in a conic, projecting from O into a circle lying on the Cyclide. Through such a generating plane, of the cone (U), there passes an infinity of tangent solids of (U), each containing another generating plane, of the opposite system. Thus, by what is said above, through the circle on the Cyclide, just obtained, there passes an infinity of spheres, each meeting the Cyclide in another circle. Any such sphere will touch the Cyclide at the two intersections of the two circles which lie upon the sphere. As the tangent solids of (U) which pass through a plane have their poles, taken in regard to Ω, upon the polar line of this plane in regard to Ω, it follows that the spheres, passing through a particular circle which lies on the Cyclide, have their centres on a line. This is then a generator of the quadric surface, in Π, obtained by projection of the quadric surface $Q(U)$. A special case arises when the particular circle consists of two lines lying on the Cyclide.

Ex. 1. In the fourfold space, we may take coordinates, (x, y, z, u, t), for which the point O is $(0, 0, 0, 0, 1)$, the tangent solid of Ω at O being $u = 0$, choosing x, y, z so that the conical sheet of lines of Ω at O is given by $u = 0$, $x^2 + y^2 + z^2 = 0$. Then the equation of Ω is of the form $-ut + x^2 + y^2 + z^2 + 2uP + du^2 = 0$, where P is a homogeneous linear form in x, y, z. If the point U be $(0, 0, 0, 1, 0)$, the equation of the cone (U) is $t^2 + 2tQ + V = 0$, where Q, V are homogeneous in x, y, z, respectively linear and quadratic. The elimination of t between these equations gives the equation of the projection of their intersection, upon the solid $t = 0$; this solid is here taken to contain the point U. If, in the eliminant, we put $x + lu$, $y + mu$, $z + nu$, respectively, for x, y, z, and choose l, m, n

properly, the equation of the projection takes the form

$$(x^2 + y^2 + z^2)^2 + u^2W = 0,$$

where W is a homogeneous quadratic form in x, y, z and u. This equation, however, can be written $(x^2 + y^2 + z^2 + \lambda u^2)^2 + u^2M = 0$, where $M = W - 2\lambda (x^2 + y^2 + z^2) - \lambda^2 u^2$. By taking, for λ, a root of a certain quintic equation, the equation $M = 0$ represents a quadric cone; thus, for such λ, we can suppose $M = XY - Z^2$, where X, Y, Z are homogeneous linear forms in x, y, z and u. Then the equation of the Cyclide becomes $CD + u^2XY = 0$, where C, D are of the forms $x^2 + y^2 + z^2 + \lambda u^2 \pm uZ$. This is satisfied by $C = 0$, $X = 0$, and by $C = 0$, $Y = 0$. The equation $C = 0$ represents a sphere, and $Z = 0$ is any plane through the vertex of the cone $M = 0$, of which $X = 0$, $Y = 0$ are tangent planes. There are, therefore, ∞^2 such spheres.

Ex. 2. The Cyclide can be generated as the locus of a circle which is the intersection of a varying sphere passing through a fixed circle of the surface, with a corresponding varying sphere passing through another fixed circle of the surface.

Ex. 3. The Cyclide, we have seen, is the envelope of a varying sphere, with centre at a varying point, say H, of a fixed quadric surface, Q', the sphere cutting at right angles a fixed sphere, S. The varying sphere touches the Cyclide in two points. These two points of contact are, in fact, the limiting points of the system of coaxial spheres defined by the fixed sphere S, and the tangent plane, at H, of the quadric Q' (cf. Vol. iii, p. 74). The Cyclide is thus the locus of the limiting points, so defined by the fixed sphere S and the varying tangent plane of the quadric surface Q'. To prove this, consider the figure in the space of four dimensions. Let L, L' be the two points where a generator of the cone (U) meets Ω; take the polar plane, in regard to Ω, of this line LL'. This plane lies on the polar solid, in regard to Ω, of every point of the line. In particular, it lies in the principal solid which is the polar of the vertex U. The plane touches the quadric surface $Q(U)$. The plane meets Ω in a conic, which we denote by λ; this conic is, therefore, on the section of Ω by the polar solid of U. As the plane is in the tangent solid of Ω at L, the conic λ is on the conical sheet of lines of Ω through L. When we project, from O, this conical sheet becomes a quadric cone, in Π, whose vertex is the projection of L. This cone contains the conic which is the projection of λ; and also, as in preceding cases, contains the conic ω. The cone, therefore, contains the section of a *principal sphere*, S, by a tangent plane of the quadric surface, Q', (arising by projection of $Q(U)$), and contains ω. The vertex of this cone, the projection of L, is, therefore, a limiting point of the coaxial system of spheres defined by the principal sphere, and the tangent plane. By definition, L is on the quartic surface Γ, and projects into one of the points of contact of the Cyclide with the sphere obtained from the section of Ω by the tangent solid, along LL', of the cone (U).

If, in the threefold space, the quadric surface Q' and the sphere S be, respectively,

$$a^{-1}x^2 + b^{-1}y^2 + c^{-1}z^2 + d^{-1}t^2 = 0,$$

$$x^2 + y^2 + z^2 + 2fxt + 2gyt + 2hzt + kt^2 = 0,$$

the Cyclide is

$$4t^2\left[a\,(x+ft)^2 + b\,(y+gt)^2 + c\,(z+ht)^2\right] + d\,(x^2 + y^2 + z^2 - kt^2)^2 = 0.$$

Ex. 4. The planes of the circles, in which the Cyclide is met by the enveloping spheres, are tangent planes of one of five quadric cones. These are often called the *cones of Kummer*.

The five generations of the Cyclide, with confocal quadric surfaces. The Cyclide has, at any point of its double curve, ω, two tangent planes, one touching each sheet. We prove, now, that all these planes touch the quadric surface, Q', of the solid Π, which is obtained by projection, from O, of the quadric $Q\,(U)$. We may call this quadric surface a *principal quadric* of Π; and call the sphere S, obtained by projection of the section of Ω by the polar solid of the vertex U, a *principal sphere*. Every generating line, of the conical sheet of lines of Ω at O, meets the point-cone (U) in two points, say K and K'. The locus of the point K is the curve, in the tangent solid of Ω at O, which is the intersection of the conical sheet of lines of Ω at O, with the quadric surface in which the cone (U) meets the tangent solid of Ω at O. This curve is of the fourth order. Consider the tangent solid of Ω at the point K. This solid passes through O, and meets the tangent solid of Ω at O in a plane, which touches the conical sheet of lines through O (p. 38, above). This plane, therefore, contains a tangent line of the conic ω. The point K lies on the surface Γ. The tangent plane of Γ at this point,—being the intersection of the tangent solid of (U) at this point with the tangent solid of Ω at this point,—projects, from O, into a tangent plane of the Cyclide. This plane is the intersection of Π with the solid, through O, containing the tangent plane of Γ at K; this solid is no other than the tangent solid of Ω at K. Thus, the tangent plane of Γ, at K, projects into a tangent plane, of the Cyclide, at the point where OK meets the conic ω. The other tangent plane of the Cyclide, at this point, arises, in a similar way, from K'.

Next consider the line KU. It lies on the cone (U), and is the intersection of two generating planes of this cone. Its polar plane, in regard to Ω, is, by definition, a tangent plane of the quadric surface $Q(U)$. This plane also lies in the tangent solid of Ω at K, because the line KU contains the point K. Thus the plane, into which the tangent plane of Γ, at K, projects, is also a tangent plane of the principal quadric Q', of Π. The point of contact is the projection of the point, of the quadric surface $Q(U)$, which is the pole, in regard to Ω, of the tangent solid of the cone (U) along KU.

It is thus shewn that the tangent plane of Γ, at K, projects into a tangent plane, of the Cyclide, at a point of the conic ω, this plane also touching the quadric Q'.

In the general case, there are, beside the cone (U), four other quadric point-cones containing the surface Γ; to each cone corresponds a *principal* solid, the polar of its vertex in regard to Ω, containing the other four vertices. The sections of Ω by these principal solids project into five principal spheres in the solid Π. To each cone will also correspond a quadric surface, such as $Q(U)$, lying in the corresponding principal solid, this being the polar, in regard to Ω, of the aggregate of the tangent solids of the cone; the quadric surfaces, so arising, project into principal quadric surfaces in Π, such as Q'. From the conjugate relation of the vertices of the cones in the fourfold space, it is clear that, in the solid Π, the centres of four of the principal spheres form a self-polar tetrad for the remaining sphere, and for the corresponding principal quadric surface; also, the join of any two of these centres is at right angles (in regard to the conic ω, as Absolute conic) to the plane containing the other three; while, further, any two of the principal spheres cut at right angles (p. 40, above). Likewise, by what we have just proved, the five principal quadrics belong to a confocal system, being all touched by the tangent planes of the Cyclide at the points of the conic ω. This conic is then one of the four focal conics of these confocal quadric surfaces (cf. Vol. III, p. 92). The Cyclide is the envelope of a sphere, with its centre on any one of the principal quadric surfaces, described to meet the corresponding principal sphere at right angles.

Ex. 1. A confocal system of quadric surfaces, in space of three dimensions, has *four* focal conics. Thus, by what has been shewn, the tangent solids of Ω, at the points such as K, should meet the solid Π in planes containing, not only the tangent lines of the conic ω, but also the tangent lines of three other conics. Reciprocating in regard to Ω, this statement is, that the joins of an arbitrary point, P, to the points K, lie on the planes of four quadric *line-cones* (see above, p. 121). In fact, K describes a quartic curve in the tangent solid of Ω at O, and, thus, lies on four quadric conical sheets in this solid. If V be the vertex of one of these, the line PK evidently lies in the plane VPK, which describes a line-cone with VP as axis. This proves the statement in its reciprocal form.

Ex. 2. Let U, T be the vertices of two quadric point-cones (U) and (T), which contain the surface Γ. If a generating line, UP, of (U), be such that the tangent solid of (U), along UP, touches Ω, the plane UTP meets the polar plane of the line TU, in regard to Ω, upon a certain conic. The same conic is obtained from the cone (T) by a similar definition. Denote this conic by γ.

The angle between two enveloping spheres of the Cyclide, of different systems, arising, say, from tangent solids of the cones (U) and (T), respectively, is equal to the interval between these two tangent solids, measured in regard to Ω (p. 39, above); or, equal to the interval between the poles of these solids, in regard to Ω. Let these solids touch (U) and (T) along the lines UK and TL, respectively; and let the planes UTK and TUL meet the polar plane of the line UT, in regard to Ω, in the points K' and L', respectively. The angle in question, between the corresponding enveloping spheres of the Cyclide, is equal to the interval $K'L'$, measured in regard to the conic γ. (Cf. Jessop, *Quartic Surfaces*, 1916, p. 106.)

Ex. 3. It has been seen that the Cyclide may be regarded as the inverse of a cubic surface containing the conic ω; and, conversely, that such a cubic surface is obtained by inversion of the Cyclide when the centre of inversion lies thereon. By taking the vertex of the harmonic inversion, in the space of four dimensions, in regard to Ω, to be at the vertex, U, of one of the point-cones, we see that the Cyclide inverts into itself when the centre of inversion is the centre of any one of the five principal spheres. When the centre of inversion is arbitrary, a Cyclide inverts into another Cyclide.

Ex. 4. For a quartic curve, in threefold space, which is the intersection of a quadric surface with a quadric cone, there are four points of the curve of which the tangent line is a generator of the cone; these are the points of the curve lying on the polar plane of the vertex of the cone, taken in regard to the quadric surface. By supposing the curve to be that considered above (p. 177), the intersection of (U) with the conical sheet of lines of Ω through O, we see that there are four points K which coincide with the associated points, K'. These will lie on the plane which is the intersection of the polar solids of O in regard to all quadrics containing the quartic surface Γ. The tangent solids of Ω at the four points K will meet in the polar line of this plane, in regard to Ω; and this line passes through O. Any such tangent solid contains the tangent plane of Γ at the corresponding point K. Thus, on projection, we see that there are four points of the conic

ω at which the two tangent planes of the Cyclide coincide with one another; and that the four tangent planes at these points have a point common.

Taking, again, the plane which is the intersection of the polar solids of O in regard to all the quadrics containing Γ, we may consider the locus of the polar lines of this plane, in regard to these quadrics. This is a cubic cone, with O as vertex. Hence, for the Cyclide, we can infer a cubic curve containing the following nine points:—the five points obtained from the vertices of the cones such as (U), these being the centres of the principal spheres; the three intersections of the opposite pairs of joining lines, of the four points of the conic ω at which the tangent planes of the Cyclide coincide; and the point of intersection of the tangent planes of the Cyclide at these four points (Segre, *Math. Annal.* xxiv, 1884, p. 330).

Further, considering the intersection, with Γ, of the polar solids of O, in regard to the quadrics containing Γ, and the intersection of such a solid with Ω, we can prove, for the Cyclide, that there are ∞^1 quadric surfaces, each touching the Cyclide along a quartic curve, all passing through the four specified points of the conic ω, there being one of these quartic curves through any point of the Cyclide. Each of these curves is the intersection of the Cyclide with a quadric surface passing through ω, say with a sphere; these spheres have a common centre, the point at which the tangent planes of the Cyclide, at the four specified points of ω, co-intersect (*loc. cit.* p. 337).

For the Cyclide expressed by

$$(x^2 + y^2 + z^2)^2 + (ax^2 + by^2 + cz^2)\,t^2 + 2\,(fx + gy + hz)\,t^3 + dt^4 = 0,$$

the tangent planes coincide at any one of the four points $(p, q, r, 0)$, where $p^2 = b - c$, $q^2 = c - a$, $r^2 = a - b$; and the tangent planes at these four points intersect in $(0, 0, 0, 1)$.

Confocal Cyclides. An enveloping sphere of the Cyclide was obtained by projecting the section, of Ω, by a solid which touches one of the point-cones (U). Of such solids, however, there is an infinity (∞^1) which also touch Ω; and the locus of the points of contact of these solids with Ω is an important curve. We denote this curve by q. Consider, in the principal solid which is the polar of the vertex U in regard to Ω, the quadric surface in which this polar solid meets Ω; this quadric surface we denote by $[\Omega]$; and, also, the quadric surface in which this solid meets the cone (U), denoting this quadric surface by $[U]$. The quadric surface $Q\,(U)$, in this solid, is the polar reciprocal of $[U]$ in regard to $[\Omega]$. A tangent solid of Ω which touches the cone (U), as it passes through the point U, must touch Ω at a point of the quadric surface $[\Omega]$, and must also touch the quadric surface [U]. This tangent solid thus meets the principal solid, which is the polar of the point U, in regard to Ω, in a common tangent plane of the quadric surfaces [Ω], [U]. The curve q is the locus of the points of contact, of these common tangent planes, with [Ω]; it is a quartic curve, being the intersection of [Ω] with $Q(U)$. When we project from O, on to the solid Π, the quadric surface [Ω] becomes a principal sphere, and the quadric surface $Q(U)$ becomes a principal quadric surface, Q'. Thus the curve q projects into the intersection of these. Such a curve, obtained by projection of q, of which there are five in general, may be called a *focal curve* of the Cyclide. The section of Ω, by a tangent solid of the cone (U) which touches Ω, is a conical sheet of lines, of which the generators meet the generators of the conical sheet of lines of Ω through O (p. 38, above); thus, the projection from O, on to the solid Π, of this section, is a quadric cone passing through the conic ω; the vertex of this cone is a point of the quartic focal curve, q', obtained by projection of q. Such a cone, when ω is regarded as the Absolute conic, is often called a *point-sphere*. The focal curve q' can then be said to be the locus of the vertices of point-spheres whose section with the Cyclide consists of two circles (which touch one another).

It can be shewn that if, upon a sphere, of a threefold space Π, that is, a quadric surface passing through a given conic, ω, of this space, there be given a quartic curve, q', the intersection of the sphere with another quadric surface; and, also in this solid Π, there be given a point, H'; then, three Cyclides can be constructed, containing the point H', all having the curve q' as focal curve. Further, that the tangent planes of any two of these Cyclides, at any point common to both, are at right angles; that is, that these tangent planes meet the plane of the conic ω in lines which are conjugate to one another in regard to this conic. It can then be shewn, further, that only these three Cyclides are possible under the given conditions. The corresponding result for curves in a plane is proved above, p. 95. For, in a fourfold space containing the solid Π, take a point, O, not itself in Π; consider the conical sheet of lines joining O to the given conic ω. A quadric, Ω, can be described containing this conical sheet, this being any quadric whose section, by the solid which is defined by O and the plane of ω, consists of this conical sheet. Next, consider the quadric point-cone which is the aggregate of the lines joining O to the points of the given sphere in Π. This will meet the quadric Ω in a quartic surface; part of this, however, is the quadric surface constituted by the conical sheet $O\omega$; the remaining part of the intersection is then another quadric surface. Thus, assuming that a quadric *surface*, in fourfold space, necessarily lies in a solid, we

can regard the given sphere, in Π, as the projection, from O, of
the section of Ω by a solid; denote this solid by Φ. The curve
q', on the given sphere in Π, is the projection from O of a curve,
say q, lying in the section, with Ω, of the solid Φ; this section is
a quadric surface, and the curve q, like q', is a quartic curve.
Let the pole, in regard to Ω, of the solid Φ, be the point U,
supposed not to lie on Ω; also, let the line OH' meet Ω again in
the point H. The polar plane, in regard to Ω, of the line UH,
will lie in the solid Φ, which is the polar solid of U; denote this
polar plane of UH by η. We know that, in the solid Φ, there can
be drawn, through the curve q, three quadric surfaces to touch the
plane η (Vol. III, p. 120); the polar reciprocals of these quadric
surfaces, in regard to Ω, are three quadric point-cones, with vertex
at U; these cones touch the tangent solids of Ω at the points of
the curve q (which lies on Ω), and contain the line UH. The
intersection of Ω, with any one of these three point-cones, projects
from O, on to Π, into a Cyclide having the given curve q' as focal
curve, and passing through H'. This proves the existence of three
such Cyclides.

We next prove that any two of these cut at right angles (in
regard to ω, as Absolute conic), at their common point H':—The
plane η, polar of the line UH in regard to Ω, is the intersection of
the solid Φ with the tangent solid of Ω at H. Thus the plane η
contains the conic, in which the solid Φ is intersected by the
conical sheet of lines of Ω which pass through H; this conic is
also on the quadric surface (Ω, Φ), upon which the curve q lies;
denote this conic by h. It is known that the three points of contact
with the plane η, of quadric surfaces, in the solid Φ, which contain
the curve q and touch η, form a self-polar triad in regard to the
conic section of η with any quadric surface, in Φ, containing q;
these three points of contact are thus a self-polar triad in regard
to the conic h, which is the intersection of the quadric surface
(Ω, Φ) with the tangent solid of Ω at H. The polar solids, in
regard to Ω, of these three points of contact, are the tangent
solids, at H, of the three point-cones, of vertex U, above drawn
through the line UH; the polar planes, in regard to Ω, of the
lines of the conical sheet, of lines of Ω through H, are the tangent
planes of this same conical sheet. Thus, reciprocating in regard
to Ω, the tangent planes at the point H, of the three surfaces in
which Ω is met by the three point-cones above drawn through UH,
are conjugate in pairs in regard to the conical sheet of lines of Ω
through H. On projecting from O, since every line of Ω through
H meets a line of Ω through O, we obtain three Cyclides through
the point H', of which every two have their tangent planes, at H',
at right angles in regard to the conic ω, regarded as Absolute

conic. By a similar argument every two of these surfaces cut at right angles at every one of their common points.

And, as in the corresponding case of a quartic curve in a plane, it was proved (p. 95, above), that there were only two such curves, so it can be shewn here that there are only three Cyclides subject to the given conditions.

Ex. 1. Let the Cyclide be the projection of the intersection of the quadric Ω given by $x^2 + y^2 + z^2 + t^2 + u^2 = 0$, with a quadric given by $ax^2 + by^2 + cz^2 + dt^2 + eu^2 = 0$. Then the coordinates, x_1, y_1, z_1, t_1, u_1, of a point of Ω whereat the tangent solid touches the point-cone (U), of vertex $(0, 0, 0, 0, 1)$, containing the intersection of these quadrics, are such that

$$u_1 = 0, \quad x_1^2 + y_1^2 + z_1^2 + t_1^2 = 0,$$
$$(a - e)^{-1} x_1^2 + (b - e)^{-1} y_1^2 + (c - e)^{-1} z_1^2 + (d - e)^{-1} t_1^2 = 0.$$

Thus the curve q, and the focal curve, q', obtained by projecting this, depend only on the differences of a, b, c, d, e. A system of Cyclides with common focal curves are then those given by the projection of the intersection of

$$x^2 + y^2 + z^2 + t^2 + u^2 = 0, \quad (a+\lambda)^{-1} x^2 + (b+\lambda)^{-1} y^2 + \ldots + (e+\lambda)^{-1} u^2 = 0,$$

where λ is arbitrary. The condition that two such Cyclides, given by the values, λ_1, λ_2, of λ, should cut at right angles at the common point, $(x_1, y_1, \ldots, u_1)$, is (p. 39, above) that the tangent solids $(a + \lambda_1)^{-1} xx_1 + \ldots = 0$, $(a + \lambda_2)^{-1} xx_1 + \ldots = 0$ should be conjugate in regard to Ω. This condition is

$$(a + \lambda_1)^{-1} (a + \lambda_2)^{-1} x_1^2 + \ldots = 0;$$

it is a consequence of the two equations

$$(a + \lambda_1)^{-1} x_1^2 + \ldots = 0, \quad (a + \lambda_2)^{-1} x_1^2 + \ldots = 0.$$

The condition that a Cyclide, of the confocal system obtained, should pass through an arbitrary point, is a cubic equation for λ.

Ex. 2. Through any point, H, of the quartic surface, Γ, defined as the intersection of the quadric Ω, and another quadric, Ω', a line may be drawn which is the polar, in regard to Ω, of the tangent plane of Γ at H. This line is the locus of the poles, in regard to Ω, of the tangent solids, at H, of all the quadrics (say $\theta\Omega + \Omega' = 0$) which contain Γ. The line lies in the tangent solid of Ω at H; and is the polar line, in regard to the conical sheet of lines of Ω through H, of the tangent plane of Γ at H, this plane, and the conical sheet, being both in the tangent solid of Ω at H. By projection of Γ, from a point O, of Ω, into a Cyclide of the solid Π, this line becomes the normal of the Cyclide at the point arising from H; namely, the line which is conjugate to the tangent plane of the Cyclide with respect to the Absolute conic ω. If H be $(x_0, y_0, z_0, t_0, u_0)$, and Ω be $x^2 + \ldots + u^2 = 0$, while Ω' is

$$ax^2 + \ldots + eu^2 = 0,$$

the general point of the line in question is $(\xi, \eta, \zeta, \tau, \nu)$, where $\xi = (a + \theta) x_0, \ldots, \nu = (e + \theta) u_0$, the θ being variable. The line, as we see by taking polars in regard to Ω, touches all the quadrics which are the polar reciprocals, in regard to Ω, of the quadrics $\lambda\Omega + \Omega' = 0$; that is, it touches all the quadrics

$$(a + \lambda)^{-1} x^2 + \ldots + (e + \lambda)^{-1} u^2 = 0;$$

except two of these, upon which it lies entirely. These two are those for which λ satisfies the equation

$$(a + \lambda)^{-1} x_0^2 + \ldots + (e + \lambda)^{-1} u_0^2 = 0,$$

which is a quadratic in virtue of

$$x_0^2 + \ldots + u_0^2 = 0, \quad ax_0^2 + \ldots + eu_0^2 = 0.$$

The tangent solid of Ω at H touches every one of these polar reciprocal quadrics, at points of this line. In particular, however, when λ has one of the five values $-a, \ldots, -e$, the line meets the corresponding locus, now become a quadric surface, only once; in the point in which the line meets the principal solid in which this quadric surface lies.

The aggregate of all the points of all such lines, when $(x_0, y_0, \ldots, u_0)$ becomes, in turn, all points of the surface Γ, is a ruled locus of ∞^3 points. By substitution of $x_0 = (a + \lambda)^{-1}\xi$, etc., in the equations $x_0^2 + \ldots + u_0^2 = 0, ax_0^2 + \ldots + eu_0^2 = 0$, or by the geometrical property of the lines, we see that the equation of this ruled locus can be formed as the discriminant, in regard to λ, of the equation

$$(a + \lambda)^{-1}\xi^2 + \ldots + (e + \lambda)^{-1}\nu^2 = 0 ;$$

so that the locus is of the twelfth order. Thus, on projection, we infer that twelve normals of a Cyclide pass through an arbitrary point. The line of the ∞^3 ruled locus, through a point, $(x_0, \ldots, u_0)$, on the curve in which the surface Γ meets the quadric

$$(a + \lambda)^{-1}\xi^2 + \ldots + (e + \lambda)^{-1}\nu^2 = 0,$$

lies entirely on this quadric, as already remarked. In particular, for $\lambda = \infty$, the lines of this locus which meet Γ in the curve given by $x_0^2 + \ldots + u_0^2 = 0, ax_0^2 + \ldots + eu_0^2 = 0, a^2x_0^2 + \ldots + e^2u_0^2 = 0$, lie entirely on the quadric Ω; such lines form a surface, lying on Ω, of order $2.12 - 2.4$, or sixteen. The lines of this surface are obtainable, also, by remarking, first, that, in the tangent plane of Γ at any point P, there are two lines lying on Ω, the intersection of this tangent plane with the conical sheet of lines of Ω through P;

and noticing, then, that, when P is on the curve whose equations have been given, the tangent plane of Γ touches this conical sheet along a line of points, with coordinates $(a + \lambda) x_0, \dots, (e + \lambda) u_0$. Thus, the tangent plane of Γ, at any point of this curve, is also a tangent plane, at another point, of the surface in which Ω is met by any one of the quadrics $(a + \lambda)^{-1} x^2 + \dots + (e + \lambda)^{-1} u^2 = 0$. For, this tangent plane of Γ lies in the tangent solid of Ω at every point of the line whose points have coordinates $(a + \lambda) x_0, \dots, (e + \lambda) u_0$; and lies in the tangent solid of the quadric

$$(a + \lambda)^{-1} x^2 + \dots + (e + \lambda)^{-1} u^2 = 0,$$

because this, being the tangent solid of Ω at the point $(x_0, \dots, u_0)$ of the curve, touches the line at this point. The curve in question,

$$(x_0^2 + \dots + u_0^2 = 0, \ a x_0^2 + \dots + e u_0^2 = 0, \ a^2 x_0^2 + \dots + e^2 u_0^2 = 0),$$

lies also on the polar reciprocal of the quadric $x^2 + \dots + u^2 = 0$, in regard to the quadric $a x^2 + \dots + e u^2 = 0$. On projection, the focal curves of the Cyclide appear as lying doubly on the developable surface formed by the tangent planes of the Cyclide at points of the conic ω, these planes touching the confocal Cyclides.

We may, however, go further. Of the ∞^1 lines, forming the surface of order sixteen, spoken of, lying on Ω, there are sixteen lines which lie on Γ. For the line of points of coordinates

$$(a + \lambda) x_0, \dots, (e + \lambda) u_0,$$

will lie on $a x^2 + \dots + e u^2 = 0$, if, beside the three equations for $(x_0, \dots, u_0)$, we also have $a^3 x_0^2 + \dots + e^3 u_0^2 = 0$. The four equations have sixteen solutions; if we denote the quintic polynomial, in t, whose roots are a, b, c, d, e by $f(t)$, the four equations are satisfied by $x_0^2 f'(a) = y_0^2 f'(b) = \dots = u_0^2 f'(e)$. The curve, of order eight, for which $x_0^2 + \dots = 0$, $a x^2 + \dots = 0$, $a^2 x_0^2 + \dots = 0$, is expressed, in terms of a parameter θ, by

$$x = (\theta + a)^{\frac{1}{2}} [f'(a)]^{-\frac{1}{2}}, \dots, \ u = (\theta + e)^{\frac{1}{2}} [f'(e)]^{-\frac{1}{2}};$$

the points of any one of the sixteen lines of the surface Γ are given by $x = (\lambda + a) [f'(a)]^{-\frac{1}{2}}, \dots, \ u = (\lambda + e) [f'(e)]^{-\frac{1}{2}}$, for varying values of λ. Here $f'(x)$ denotes df/dx.

More generally, in space of n dimensions, in which $x_1, \dots, x_{n+1}$ are the coordinates, we may consider the surface given by the $(n-2)$ equations, $a_1^r x_1^2 + \dots + a_{n+1}^r x_{n+1}^2 = 0$, for $r = 0, 1, \dots, (n-3)$. Upon this there is a curve, of order 2^{n-1} and genus $2^{n-2}(n-3) + 1$, given by these equations together with

$$a_1^{n-2} x_1^2 + \dots + a_{n+1}^{n-2} x_{n+1}^2 = 0,$$

of which the points have coordinates of the forms

$$(\theta + a_1)^{\frac{1}{2}} [f'(a_1)]^{-\frac{1}{2}}, \dots,$$

where $f(t)$ is the polynomial having $a_1, ..., a_{n+1}$ for roots. This leads to the consideration of 2^n lines of the surface, each consisting of points of coordinates such as $x_s = (\lambda + a_s)[f'(a_s)]^{-\frac{1}{2}}$. For the case $n = 3$, cf. Vol. III, p. 92. The case $n = 5$ arises in Chap. VII below.

Ex. 3. It is not a difficult matter to obtain *the lines of curvature upon a Cyclide*. We do this by a method which is, in part, applicable to any surface. The question is, however, of special interest here because, as will be seen in Chap. VII (below, p. 231), these lines are in correspondence with the inflexional lines of a Kummer surface.

At any point of a surface, in space of three dimensions, two lines can be drawn, in the tangent plane at this point, which meet the surface in three coincident points there. These are the *inflexional* directions. If the point be $(x_0, ...)$, and the surface be $f(x, ...) = 0$, the lines are found by combining the tangent plane with the polar quadric of the point, given by $D^2 f_0 = 0$, where D is the operator $x\partial/\partial x_0 + ...$, and f_0 the value of f at $(x_0, ...)$. If any sphere be drawn to touch the surface at the point, its inflexional directions give two other lines lying in the tangent plane of the surface. The directions of the lines of curvature of the surface, at the point, are along the two lines, in the tangent plane, which are the double rays of the involution determined by the two pairs of inflexional directions which we have spoken of. In fact, the curve of intersection of the sphere with the surface has a double point at the point of contact, whose tangent lines lie in the tangent plane, and form a pair of the involution in question; there are two such spheres for which the curve of intersection has a cusp at the point; the two cuspidal tangents are the directions of the lines of curvature.

We prove these statements by regarding the surface under consideration, in the threefold space, Π, as obtained by projection, upon Π, of the intersection of a quadric (∞^3), Ω, lying in a space of four dimensions, which contains Π, with another (∞^3) locus, Ω', lying in this fourfold space. We shall speak of Ω' also as a quadric, so that the direct application is to the Cyclide. The centre of projection, O, is on Ω; the point of the surface (Ω, Ω'), which we consider, is denoted by P, and its coordinates by (x_0, y_0, z_0, t_0, u_0); the tangent solids of Ω and Ω' at P are denoted by T and T'. We take solids, $\theta T + T'$, passing through the tangent plane, (T, T'), of the surface (Ω, Ω'), at the point P. These solids intersect Ω in surfaces which project, from O, into *spheres* in the space Π, the Absolute conic being the intersection of Π with the conical sheet of lines of Ω through O. There are two lines through P, the intersection of the plane (T, T') with the cone (T, Ω), in which the solid $\theta T + T'$ meets the conical sheet of lines of Ω through P; these lines meet the conical sheet of lines of Ω through O, and project into the generators, or inflexional lines, of the sphere into which the surface $(\theta T + T', \Omega)$ projects, at the point which is the projection of P. These two lines lie in the plane (T, T'), and are independent of θ. The solid $\theta T + T'$ meets the surface (Ω, Ω') in a curve; Ω' being a quadric, this curve, lying on the quadric $\theta\Omega + \Omega'$, lies on the conical sheet $(\theta T + T', \theta\Omega + \Omega')$, the solid $\theta T + T'$ being the tangent solid of $\theta\Omega + \Omega'$; the curve may be given as the intersection of this conical sheet, $(\theta T + T', \theta\Omega + \Omega')$, with the quadric Ω. Thus, the curve has a double point at P, whose tangent lines are the two in which T meets the conical sheet; that is, these tangent lines are the two lines in which the quadric $\theta\Omega + \Omega'$ is met by the plane (T, T'). For different values of θ these form an involution of pairs of lines in this plane, intersecting at P. They project, from O, into the tangent lines of the curve in which the surface, obtained by the projection of (Ω, Ω'), is met by a sphere touching this surface at the projection of P. For the particular value of θ for which the solid $\theta T + T'$ contains the point O, this sphere is replaced by the tangent plane of the surface obtained by projection of (Ω, Ω'), and the tangent lines become the inflexional directions of this surface. For the value $\theta = \infty$, we have the two lines, first considered, in which the plane (T, T') meets Ω, which project into the common generators of all the spheres. For five particular values of θ, the quadric $\theta\Omega + \Omega'$ becomes a cone, and we have, on projection, the tangent lines of a pair of conics of the Cyclide, through the point into which P projects.

There are, also, two values of θ for which the tangent plane (T, T') meets $\theta\Omega + \Omega'$ in two coincident lines, namely, the plane $(T, \theta T + T')$ touches the conical sheet $(\theta T + T', \theta\Omega + \Omega')$. For, we know that the tangent planes of this cone are the intersections, of $\theta T + T'$, with the tangent solids of $\theta\Omega + \Omega'$ at points of this conical sheet (p. 38, above). Thus, for the plane $(T, \theta T + T')$ to touch the conical sheet, it must be identical with a plane $(K, \theta T + T')$, where K is the tangent solid of $\theta\Omega + \Omega'$ at a point, $(x_1, y_1, \ldots)$, of this conical sheet. Now, if Ω, Ω' be, respectively, $x^2 + \ldots + u^2 = 0$, $ax^2 + \ldots + eu^2 = 0$, the condition, that the planes

$$[(a + \theta)\, xx_1 + \ldots = 0, \quad (a + \theta)\, xx_0 + \ldots = 0],$$
$$[(a + \theta)\, xx_1 + \ldots = 0, \quad xx_0 + \ldots = 0]$$

should be identical, is the existence of equations of the form $(a + \theta)\, x_1 = \rho\, (a + \psi)\, x_0$, for each coordinate, for proper values of ψ and ρ; or, if $x_2 = x_1 - \rho x_0$, etc., that $(a + \theta)\, x_2 = \rho\, (\psi - \theta)\, x_0$, etc., or, say $(a + \theta)\, x_2 = \sigma x_0$. The condition that $(x_1, \ldots)$, and hence also

$(x_2, \ldots)$, should lie on the conical sheet, requiring

$$(a + \theta) x_0 x_2 + \ldots = 0, \quad (a + \theta) x_2{}^2 + \ldots = 0,$$

is satisfied if $x_0{}^2 (a + \theta)^{-1} + \ldots = 0$, in virtue of $x_0{}^2 + \ldots = 0$. As we also have $a x_0{}^2 + \ldots = 0$, this equation for θ is only a quadratic; denote its roots by λ and μ. Then the curve, in which the surface (Ω, Ω') is met by either of the two solids $\lambda T + T'$, $\mu T + T'$, has a cusp at P. The two cuspidal tangents give, on projection, the cuspidal tangents of two curves in which the Cyclide is met by two spheres. These we define as the directions of the lines of curvature.

The surface (Ω, Ω'), if $f(t)$ be the quintic polynomial whose roots are $a, b, \ldots, e$, has the coordinates of its points $(x, y, \ldots)$ represented by equations $x^2 = (a + \theta)(a + \phi)/f'(a), \ldots, u^2 = (e + \theta)(e + \phi)/f'(e)$; the values of θ, ϕ appropriate to the point $(x, y, \ldots)$ are the roots of the quadratic equation $x^2 (a + \theta)^{-1} + \ldots + u^2 (e + \theta)^{-1} = 0$. If we pass, along the curve of the surface for which θ is constant, from $(x, y, \ldots)$ to $(x + dx, y + dy, \ldots)$, we have $dx = \frac{1}{2} x \, d\phi (a + \phi)^{-1}$, etc.; namely, the tangent line of this curve is the line from $(x, y, \ldots)$ to the point $(x (a + \phi)^{-1}, y (b + \phi)^{-1}, \ldots)$. This tangent line is then, by what we have proved above, the cuspidal tangent of the intersection of (Ω, Ω') with the solid $\phi T + T'$, where T, T' are the tangent solids of Ω and Ω' at the point considered. Thus, the curve along which θ is constant has its tangent line along a direction of curvature, at every point, but the *osculating* solid (and consequent *sphere*) has a parameter ϕ which varies from point to point of this curve.

This result, after what is proved above, is in accord with a theorem, associated with the name of Dupin, that if, in space of three dimensions, three surfaces be such that every two of them cut at right angles, in regard to the Absolute conic, at all their common points, then, at a point common to all the surfaces, the tangent line of the common curve of any two of these surfaces, is the tangent line of a line of curvature on each. In particular, a Cyclide is intersected in a line of curvature by any one of the five principal spheres.

Ex. 4. It was remarked that the line joining a point, $(x, y, \ldots)$, of the surface Γ, or (Ω, Ω'), to a point of coordinates $(a + \psi) x$, $(b + \psi) y, \ldots$, becomes, on projection, a normal of the Cyclide. This line is, evidently, on projection, at right angles to the line, considered in Ex. 3, joining $(x, y, \ldots)$ to the point of coordinates $(a + \psi)^{-1} x$, $(b + \psi)^{-1} y, \ldots$. Consider two points $(x, y, \ldots)$, $(x + dx, y + dy, \ldots)$, of Γ, on the curve given, for constant θ, by the equation $(a + \theta)^{-1} x^2 + \ldots + (e + \theta)^{-1} u^2 = 0$, these points corresponding to values ϕ, $\phi + d\phi$, of the second parameter, ϕ, determining points of Γ, as in Ex. 3. Corresponding to the first point, take the point of coordinates $(a + \phi) x$, $(b + \phi) y, \ldots$; this is, in fact, the pole, in regard to Ω, of the tangent solid of $\phi\Omega + \Omega'$ at the point $(x, y, \ldots)$, and projects into the centre of the sphere into which the surface $(\Omega, \phi T + T')$ projects. Corresponding to the second point, take the point of coordinates $(a + \psi)(x + dx)$, $(b + \psi)(y + dy), \ldots$, where $\psi = \phi - \tfrac{1}{2}d\phi$. Since $dx = \tfrac{1}{2}x d\phi(a + \phi)^{-1}$, we have

$$(a + \phi - \tfrac{1}{2}d\phi)(x + dx) = (a + \phi - \tfrac{1}{2}d\phi)[1 + \tfrac{1}{2}(a + \phi)^{-1}d\phi]x$$
$$= (a + \phi)x - \tfrac{1}{2}d\phi dx.$$

Thus, on projection, we have the result often expressed by saying that the normals, at two consecutive points of a line of curvature, intersect one another. These normals, that is, form a developable surface, whose *cuspidal edge* (Vol. III, pp. 131, 132), or *edge of regression*, is the locus of the centres of the osculating spheres.

Ex. 5. By considering when the two cuspidal directions obtained in Ex. 3 can coincide, prove that the lines of the surface Γ lie on the quartic locus, obtained as the discriminant, in regard to θ, of the quadratic equation $(a + \theta)^{-1}x^2 + \ldots = 0$, whose equation is

$$(p_1\Omega_2 + \Omega_3)^2 = 4\Omega_2(p_2\Omega_2 + \Omega_4),$$

where

$$\Omega_r = a^r x^2 + \ldots + e^r u^2, \quad p_1 = a + \ldots + e, \quad p_2 = ab + ac + \ldots + de.$$

Ex. 6. By inversion of a surface in threefold space, a line of curvature is changed into a line of curvature of the inverse surface.

Ex. 7. Consider, in threefold space, whose Absolute conic is ω, a curve, γ, lying upon a surface of this space. The tangent lines of this curve γ meet the plane of the Absolute conic in points, T, describing a curve, τ, of this plane. The normals of the surface, at points of the curve γ, that is, the lines conjugate to the tangent planes in regard to the Absolute conic, meet the Absolute plane in points, N, describing a curve, ν, of this plane. If, for every point of the curve γ, the tangent line of the curve ν, at the point N, passes through the corresponding point, T, of the curve τ, then the curve γ is a line of curvature of the surface. The condition may be expressed briefly by saying that the curve defined by the normals is a curve of pursuit of the curve defined by the tangents.

Ex. 8. Two surfaces, in threefold space, are such as to cut at right angles at every point of their curve of intersection, which is known to be a line of curvature on one of the surfaces. Considering, in the Absolute plane, the curve enveloped by the tangent planes of one of the surfaces, at the points of their common curve, and the like curve enveloped by the tangent planes of the other surface, at the points of this curve, shew that these curves are polar reciprocals of one another in regard to the Absolute conic.

Ex. 9. If a line of curvature of a surface, in threefold space, be

a plane curve, prove that the tangent plane of the surface at points
of the curve is at a constant inclination to the plane of the curve
(with respect to the Absolute conic). Deduce this by proving that,
if the normals of a plane curve, defined as lines conjugate to the
tangent lines with respect to an Absolute *conic* of the plane, all
meet in a point, then the curve is the locus of a point which is at
a constant *interval* from the point of intersection of the normals,
the interval being defined by the Absolute conic (as in Vol. ii,
pp. 168 ff.).

Ex. 10. If the tangent plane of a surface, in threefold space, at
a point (x, y, z, t), be expressed by $lX + mY + nZ + pT = 0$, an
inflexional line of the surface satisfies the differential equation

$$dx\,dl + dy\,dm + dz\,dn + dt\,dp = 0.$$

If l, m, n, l', m', n' be the line coordinates of the normal of the surface, a line of curvature satisfies the differential equation

$$dl\,dl' + dm\,dm' + dn\,dn' = 0.$$

More generally, while the lines of a ruled surface, in space of three
dimensions, are represented, in space of five dimensions, by the
points of a curve lying on the fundamental quadric Ω (p. 40, above),
the lines of a developable surface, or the tangents of a curve, in the
threefold space, are represented by a curve on Ω of which all the
tangent lines also lie on Ω.

The matters discussed in these Examples arise again below, in
Chap. vii (p. 232). For a further generalisation see Darboux,
Théorie...des Surfaces, Livre vii, p. 485, § 840.

In regard to some particular cases. The Dupin Cyclide.
In considering the quartic surface, in the space of four dimensions,
as the intersection of two quadrics, we have generally assumed that
the two quadrics are general, so that there are five quadric point-
cones passing through the surface, whose vertices form a self-polar
pentad for the quadrics. In this case the equations of the two
quadrics can be supposed to be of the forms, $x^2 + ... + u^2 = 0$, and
$ax^2 + ... + eu^2 = 0$, in which no two of the coefficients $a, b, ..., e$ are
equal. And we have given the centre of projection, by which we
pass to the Cyclide, a quite general position. In this Volume we
do not enter into the discussion of the possibilities which can arise
in less general cases; Segre (*Math. Annal.* xxiv, p. 440) enumerates
more than seventy. We make, however, some remarks, with the
aim of indicating the nature of the special cases; and, in particular,
on account of its historical and intrinsic interest, we shew how
Dupin's Cyclide can be obtained.

(*a*) It may happen that, through a surface, in fourfold space,
which is defined by the intersection of two quadrics, which are cones,

there passes no quadric which is not a cone. In other words, if $U = 0$, $V = 0$ be the given quadric cones, it may happen that the discriminantal determinant, of five rows and columns, of the quadric $U + \lambda V = 0$, vanishes for all values of λ. To illustrate this possibility, suppose $U = y^2 + z^2 + u^2 - tu$, $V = yz - xu$, each of which, as depending on four linear functions of the coordinates only, belongs to a cone. We then have

$$U + \lambda V = y^2 + \lambda yz + z^2 + u(u - t - \lambda x);$$

for every value of λ, this likewise depends on four linear functions of the coordinates only, and also belongs to a cone.

The projection of the surface (U, V), in this case, from the point $(0, 0, 0, 0, 1)$, gives, in the space (x, y, z, t), the surface whose equation is $y^2z^2 + z^2x^2 + x^2y^2 - xyzt = 0$. This is *Steiner's quartic surface* (Vol. III, p. 222); its properties can then be deduced by considering the surface (U, V). The cones $U = 0$, $V = 0$ have a common tangent solid at every point of the line joining their vertices, and the projection is made, on to this solid, from a point on one of the cones.

The Steiner surface may also be obtained by projecting the Veronese surface, in space of five dimensions (p. 52, above), by means of planes passing through a line which does not meet this surface.

(*b*) When, beside cones, there is at least one general quadric containing the surface, say $U = 0$, the discriminant of $U + \lambda V$, not vanishing identically in regard to λ, vanishes for, at most, five values of λ. There are then other quadrics which are not cones, beside $U = 0$, containing the surface. Let $V = 0$ be such a quadric. Then, two cases are possible. Either, there is *not* a point, common to the quadrics $U = 0$, $V = 0$, at which these have the same, definite, tangent solid; or there is such a point, or several such. We sketch now, incompletely, a proof that, in the former of these cases, the equations of $U = 0$, $V = 0$ can be supposed to be of the respective forms $x^2 + y^2 + \ldots + u^2 = 0$, $ax^2 + by^2 + \ldots + eu^2 = 0$, in which no two of the coefficients $a, b, \ldots, e$ are equal. This will then be the general case considered above. In fact, under the hypothesis made, there will be at least one point, not lying on $U = 0$, nor on $V = 0$, for which the five ratios, respectively of $\partial U/\partial x$, $\partial U/\partial y$, ..., $\partial U/\partial u$ to $\partial V/\partial x$, $\partial V/\partial y$, ..., $\partial V/\partial u$, are equal. If we change the notation so that this point becomes $(0, 0, 0, 0, 1)$, and its polar solid, in regard to either $U = 0$, $V = 0$, which is the same for both, becomes $u = 0$, then the equations of the two quadrics take the forms $u^2 + \phi = 0$, $eu^2 + \psi = 0$, where ϕ, ψ are quadratic forms in x, y, z, t only. The quadric surfaces, $\phi = 0$, $\psi = 0$, regarded as lying in the solid $u = 0$, will not have a common point at which their tangent planes coincide; for then, the solid joining this plane to the point $(0, 0, 0, 0, 1)$

would be a common tangent solid of the two original quadrics, at a common point. And the discriminant of $\lambda\phi + \psi$ is not identically zero. We can therefore apply a similar argument to $\phi = 0$, $\psi = 0$, reducing them to the respective forms $t^2 + \xi = 0$, $dt^2 + \eta = 0$, where ξ, η are quadratic forms in x, y, z, only. And so on. Thereby U, V are reduced, respectively, to forms $x^2 + \ldots + u^2$, $ax^2 + \ldots + eu^2$. Herein, however, we cannot have, for example, $d = e$; since, then, the quadrics $U = 0$, $V = 0$ would have the common tangent solid, $t + iu = 0$, at the common point $(0, 0, 0, 1, i)$.

(c) Suppose now that $U = 0$, $V = 0$ have a common point, which we may take to be $(0, 0, 0, 0, 1)$, and, thereat, a definite common tangent solid, which we may take to be $t = 0$. Their equations are then capable of the forms $ut + \phi = 0$, $ut + \psi = 0$, where ϕ, ψ are quadratic forms in x, y, z, t, only. The surface of intersection of the quadrics thus lies on $\phi - \psi = 0$. This is, then, either a point-cone whose vertex is $(0, 0, 0, 0, 1)$, or a line-cone whose axis passes through this point, or a pair of solids, or a single repeated solid. In all these cases, the projection of the quartic surface (U, V), from this point, upon any solid, is a *quadric* surface, of a general or particular kind. Thus the Cyclide arising by projection of the surface (U, V) from a general point, say, of $U = 0$, is one which is the inverse, in the solid upon which we project, of a quadric surface, of general or of particular kind. It is clear that there are many cases.

(d) In what has preceded, we have spoken of projecting the quartic surface of fourfold space, into a Cyclide, from a centre of projection lying on a particular quadric which contains the surface; and have used this quadric to define inversion in the space of the Cyclide. In general, as a single quadric of the form $U + \lambda V = 0$ passes through an arbitrary point of the fourfold space, this is equivalent to defining the Cyclide as arising by projection from an arbitrary point of the space. If we do this, however, special cases may arise. For instance, it may happen that the quadric, of the family $U + \lambda V = 0$, which passes through the centre of projection, is a cone. Then it is easy to see that the surface of the fourth order, obtained by projection, in place of having a double conic of general kind, has two double lines, which may coincide. Or, again, it may happen, when the surface in the fourfold space lies on a line-cone, that the tangent solid, at the centre of projection, of the quadric, of the family, which passes through this point, is a tangent solid of this line-cone. In this case, while there may exist a double conic for the quartic surface obtained by projection, it will be such that the tangent planes of this surface coincide at every point of this conic. The double conic then becomes a *cuspidal* conic.

(*e*) For historical and intrinsic reasons, some account may be given of the case in which the quartic surface, in fourfold space, lies on one, and, more particularly, on two quadric line-cones. The former arises when the surface is given by two equations of the forms $x^2 + y^2 + \ldots + u^2 = 0$, $ax^2 + by^2 + eu^2 = 0$; the latter arises when, further, $a = b$, the second line-cone containing the surface then being $a(z^2 + t^2) + (a - e) u^2 = 0$. The latter surface projects into what has been called *Dupin's Cyclide* (Dupin, *Applications de géométrie et de méchanique*, 1822, p. 200). This surface is in point correspondence with the *Tore* (or *Anchor Ring*); and either can be inverted into the other, or into a quadric cone of revolution. Beautiful stereoscopic diagrams of forms of Dupin's Cyclide are given in Maxwell's *Scientific Papers*, Vol. II, p. 158.

Consider first the case when the quartic surface in fourfold space lies upon only one quadric line-cone. This cone, then, arises in place of two of the five point-cones existing in the general case. The axis of the line-cone meets a general quadric of the family, say the quadric Ω, in two points, say M and N. On projection from a point, O, of Ω, these become two double points of the resulting Cyclide. A tangent solid of the line-cone, which touches this at all points of a plane, meets Ω in a quadric surface; on projection, we have a sphere touching the Cyclide in the points of a circle, passing through the two double points. The centres of such spheres lie on a conic; this is the projection of the conic which is the polar reciprocal of the line-cone, in regard to Ω; the plane of this conic is the projection of the polar plane, in regard to Ω, of the axis of the line-cone. The two conical sheets, of lines lying upon Ω, passing through M and N, respectively, intersect in a conic, lying in the polar plane of the axis of the line-cone; this plane is in the polar solid, in regard to Ω, of every point of this axis. Thus, on projection, there is a circle, lying in the plane which bisects, at right angles, the line joining the two double points of the Cyclide. Every sphere passing through this circle has its centre on this joining line, and the Cyclide inverts into itself in regard to this sphere. In particular, there is one such sphere which reduces, beside the Absolute plane, to a plane; so that there is a plane in regard to which the Cyclide is symmetrical.

It is possible to find points, H, not lying on Ω, such that the points, M', N', in which the lines HM, HN meet Ω again, lie on the tangent solid of Ω at the point O. For, let HO meet Ω again in O'; if M' be a point on the tangent solid at O, the line OM' lies on Ω; the plane, HMO, meeting Ω in the line OM', meets Ω in another line; thus, MO' is on Ω, or O' lies on the tangent solid at M; and conversely. Similarly, if N' be on the tangent solid at O, the point O' lies on the tangent solid at N. Take, then, for O', any point of the conic, on Ω, which is common to the tangent solids at M and N; for H, take any point of the line OO'. This point, H, can be used for the centre of the harmonic inversion, in the fourfold space, which gives rise to inversion, in regard to a sphere, in the threefold space upon which we project. Thereby it is shewn that the Cyclide, obtained by projection of the quartic surface which is the intersection of Ω with the quadric line-cone having MN for axis, can be inverted, in the threefold space, into a Cyclide having two double points on the Absolute conic, the centre of inversion being any point of a certain circle. It is a simpler remark that a point H, not on Ω, can be taken on the line OM; so that the Cyclide can also be inverted into a quadric cone.

Ex. 1. Consider the quartic surface given by the equations $x^2 + y^2 + \ldots + u^2 = 0$, $ax^2 + by^2 + \ldots + eu^2 = 0$; denote the former quadric by Ω; let O, of coordinates $(x_0, y_0, \ldots u_0)$, be any point of Ω. Also, let A, B, C, D, respectively, denote $a - e, b - e, c - e, d - e$, and A', B', C', E', respectively, denote $a - d, b - d, c - d, e - d$, so that $E' + D = 0$. The quartic surface thus lies on the two point-cones

$$Ax^2 + By^2 + \ldots + Dt^2 = 0, \quad A'x^2 + B'y^2 + \ldots + E'u^2 = 0.$$

We have, identically,

$$(Ax^2 + By^2 + Cz^2 + Dt^2)(Ax_0^2 + By_0^2 + Cz_0^2 + Dt_0^2)$$
$$- (Axx_0 + Byy_0 + Czz_0 + Dtt_0)^2$$

equal to

$$[l'\,(BC)^{\frac{1}{2}} + l\,(AD)^{\frac{1}{2}}]^2 + [m'\,(CA)^{\frac{1}{2}} + m\,(BD)^{\frac{1}{2}}]^2 + [n'\,(AB)^{\frac{1}{2}} + n\,(CD)^{\frac{1}{2}}]^2,$$

where $l = tx_0 - t_0x$, $l' = yz_0 - y_0z$, etc. Let this last be denoted by U. We thus infer that, when $(x, y, \ldots t)$ lies on the quartic surface, we have

$$Axx_0 + Byy_0 + Czz_0 + Dtt_0 = (- U)^{\frac{1}{2}}.$$

If V denote what U becomes when, for A, B, C, D, are put, respectively, A', B', C', E', and, for l, m, n, are put, respectively, $ux_0 - u_0x$, $uy_0 - u_0y$, $uz_0 - u_0z$, we similarly have, on the quartic surface,

$$A'xx_0 + B'yy_0 + C'zz_0 + E'uu_0 = (- V)^{\frac{1}{2}}.$$

Thus, if T denote $xx_0 + yy_0 + zz_0 + tt_0 + uu_0$, we have, on the quartic surface,

$$(d - e)\,T = (- U)^{\frac{1}{2}} - (- V)^{\frac{1}{2}}.$$

In this equation, $T = 0$, $l' = 0$, $l = 0$, etc., $ux_0 - u_0x = 0$, etc., all represent solids passing through the point O. Thus this equation is, effectively, an irrational form of the equation of the general Cyclide which is obtained by projecting the quartic surface from the point O.

Ex. 2. For the case when the quartic surface is given by a general quadric, Ω, and a *line-cone*, we may proceed as in Ex. 1. We may also proceed as follows: Let $t = 0$ be the tangent solid of the quadric Ω at the point from which the projection, to the space of three dimensions, is to be made; this point we now denote by U. Let $u = 0$ be the polar solid of U in regard to the line-cone; this solid contains the axis of the cone. Let the point in which $t = 0$ meets the axis of the line-cone be Z, and let $z = 0$ be the polar solid of Z in regard to Ω; thus $z = 0$ is undefined when $t = 0$ contains the axis of the line-cone. The solid $z = 0$ contains the point U; and contains the polar plane, in regard to Ω, of the axis of the line-cone. We define two other solids, $x = 0$, $y = 0$, as follows: Consider the plane $u = 0$, $z = 0$; this meets the quadric Ω in a conic; in this plane take the pole of the line $u = z = t = 0$, in regard to this conic, say the point T; pairs of lines drawn from T, conjugate in regard to this conic, meet the line $u = z = t = 0$ in pairs of points in involution. The line-cone meets the plane $u = 0$, $z = 0$ in two lines intersecting in the point in which the axis of the cone, which lies in $u = 0$, meets this plane; let this point be $(x_0, y_0, 0, 1, 0)$. We can then choose $x = 0$, $y = 0$, vanishing at U, so that the solids $x - x_0 t = 0$, $y - y_0 t = 0$ contain the axis of the cone, and meet the plane $z = 0$, $u = 0$ in two lines harmonic in regard to the lines in which the line-cone meets this plane; while, at the same time, the three lines, of the plane $z = 0$, $u = 0$, given by $x = 0$, $y = 0$, $t = 0$, are a self-polar triad in regard to the conic in which this plane meets Ω.

With this choice of coordinates we may suppose the equations of Ω, and of the line-cone, to be, respectively,

$$x^2 + y^2 + z^2 - r^2 t^2 - 2tu = 0, \quad a^2 (x - x_0 t)^2 + b^2 (y - y_0 t)^2 - u^2 = 0,$$

where a, b, r are certain constants. By elimination of u we have the equation of the Cyclide,

$$(x^2 + y^2 + z^2 - r^2 t^2)^2 = 4t^2 [a^2 (x - x_0 t)^2 + b^2 (y - y_0 t)^2].$$

This represents a surface, symmetrical in regard to $z = 0$, which is the envelope of the sphere

$$(x - ta \cos \theta)^2 + (y - tb \sin \theta)^2 + z^2 = t^2 [(x_0 - a \cos \theta)^2 + (y_0 - b \sin \theta)^2 + r^2 - x_0^2 - y_0^2];$$

the centre of this sphere is $(a \cos \theta, b \sin \theta, 0, 1)$, lying on a certain conic; and the sphere passes through the points

$$[x_0, y_0, \pm (r^2 - x_0^2 - y_0^2)^{\frac{1}{2}}, 1].$$

The surface inverts into itself, in regard to the sphere,

$$x^2 + y^2 + z^2 - 2xx_0t - 2yy_0t - 2zht + r^2t^2 = 0,$$

whatever h may be.

If we take λ_1, λ_2, λ_3 as the roots of $f(\lambda) = 0$, where

$$f(\lambda) = (a^2 - \lambda)^{-1} a^2 x_0^2 + (b^2 - \lambda)^{-1} b^2 y_0^2 - r^2 - \lambda,$$

and put, to define ξ_1, ξ_2, ξ_3, ξ_4,

$$x - x_0t = x_0 \sum_{k=1}^{3} (a^2 - \lambda_k)^{-1} \lambda_k \xi_k,$$

$$y - y_0t = y_0 \sum_{k=1}^{3} (b^2 - \lambda_k)^{-1} \lambda_k \xi_k, \quad u = \sum_{k=1}^{3} \lambda_k \xi_k,$$

together with $t = \xi_1 + \xi_2 + \xi_3 + \xi_4$, we find

$$a^2 (x - x_0t)^2 + b^2 (y - y_0t)^2 - u^2 = \sum_{k=1}^{3} \lambda_k^2 f'(\lambda_k) \xi_k^2,$$

$$z^2 + x^2 + y^2 - r^2t^2 - 2tu = z^2 + \sum_{k=1}^{3} \lambda_k f'(\lambda_k) \xi_k^2 + (x_0^2 + y_0^2 - r^2) \xi_4^2,$$

where $f'(\lambda)$ denotes the derivative of $f(\lambda)$.

Ex. 3. The equations when the quartic surface, in the fourfold space, is given by the intersection of *two* line-cones may be obtained as a particular case of Ex. 2. Two of the roots of $f(\lambda) = 0$ will, in fact, become equal to b^2, if $y_0 = 0$ and $a^2x_0^2 = (r^2 + b^2)(a^2 - b^2)$. In general, if a^2, b^2, r^2, $r^2 - x_0^2 - y_0^2$ are positive real quantities $(a^2 > b^2)$, the roots of $f(\lambda) = 0$ are in the intervals of $-\infty$, 0, b^2, a^2. Suppose the two particular conditions satisfied; introduce μ instead of r, given by $\mu^2 = r^2 + b^2$, and put $c^2 = a^2 - b^2$. With $C = (\mu^2 - c^2)^{\frac{1}{2}}$, $A = (\mu^2 - a^2)^{\frac{1}{2}}$, the reduction to the general formulation is given by putting

$$bCX = a\mu x - \mu^2 ct - cu, \quad bACT = acx - \mu c^2 t - \mu u, \quad bAU = \mu cx + a(b^2 - \mu^2)t - au,$$

which lead to

$$(ax - \mu ct)^2 + b^2y^2 - u^2 = b^2[y^2 + X^2 - A^2T^2],$$
$$b^2z^2 - (cx - \mu at)^2 + (u - b^2t)^2 = b^2[z^2 + C^2T^2 - U^2],$$

and hence

$$x^2 + y^2 + z^2 + (b^2 - \mu^2)t^2 - 2tu = y^2 + z^2 + X^2 + b^2T^2 - U^2.$$

The quartic surface in fourfold space is thus the intersection of two line-cones. The Cyclide obtained by projection has, therefore, four double points; its equation is capable, among others, of the forms

$$[x^2 + y^2 + z^2 + (b^2 - \mu^2)t^2]^2 = 4t^2[(ax - \mu ct)^2 + b^2y^2],$$
$$[x^2 + y^2 + z^2 - (b^2 + \mu^2)t^2]^2 = 4t^2[(cx - \mu at)^2 - b^2z^2].$$

A particular irrational form (noticed by Kummer, *Berlin. Monatsber.*, 1863), which may be found as in Ex. 1, is

$$b^2t + [(ax - \mu ct)^2 + b^2y^2]^{\frac{1}{2}} - [(cx - \mu at)^2 - b^2z^2]^{\frac{1}{2}} = 0.$$

The polar reciprocals of the line-cones, with respect to the general quadric, Ω, containing the surface, which is found above, expressed by five squares, are given by $U = 0$, $z = 0$, $a^{-2}x^2 + b^{-2}y^2 = t^2$, and by $X = 0$, $y = 0$, $c^{-2}x^2 - b^{-2}z^2 = t^2$. On projection, to the space (x, y, z, t), we have two conics, in the planes $z = 0$, $y = 0$, so related that their common tangent planes touch the Absolute conic $t = 0$, $x^2 + y^2 + z^2 = 0$. Reciprocally, this is the statement that, in the fourfold space, the two line-cones meet the solid $t = 0$ in two quadric cones (given by $a^2x^2 + b^2y^2 - u^2 = 0$, $c^2x^2 - b^2z^2 - u^2 = 0$) whose common curve is on Ω, as well as on $t = 0$. If we denote the irrational form, noticed, by $p + (qr)^{\frac{1}{2}} - (sv)^{\frac{1}{2}} = 0$, the Absolute conic is given by $p = 0$, $qr - sv = 0$; the double points are given by $q = r = p^2 - sv = 0$ and $s = v = p^2 - qr = 0$, that is by

$$ax_0 - \mu ct_0 = y_0 = az_0 \pm b(\mu^2 - a^2)^{\frac{1}{2}} t_0 = 0$$

and
$$cx_1 - \mu at_1 = z_1 = cy_1 \pm b(c^2 - \mu^2)^{\frac{1}{2}} t_1 = 0,$$

for which $c^{-2}x_0^2 - b^{-2}z_0^2 = t_0^2$, $a^{-2}x_1^2 + b^{-2}y_1^2 = t_1^2$. The Cyclide is obtainable as the envelope of spheres

$$(x - ta\cos\theta)^2 + (y - tb\sin\theta)^2 + z^2$$
$$= t^2[(t_0^{-1}x_0 - a\cos\theta)^2 + b^2\sin^2\theta + t_0^{-2}z_0^2],$$

which have their centres on the conic $z = 0$, $a^{-2}x^2 + b^{-2}y^2 = t^2$ and pass through the first pair of double points. It is likewise obtainable as the envelope of spheres, with centres on the conic $y = 0$, $c^{-2}x^2 - b^{-2}z^2 = t^2$, passing through the second pair of double points. Any sphere of the former system touches any sphere of the latter system. It is clear, in fact, in the fourfold space, that, at a common point of the two line-cones defining the surface, the tangent solid of Ω is met in the same plane by the tangent solids of both the line-cones. Thus, for the Cyclide, taking two spheres of either system, the Cyclide can be generated as the envelope of a sphere, with centre on a certain conic, which touches two fixed spheres; or, again, as the envelope of one of the four systems of spheres which touch three given spheres,—a simple result of inversion from a common point of the three spheres. Further, the circles in which the Cyclide is touched by the enveloping spheres of the two systems, are easily seen to be *lines of curvature* (p. 186, above). Dupin's Cyclide may, indeed, be defined by this property of having two systems of lines of curvature which are circles. Also, we notice, any normal of the Cyclide meets the two conics

$(a^{-2}x^2 + b^{-2}y^2 - t^2 = 0,\ z = 0),\quad (c^{-2}x^2 - b^{-2}z^2 - t^2 = 0,\ y = 0)$; and conversely. This Cyclide is the only surface, in threefold space, of which the normals meet two curves; if the surface be a wavefront this is an optical theorem (Maxwell, *Papers*, II, p. 144, 1867). In the case of the general Cyclide, we have defined a line, in the fourfold space (p. 183, above), which, on projection, becomes the normal; in the present case this is the line whose points, for varying ψ, have coordinates

$$\psi x - \mu act,\ (\psi - c^2)y,\ (\psi - a^2)z,\ (\psi - a^2)t + u,$$
$$(\psi - c^2 - \mu)u + \mu c\,(ax - \mu ct),$$

where (x, y, z, t, u), arising for $\psi = \infty$, is the foot of the normal.

Ex. 4. Prove that the Dupin Cyclide is obtainable by rationalising the equation

$$\psi_1^{\frac{1}{2}} + \psi_2^{\frac{1}{2}} + \psi_3^{\frac{1}{2}} = \mu,$$

where ψ_1, ψ_2, ψ_3 are the three roots of the equation $C(\psi) = 0$, in which

$$C(\psi) = \psi^{-1}x^2 + (\psi - c^2)^{-1}y^2 + (\psi - a^2)^{-1}z^2 - t^2\,;$$

the equation of the Cyclide so obtained, one factor of the complete result of rationalisation, appears in the form

$$[x^2 + y^2 + z^2 + t^2(c^2 + a^2 - \mu^2)]^2$$
$$= 4t^2\,[(act - \mu x)^2 + x^2(a^2 + c^2 - \mu^2) + a^2y^2 + c^2z^2].$$

This result is obtained by Maxwell from optical considerations. It can be shewn that, if w denote the interval, measured in regard to the quadric surface $C(\theta) = 0$, between two points of a line which touches the two confocal quadric surfaces $C(p) = 0$, $C(q) = 0$, then

$$2w\frac{(\theta - p)(\theta - q)}{\Theta} = \sum_{k=1}^{3}\int\frac{(\psi_k - p)(\psi_k - q)}{(\psi_k - \theta)\Psi_k}\,d\psi_k,$$

where $\Theta^2 = 4\theta(\theta - c^2)(\theta - a^2)(\theta - p)(\theta - q)$, or, say, $\Theta^2 = F(\theta)$, and $\Psi_k^2 = F(\psi_k)$. When $p = c^2$, $q = a^2$, which corresponds to a line meeting the two focal conics, as in Ex. 3, this gives

$$\exp.\ (2w) = \prod_{k=1}^{3}(\theta^{\frac{1}{2}} + \psi_k^{\frac{1}{2}})/(\theta^{\frac{1}{2}} - \psi_k^{\frac{1}{2}});$$

and if, herein, we suppose θ to increase indefinitely, putting $\theta^{\frac{1}{2}}w = \mu$, this will be found to lead to Maxwell's result quoted. (Cf., also, Darboux, *Théorie...des Surfaces*, Livre IV, pp. 297, 308; and a note, *Proc. Camb. Phil. Soc.*, XX, 1921, p. 129.)

Ex. 5. It was shewn (p. 193) that, in the fourfold space, two given points of the general quadric Ω, can be projected into two other points of Ω lying on the tangent solid at a given point, O, of Ω. It was seen that the centre of such projection is any point (not on Ω), of the conical sheet joining O to the conic in which the polar plane, of the joining line of the two given points, meets Ω. Such a projection changes a line-cone into a line-cone; thus a quartic surface, in the fourfold space, given by the intersection, lying on Ω, of two quadric line-cones, is changed, by such projection, into a quartic surface given by two such line-cones, of which, however, two double points lie on the tangent solid of Ω at O. Such a quartic surface projects, from O, on to any solid, into a *Tore*, or *Anchor Ring*. When this is proved, it will follow that the Dupin Cyclide can be inverted into a Tore. The necessary centre of inversion, in the threefold space, is any point of either of two circles; one of these circles is the intersection of the cones (or *point-spheres*) which join a pair of the double points of the Dupin Cyclide to the Absolute conic.

Suppose, in fact, that Ω contains the intersection of one quadric line-cone, whose axis meets Ω in two points, A and B, with another quadric line-cone, whose axis meets Ω in C and D. Thus CDA is a plane of the line-cone (CD); it can be seen easily that the tangent solid of Ω at A is likewise a tangent solid of the line-cone (CD), touching it along the plane CDA. Thus, the polar plane of the line AB, in regard to Ω, passes through the line CD. Now suppose that A and B are in the tangent solid of Ω at a point O, so that the polar plane of the line AB is the plane OCD. Then, projecting from O, one of the two focal conics of the resulting Cyclide degenerates into the line which is the projection of CD. The other focal conic, the projection of the polar reciprocal, in regard to Ω, of the line-cone (CD), becomes a circle. For, this requires that the polar reciprocal of this line-cone has two points, lying on Ω, which are on the tangent solid at O; or, that there are two tangent planes of this line-cone which touch Ω and pass through O; by what has been said, these are the tangent planes of Ω at A and B. The proof can, then, be completed easily; and it can be shewn, further, that the tangent planes of the Tore, at either of the two double points which lie on the Absolute conic, coincide. Each of these points is itself a coincidence of two of the four points of the Absolute conic, generally existing (p. 180), at which the tangent planes of the Cyclide are the same.

If, in the equations of Ex. 3, we suppose $c = 0$ $(a = b)$, the double points of the Cyclide are given by $t_1 = z_1 = x_1{}^2 + y_1{}^2 = 0$, and

$$x_0 = y_0 = z_0 \pm (\mu^2 - a^2)^{\frac{1}{2}} t_0 = 0,$$

and the equation of the surface becomes

$$[x^2 + y^2 + z^2 - (\mu^2 - a^2) t^2]^2 = 4t^2 a^2 (x^2 + y^2).$$

To obtain such a surface by inversion of the original, we are to take the centre of inversion at any point of the circle of intersection of the cones

$$[cx - \mu at]^2 + [cy \pm b(c^2 - \mu^2)^{\frac{1}{2}} t]^2 + c^2 z^2 = 0,$$

which lies in $y = 0$. Similarly, taking the other pair of double points, a centre of inversion may be taken at any point of the circle $z = 0$, $(ax - \mu ct)^2 + a^2 y^2 + b^2 (\mu^2 - a^2) t^2 = 0$.

By taking the inversion from the point $cx = (\mu a + bC \cos \theta) t$, $cy = bC \sin \theta t$, where $C = (\mu^2 - c^2)^{\frac{1}{2}}$, it is found that the Tore obtained is given by the equation

$$[\xi^2 + \eta^2 + \zeta^2 + R^2 t^2]^2 = m^2 (\xi^2 + \eta^2) t^2,$$

where

$$m^2 = k^4 c^4 C^{-2} (b\mu + aC \cos \theta)^{-2}, \quad 4R^2 = -A^2 m^2 b^{-2}, = (a^2 - \mu^2) m^2 b^{-2},$$

and k^2 is arbitrary. If

$$\sigma = \tfrac{1}{2} ck^2/bC, \quad \rho^2 = t^2 k^2/(x_1^2 + z_1^2 + y^2), \quad p = \mu a + bC \cos \theta,$$
$$q = \mu b + aC \cos \theta,$$

the actual formulae are

$$x_1 = x - tc^{-1} p, \quad z_1 = z - tc^{-1} bC \sin \theta, \quad \eta = \rho^2 y,$$
$$\xi = \rho^2 (x_1 \cos \theta + z_1 \sin \theta) + \sigma t, \quad \zeta = \rho^2 (x_1 \sin \theta - z_1 \cos \theta) + \sigma t a C q^{-1} \sin \theta.$$

The double points of the Tore are given by $(0, 0, \pm iR, 1)$ and $(1, \pm i, 0, 0)$.

The inversion of a Dupin Cyclide into a Tore is given by Darboux (*Sur une classe remarquable de courbes et de surfaces algébriques*, Paris, 1873, p. 242), with reference to Mannheim (*Nouv. Ann. de Math.*, 1860, p. 67). Darboux operates in three-fold space, with spheres, saying (*loc. cit.*, p. 164): *Comme on n'a pas d'espace à quatre dimensions, les méthodes de projection ne s'étendent pas à la géométrie de l'espace.*

Ex. 6. The locus of the pole, for inversion, transforming any two spheres into spheres which are equal, consists of two spheres coaxial with the first, with centres at their centres of similitude. The locus of the pole, for inversion, transforming three spheres, with *directed* radii, into three spheres of equal *and like* radii, is the circle common to the two point-spheres whose centres are the intersections of the orthogonal circle and axis of similitude of the three original spheres (Darboux, *loc. cit.*, pp. 243, 244). With different signs for the three original radii, different axes of similitude arise.

Ex. 7. Determine the character of the curve of intersection of the enveloping cones, to a quadric in fourfold space, drawn from three arbitrary points.

Prove that, in threefold space, the envelope of a sphere touching three given spheres, $S_1 = 0$, $S_2 = 0$, $S_3 = 0$, is

$$(\mu_{23}S_1)^{\frac{1}{2}} + (\mu_{31}S_2)^{\frac{1}{2}} + (\mu_{12}S_3)^{\frac{1}{2}} = 0,$$

where, for example, $\mu_{23} = 0$ is the condition of contact of the two latter spheres.

Ex. 8. With the notation of Ex. 3, putting ρ^2 for $(X^2 + Y^2 + Z^2)^{-1}$, the equations

$$aCx = \rho^2(\mu bX - cAZ)\,t^2 + \mu cCt, \quad aCz = \rho^2(cAX + \mu bZ)\,t^2 - bACt,$$

with $y = \rho^2 Yt^2$, transform the Dupin Cyclide into the cone of revolution

$$(X - \tfrac{1}{2}ct/Cb)^2 + Y^2 = b^{-2}A^2(Z - \tfrac{1}{2}\mu t/CA)^2.$$

Ex. 9. The Dupin Cyclide may be regarded as the envelope of its tangent planes, and its equation expressed tangentially. If we express tangentially one of the enveloping spheres, whose centre is on a focal conic, the equation of a tangent plane being written $x\xi + y\eta + z\zeta + t\tau = 0$, and find the envelope of this sphere as its centre moves, we find, for the two focal conics, the equations

$$(a\xi - c\omega)^2 + b^2\eta^2 - (\tau + \mu\omega)^2 = 0, \quad (c\xi - a\omega)^2 - b^2\zeta^2 - (\tau + \mu\omega)^2 = 0,$$

where $\omega^2 = \xi^2 + \eta^2 + \zeta^2$. The enveloping spheres, whose centres are at the points of the focal conics which lie on $t = 0$, give four double tangent planes, whose equations are

$$cx - \mu at \pm bz = 0, \quad ax - \mu ct \pm iby = 0.$$

In the fourfold space, we may consider the aggregate of the tangent planes of the quartic surface of intersection of the two quadric line-cones: Considering an arbitrary tangent solid of each of these line-cones, with the plane along which it touches the line-cone, these two planes have a common point; the intersection of the tangent solids is the tangent plane of the quartic surface at this point. After projection, this becomes the statement that any enveloping sphere, of one of the two systems, touches any sphere of the other system. The tangent solids, of the two line-cones, respectively given by

$$(ax - \mu ct)\cos\theta + by\sin\theta - u = 0, \quad (cx - \mu at)\cos\phi - ibz\sin\phi - (u - b^2t) = 0,$$

meet in a plane lying on the solid $x\xi + y\eta + z\zeta + t\tau = 0$, where

$$\xi = a\cos\theta - c\cos\phi, \quad \eta = b\sin\theta, \quad \zeta = ib\sin\phi, \quad \tau = -\mu(c\cos\theta - a\cos\phi) - b^2,$$

which, with $\omega^2 = \xi^2 + \eta^2 + \zeta^2$, lead to $\omega = c\cos\theta - a\cos\phi$. These satisfy the two equations given above. There are two tangent solids of either line-cone which pass through the centre of projection; these give the four double tangent planes.

Dually, in the fourfold space, we may consider the surface, in the solid $t = 0$, which is the locus of the intersection of this solid with the line joining the two points

$$(a \cos \theta,\ b \sin \theta, 0, 1, \mu c \cos \theta - \mu^2 + b^2),$$
$$(c \cos \phi, 0, - ib \sin \phi, 1, \mu a \cos \phi - \mu^2);$$

these are points of those conics which are the polar reciprocals of the line-cones, in regard to the quadric

$$x^2 + y^2 + z^2 + (b^2 - \mu^2)\, t^2 - 2tu = 0.$$

This is the line which, on projection, becomes the normal of the Cyclide; it meets $t = 0$ in the point $(\xi, \eta, \zeta, 0, -\tau)$.

Ex. 10. Prove that the Dupin Cyclide is the projection of a quartic surface, in the fourfold space, for which the coordinates of a point are, respectively, the five

$$\mu\, (a \cos \theta - c \cos \phi) - b^2 \cos \theta \cos \phi,\ b \sin \theta\, (\mu - a \cos \phi),$$
$$ib \sin \phi\, (\mu - c \cos \theta),\ c \cos \theta - a \cos \phi,\ b^2\, (\mu - a \cos \phi).$$

The first four of these give the coordinates of a point of the Cyclide. The equations of the surface are obtainable by eliminating θ and ϕ.

Prove, also, that the tangent planes of the Cyclide, at any point, $(x_0, y_0, z_0, 0)$, of its double conic, are expressed by

$$(xx_0 + yy_0 + zz_0)^2 - t^2\, (a^2 x_0^2 + b^2 y_0^2) = 0;$$

further, that the Cyclide is touched, in two points, by every tangent plane of the cone

$$(\mu x - act)^2 + (\mu^2 - c^2)\, y^2 + (\mu^2 - a^2)\, z^2 = 0,$$

the curve of contact, as this plane varies, lying on a sphere. Also, that the tangential equation of the Cyclide is capable of the form

$$[\mu^{-2}\, \xi^2 + (\mu^2 - c^2)^{-1}\, \eta^2 + (\mu^2 - a^2)^{-1}\, \zeta^2 - 2ac\, \xi T + M\tau^2]^2$$
$$- 4\mu^4\, T^2\, (\xi^2 + \eta^2 + \zeta^2) = 0,$$

where $M = \mu^2\, (\mu^2 - a^2)\, (\mu^2 - c^2)$, $T = (ac\xi + \mu\tau)/M$. Determine all the lines lying upon the Dupin Cyclide.

CHAPTER VII

RELATIONS IN SPACE OF FIVE DIMENSIONS.
KUMMER'S SURFACE

Klein's figure in three dimensions, related to a figure in five dimensions. We have studied in some detail, in Chap. v, a figure, in space of four dimensions, containing fifteen lines meeting, in threes, in fifteen points; for convenience we may describe this here as Segre's figure (cf. Stéphanos, *Compt. Rend.*, xciii, 1881, p. 634; quoted by Segre). We shall shew now that these fifteen lines are in correspondence with the joining lines of six points in space of five dimensions. The transformation utilises formulae arising in the representation of the points, and linear complexes of lines, in space of three dimensions, by use of space of five dimensions; for clearness sake we begin by referring again to this representation.

Let l, m, n, l', m', n' be the coordinates of a line, in space of three dimensions, and put $x = l - l'$, $y = m - m'$, $z = n - n'$, $u = l + l'$, $v = m + m'$, $w = n + n'$. We have seen (p. 46, above) that the lines through a point, (ξ, η, ζ, τ), of the threefold space, are represented, in the fivefold space in which (x, y, z, u, v, w) are coordinates, by the points of a plane whose three equations are $(u, v, w) = D\,(x, y, z)$, where D, written in explicit form, consists of the elements, of the matrix M given by

$$M = \begin{pmatrix} \tau^2 + \xi^2 - \eta^2 - \zeta^2, & 2\,(\xi\eta - \zeta\tau)\ , & 2\,(\zeta\xi + \eta\tau) \\ 2\,(\xi\eta + \zeta\tau)\ , & \tau^2 + \eta^2 - \zeta^2 - \xi^2, & 2\,(\eta\zeta - \xi\tau) \\ 2\,(\zeta\xi - \eta\tau)\ , & 2\,(\eta\zeta + \xi\tau)\ , & \tau^2 + \zeta^2 - \xi^2 - \eta^2 \end{pmatrix},$$

each element divided by $\xi^2 + \eta^2 + \zeta^2 + \tau^2$. For brevity, the elements of the rows of the matrix M will be denoted, respectively, by l_1, m_1, n_1; l_2, m_2, n_2; l_3, m_3, n_3, and $\xi^2 + \eta^2 + \zeta^2 + \tau^2$ by σ. In the threefold space, the equations $x = 0$, $u = 0$, ... denote six linear complexes, of which every two are conjugate; in the fivefold space these equations define a *hexad*, self-polar in regard to the quadric, Ω, whose equation is $u^2 + v^2 + w^2 = x^2 + y^2 + z^2$. The plane, in the fivefold space, representing the lines through a point of the threefold space, is a plane, of one system, lying on the quadric Ω; and there are planes of Ω of a second system, each representing the lines of a *plane* of the threefold space. For each two of the six linear complexes, or the associated polar systems, of the threefold space, there are two lines which are polars of one another for both systems; if P be any point

of the threefold space, and, on the transversal drawn from P to
these two common polar lines, P' be the point which is the harmonic
conjugate of P, in regard to the polar lines, then P' is the pole, in
either focal system, of the polar plane of P in the other focal system
(cf. Vol. III, p. 65). In the representation in space of five dimensions,
to a point and its polar plane, in regard to the focal system for
which the associated linear complex is $u = 0$, there correspond two
planes of the quadric Ω, which meet in a line of the fourfold $u = 0$;
these are harmonic inverses of one another, in regard to the fourfold
$u = 0$ and its pole in regard to Ω (p. 42, above). Thus, the representation, in the fivefold space, of the two points P, P', spoken of,
is by two planes of Ω of the same system, either obtainable from
the other by the succession of two such harmonic inversions. When
we have six linear complexes of which every two are conjugate,
there will be fifteen such involutory transformations as that from
P to P'; by these, every point of the threefold space gives rise to
fifteen others. Likewise, every plane similarly gives rise to fifteen
other planes. In the fivefold space there correspond sets of sixteen
planes of Ω, of the same system, arising from one of these planes;
the linear complexes being $x = 0$, $u = 0$, ..., as above, any point of
one of the fifteen derived planes is obtainable, from a point of the
primary plane, by change of the sign of *two* of the coordinates of
the point; thus, also, any one of the sixteen planes can be regarded
as primary. In the threefold space, there is a figure of sixteen
points and sixteen planes, arising by taking an arbitrary point and
the fifteen derived points, and then the polar planes of all the points
in regard to all the six conjugate focal systems. Only sixteen planes
arise in this way; each plane contains six of the points, the poles
of this plane in the various polar systems; and through each point
there pass its six polar planes. The thirty-two planes of the quadric
Ω, in the fivefold space, which correspond to this figure, are all
derivable from one plane of Ω, by combination of the six processes
of harmonic inversion, in which one of the points of the fundamental hexad, and its polar plane in regard to Ω, are fundamental.
Further, the six lines in which an arbitrary plane of Ω meets the
primary fourfolds, $x = 0$, $u = 0$, ..., touch a conic of this plane. For,
in the notation above suggested, these six lines are given, respectively, by

$$x = 0, \quad y = 0, \quad z = 0, \quad l_1 x + m_1 y + n_1 z = 0, \quad ..., \quad l_3 x + m_3 y + n_3 z = 0;$$

and they touch the conic in which the plane is met by the cone

$$(l_1 l_2 l_3 x)^{\frac{1}{2}} + (m_1 m_2 m_3 y)^{\frac{1}{2}} + (n_1 n_2 n_3 z)^{\frac{1}{2}} = 0.$$

This conic equally lies on the cone

$$(l_1 m_1 n_1 u)^{\frac{1}{2}} + (l_2 m_2 n_2 v)^{\frac{1}{2}} + (l_3 m_3 n_3 w)^{\frac{1}{2}} = 0.$$

The fact, in the threefold space, which corresponds to this is that, in the figure of sixteen points and planes, the six poles in any plane lie on a conic, and the six planes through any point touch a quadric cone.

Ex. 1. The reverse of the formulae given above, for passing from any point, (ξ, η, ζ, τ), of the threefold space, to a plane,

$$(u, v, w) = D(x, y, z),$$

of Ω, are, if p, q, r, p', q', r', respectively, denote

$$m_3 + n_2, \quad n_1 + l_3, \quad l_2 + m_1, \quad m_3 - n_2, \quad n_1 - l_3, \quad l_2 - m_1,$$

which do not contain l_1, m_2, n_3, these following:

$$\tau^{-1}\xi = q/r' = r/q'; \quad \tau^{-1}\eta = r/p' = p/r'; \quad \tau^{-1}\zeta = p/q' = q/p'.$$

If $\theta = pp', = qq', = rr', \ n_2^2 = \alpha, \ l_3^2 = \beta, \ m_1^2 = \gamma$, then

$$p = (\alpha + \theta)^{\frac{1}{2}} + \alpha^{\frac{1}{2}}, \quad q = (\beta + \theta)^{\frac{1}{2}} + \beta^{\frac{1}{2}}, \quad r = (\gamma + \theta)^{\frac{1}{2}} + \gamma^{\frac{1}{2}}.$$

Ex. 2. In the threefold space, the polar plane,

$$\xi X + \eta Y + \zeta Z + \tau T = 0,$$

of the point (ξ, η, ζ, τ), in regard to the quadric surface

$$X^2 + Y^2 + Z^2 + T^2 = 0,$$

contains the three points $(\tau, -\zeta, \eta, -\xi), (\zeta, \tau, -\xi, -\eta), (-\eta, \xi, \tau, -\zeta)$, of which every two are conjugate to one another in regard to the quadric surface. When the coordinates of the first of these three points are put for ξ, η, ζ, τ, in the matrix M, above, the matrix is unaltered, save for a change of sign of every element in the second and third rows of the matrix. Thus, the plane, in the fivefold space, representing this point, $(\tau, -\zeta, \eta, -\xi)$, is obtained, from the plane representing the point (ξ, η, ζ, τ), by two harmonic inversions in succession, in the fourfolds $v = 0$, $w = 0$, each associated with its pole in regard to Ω. Similarly, the second and third, of the three points named, correspond, respectively, to the pairs of harmonic inversions (w, u) and (u, v). The same plane, $\xi X + \ldots + \tau T = 0$, contains also the three points $(\tau, \zeta, -\eta, -\xi), (-\zeta, \tau, \xi, -\eta), (\eta, -\xi, \tau, -\zeta)$; these are, similarly, a self-polar triad in regard to the quadric surface $X^2 + \ldots + T^2 = 0$, and correspond, respectively, to pairs of harmonic inversions, in the fivefold space, in the pairs of fourfolds $(y, z), (z, x), (x, y)$.

If we take new coordinates (X', Y', Z', T'), given by

$$(X', Y', Z', T') = \begin{pmatrix} \tau, & -\zeta, & \eta, & -\xi \\ \zeta, & \tau, & -\xi, & -\eta \\ -\eta, & \xi, & \tau, & -\zeta \\ \xi, & \eta, & \zeta, & \tau \end{pmatrix} (X, Y, Z, T),$$

the coordinates, (X', Y', Z'), of the six points of the plane $T' = 0$, are $(1, 0, 0)$, $(0, 1, 0)$, $(0, 0, 1)$, (l_1, l_2, l_3), (m_1, m_2, m_3), (n_1, n_2, n_3), respectively. These points lie on the conic

$$l_1 m_1 n_1/X' + l_2 m_2 n_2/Y' + l_3 m_3 n_3/Z' = 0,$$

of which the general point is given by

$$X' = l_1 m_1 n_1 \theta, \quad Y' = l_2 m_2 n_2 (1 - \theta), \quad Z' = l_3 m_3 n_3 \theta (\theta - 1),$$

namely, for the respective values $\theta = 1$, 0, ∞, $-\mu\nu$, $-\nu\lambda$, $-\lambda\mu$, where $\lambda = l_2/l_3$, $\mu = m_2/m_3$, $\nu = n_2/n_3$. The six points are also given by iH, jH, kH, Hi, Hj, Hk, where $H = i\xi + j\eta + k\zeta + \tau$, the symbols i, j, k being such that $i^2 = -1$, $jk = -kj = i$, etc. (Vol. III, p. 138.)

Ex. 3. Shew that the fifteen points, in the threefold space, can be constructed from the six points iH, jH, kH, iHi, jHj, kHk, the notation being as in Ex. 2. These are the six points A, B, C, P, Q, R of Ex. 5, p. 139, above. Determine the coordinates of all the points relatively to the four iH, jH, kH, H.

Ex. 4. In the fivefold space, let the poles, in regard to Ω, of the six primary fourfolds, $x = 0, \ldots, w = 0$, be divided into three pairs. The joins of the points of a pair meet the quadric Ω in two points, which represent the pair of common polar lines of the two corresponding polar systems of the threefold space. It can be shewn that the three pairs of polar lines so obtained are the pairs of opposite joins of four points of the threefold space; for instance, the tangent fourfolds of Ω at the two points $(1, 0, 0; \pm 1, 0, 0)$ contain the other two pairs of points similarly arising.

Ex. 5. The formulae suggest the consideration, in the threefold space, of ten quadric surfaces. One of these is expressed by

$$\xi^2 + \eta^2 + \zeta^2 + \tau^2 = 0;$$

the other nine are obtained by equating to zero the respective elements, l_1, m_1, n_1, etc., of the matrix M. These quadric surfaces arise geometrically because the polar plane of a point, P, in regard to any one of these quadrics, is obtained by first passing from P to a point, P', by means of one of the fifteen involutory transformations explained above, as arising from a pair of the focal systems, and then taking the polar plane of P' in a third focal system. Or, because the generators, of either system, of any one of these quadric surfaces, consist of the lines common to three of the six fundamental linear complexes, the other system of generators being the lines common to the other three complexes. This is clear, either, directly, from the equations of the quadric surfaces, or, from the representation of a point of the threefold space by a plane of Ω in the fivefold space. For instance, from the three equations of such a plane

$$\sigma u = l_1 x + m_1 y + n_1 z, \text{ etc.},$$

any point of the quadric $m_1 = 0$, or $\xi\eta - \zeta\tau = 0$, corresponds to a plane of Ω which contains a point for which $u = 0$, $x = 0$, $z = 0$. These last three equations, however, represent a plane, in the five-fold space, which meets Ω in a conic. The points of the conic thus represent one system of generators of the quadric $m_1 = 0$. We shall associate the vertices x, y, z, u, v, w of the fundamental hexad in the fivefold space, that is, the points for which these coordinates have, respectively, the values $(1, 0, 0; 0, 0, 0)$, ..., $(0, 0, 0; 0, 0, 1)$, with the respective numbers 4, 5, 6, 1, 2, 3. Then, for instance, the plane $u = x = z = 0$, containing the points y, v, w, is associated with the triad 235. This plane contains the points of Ω lying on the joins, of pairs of points of the hexad, given by 23, 25, 35; and these points represent three pairs of generators, of the quadric surface $m_1 = 0$, each of which is a pair of common polar lines for a pair of the six polar systems. The other system of generators, of the quadric surface $m_1 = 0$, is represented by the conic of Ω which is the polar of the former, lying in the plane $y = v = w = 0$; this conic, or the plane in which it lies, is then associated with the triad 164. The quadric surface may then be represented by 235, or 164; and this notation suggests six pairs of generators belonging to the quadric surface. The ten quadrics corresponding to the elements of the matrix M, and to $\sigma = 0$, will then, respectively, have the nota-tion of the annexed scheme,

234 (156),	235 (164),	236 (145)	·
314 (256),	315 (264),	316 (245)	·
124 (356),	125 (364),	126 (345)	·
·	·	·	123 (456)

as is easily seen in the same way. Now, it is clear that, in the five-fold space, there can be drawn, through the line joining any two of the six fundamental points, four planes, each containing one other of these six points (and, thus, the lines joining this to the two first points). There are, therefore, fifteen sets, each of four of the ten quadric surfaces in the threefold space, such that these four quadrics have two generators in common. Any one of these sets of four quadric surfaces consists of those whose symbols, in the annexed scheme, lie in the same row and column with any one, other than 123 (456), of the sixteen elements of the scheme. For instance, the quadric surfaces so obtainable from the fourth element of the first row all contain the two generators represented by 23; that is, the generators given by the intersections, with Ω, of the line joining the points v, w, in the figure in fivefold space.

We may, thus, from the given scheme, obtain another scheme, of four rows and columns, with no entry for the fourth element of the fourth row, by putting, in any place, the symbol for the two generators common to the four quadric surfaces arising, in the way explained, by starting from this place. This new scheme agrees in notation with one given earlier (p. 133, above).

It may be proved that the two generators common to a set of four quadric surfaces are polar lines, each of the other, in regard to each of the six remaining quadric surfaces. Also, the squares of the quadratic functions which, equated to zero, give such a set of four quadric surfaces, are connected by a linear equation. We may also remark that any quadric surface, whatever, may be expressed as a linear function of the ten quadric surfaces; and that every two of these ten are both outpolar and inpolar to one another (cf. Vol. II, p. 149).

The six linear complexes, conjugate in pairs, were studied by F. Klein, *Math. Annal.* II. (1870), p. 198, who remarks the sets of four quadric surfaces having two common generators (*Ges. Math. Abhandl.* I. (1921), p. 63). See also Hudson, *Kummer's Quartic Surface*, 1905, p. 40, etc.

A transformation of Segre's figure, in space of four dimensions, to a figure in five dimensions. The representation, in space of five dimensions, of the figure of six linear complexes in a threefold space, has suggested a notation for the fifteen pairs of common polar lines, of two complexes, for the ten quadric surfaces, and for the fifteen sets of six joins of a tetrad of points (each associated with such a symbol as 12.34.56). This notation agrees with that previously employed (p. 114, above) for the fifteen lines, for the ten singular solids, and for the fifteen points, arising in Segre's figure. As we saw, the figure of six linear complexes arises in any tangent solid of the locus Σ, discussed in connexion with Segre's figure. We now describe an independent transformation, which, though analogous, is different in essence from the above.

We consider any threefold space, with coordinates ξ, η, ζ, τ. Therein, as we have seen, we can associate, with any point (ξ, η, ζ, τ), fifteen others; the coordinates of these are derivable from ξ, η, ζ, τ by two processes: (1), the change of the signs of two of the coordinates; (2), the interchange of two of the coordinates accompanied by the interchange of the other two. These processes may be applied to the coordinates of any one of the aggregate of sixteen points, and will give the coordinates of the other fifteen, and only these. We thus have a linear group in four (homogeneous) variables. And we have sets of sixteen points of which each set is determined by any one of its points; or, as we may say, we have an *involution* of sets of sixteen points in the threefold space. We now shew that we can determine a set of five functions, all homogeneous polynomials of the same dimension in ξ, η, ζ, τ, with the two properties; (1), that the ratios of these polynomials are the same for all the points of a set of the involution; (2), that these ratios are the same *only* for sixteen points belonging to the same set of the involution. Then we regard these five functions as homogeneous coordinates in a space of four dimensions. Thereby we obtain a correspondence, which is uniquely reversible, between points of this fourfold space, and the sets of the involution in the threefold space. But, the ratios of the five functions, being dependent upon the three ratios of ξ, η, ζ, τ, will be connected by an algebraic relation, which may be taken in a rational form. The correspondence will thus be one between the points of an ∞^3 locus, in the fourfold space, and the unrestricted sets of the involution in the threefold space. For the five functions in question we take the squares of five elements of the matrix M, above (p. 203); namely, recalling the notation used before, $l_1 = \tau^2 + \xi^2 - \eta^2 - \zeta^2$, $n_2 = 2(\eta\zeta - \tau\xi)$, etc., we take the five functions $l_1{}^2$, $m_2{}^2$, $n_3{}^2$, $n_2{}^2$, $m_3{}^2$. These are easily seen to be unaltered, in their ratios, by passing from any point to any other point of the same set of the involution. The same would not be true of the unsquared functions, l_1, m_2, etc.; for instance the ratio l_1/m_2 is changed in sign by the substitution which replaces ξ, η, ζ, τ, respectively, by τ, ζ, η, ξ. As the ratios of the five chosen functions are unaltered in passing from one point to another of the same set, it follows that when these ratios are given, the ratios of ξ, η, ζ, τ are capable of at least the sixteen values which correspond to the points of such a set of the involution. If we shew algebraically that the ratios of ξ, η, ζ, τ are capable of *only* sixteen values, it will follow that these are those of a set of the involution. Thence will follow the second, characteristic, property of the ratios of the five chosen functions; namely, that they are unaltered *only* for the points of such a set. Now, the expressions of the ratios of ξ, η, ζ, τ, in terms of the five squares $l_1{}^2$, $m_2{}^2$, $n_3{}^2$, $n_2{}^2$, $m_3{}^2$, are to be found from the equations, previously remarked (p. 205),

$$\tau^{-1}\xi = (n_1 + l_3)/(l_2 - m_1), \quad \tau^{-1}\eta = (l_2 + m_1)/(m_3 - n_2),$$
$$\tau^{-1}\zeta = (m_3 + n_2)/(n_1 - l_3),$$

by putting, beside $\quad m_3 = (m_3{}^2)^{\frac{1}{2}}, \; n_2 = (n_2{}^2)^{\frac{1}{2}}$,

$$n_1 = (-l_1{}^2 + m_2{}^2 + m_3{}^2)^{\frac{1}{2}}, \; l_3 = (-l_1{}^2 + m_2{}^2 + n_2{}^2)^{\frac{1}{2}},$$
$$l_2 = (-l_1{}^2 + m_3{}^2 + n_3{}^2)^{\frac{1}{2}}, \; m_1 = (-l_1{}^2 + n_2{}^2 + n_3{}^2)^{\frac{1}{2}};$$

the ratios $\tau^{-1}\xi$, $\tau^{-1}\eta$, $\tau^{-1}\zeta$ may thus appear to be capable of 2^5, or thirty-two, values, when the ratios of the five squares are given. But the signs are, in fact, not independent; for the product $2m_3 n_2 . n_1 l_3 . l_2 m_1$ is expressible by the five squares, in the form

$$2m_3{}^2 n_2{}^2 (m_2{}^2 + n_3{}^2) + (m_3{}^2 + n_2{}^2)(m_2{}^2 n_3{}^2 + m_3{}^2 n_2{}^2 - l_1{}^2 \sigma^2),$$

where $\sigma^2 = -l_1{}^2 + m_2{}^2 + n_3{}^2 + n_2{}^2 + m_3{}^2$; as is easily verified.

The representative character of the ratios of the five functions is thus established. To obtain the relation connecting them we introduce the precise notation

$$X = (\tau^2 + \eta^2 - \zeta^2 - \xi^2)^2, \quad X' = (\tau^2 + \zeta^2 - \xi^2 - \eta^2)^2,$$
$$Y = 4(\eta\zeta - \tau\xi)^2, \quad Y' = 4(\eta\zeta + \tau\xi)^2,$$
$$Z = -(\tau^2 + \xi^2 - \eta^2 - \zeta^2)^2, \quad Z' = -(\tau^2 + \xi^2 + \eta^2 + \zeta^2)^2;$$

then it can be verified at once that

$$X + X' + Y + Y' + Z + Z' = 0, \quad \text{and} \quad (XX')^{\frac{1}{2}} + (YY')^{\frac{1}{2}} + (ZZ')^{\frac{1}{2}} = 0.$$

The latter is the equation which may be written $l_1\sigma = m_2 n_3 - m_3 n_2$; it represents the ∞^3 locus, in the space of four dimensions, of which the points are in reversible correspondence with the sets of the involution in the threefold space. By elimination of Z' between these two equations, we have the relation connecting the five chosen functions X, X', Y, Y', Z only.

We may express the matter by speaking of the *invariants* of a linear group, of finite order, in four homogeneous variables. The reader may consult Burnside, *Theory of Groups of Finite Order* (2nd Edit., Cambridge, 1911), p. 359, where, however, the variables are not homogeneous. Also, the present writer's *Multiply-Periodic Functions* (Cambridge, 1907), p. 281. In this, in line 5, the second N should be M; and, in line 22, the words "there being" should read "there not being."

The locus expressed by these equations is none other than the locus Σ previously discussed (pp. 126, 159, above). If we put

$$U = Y + Z + X', \quad V = Z + X + Y',$$
$$W = X + Y + Z', \quad T = -(X + Y + Z),$$

these corresponding, respectively, to the squares m_1^2, n_1^2, $-l_2^2$, $-l_3^2$, this locus has fifteen double lines, in each of which there vanish four of the ten functions X, Y, Z, X', Y', Z', U, V, W, T. The fifteen sets of four of these are those in the same row and column with the elements, respectively, other than Z', of the adjoined scheme. On comparison with a scheme previously given (p. 133, above), we thus have a notation, by two numbers, for each of the fifteen lines; and also a notation, by three numbers, for each of the ten solids expressed by the equations $X = 0, \ldots, T = 0$.

$$\begin{array}{cccc} Z, & U, & V & \cdot \\ W, & X, & Y & \cdot \\ T, & Y', & X' & \cdot \\ \cdot & \cdot & \cdot & Z' \end{array}$$

In general, as has been said, to a point of the locus Σ, now found, there corresponds a set of sixteen points of the threefold space (ξ, η, ζ, τ). But the fifteen lines of Σ are exceptional. Each of these corresponds to the points common to four quadric surfaces, of which three are linearly independent; but these four are, in each case, such that they have two lines in common. Thus each of the fifteen lines of the fourfold space corresponds to *two* lines of the threefold space. However, as we have seen above (p. 207), when we represent the lines of the threefold space (ξ, η, ζ, τ) by the points of a quadric Ω, in fivefold space, the representation of the two lines, common to one of the sets of four quadric surfaces, is by two particular points of Ω; namely, the two intersections of Ω with the join of two of the points of the fundamental hexad, self-polar in regard to Ω.

Thus, by passing, through the threefold space, from the fourfold space to the space of five dimensions, every one of the fifteen lines of the Segre figure corresponds to the join of two of the six fundamental points of the fivefold space.

But there is more. The fifteen lines of the Segre figure contain six sets each of five associated lines, (each line being used twice), (p. 114, above). Such a set is that whose notation is $i1$, $i2$, $i3$, $i4$, $i5$, $i6$ (with omission of ii). Wherefore, by the number notation we have used for the points of the hexad, in fivefold space, we see that a set of five associated lines, of Segre's figure, is represented by the joins of one point of the hexad to the other five points. All the associated sets are thus accounted for.

Relation of the transformation to the theory of Kummer's surface. We may also apply the transformation by which we have passed from the locus Σ, in the space (X, Y, Z, X', Y', Z'), to the space (ξ, η, ζ, τ), to another locus. We saw that the locus Σ intersects any one of its tangent solids in a Kummer's quartic surface (p. 138, above), having a double point at the point of contact of the tangent solid, and also a double point at each of the fifteen points where the tangent solid is met by the double lines of Σ. The equation of the tangent solid involves the coordinates $X, Y, \ldots$ linearly, and, in the transformation, each of these is a quartic function of the coordinates ξ, η, ζ, τ. Thus the Kummer's surface corresponds to a quartic surface in the space (ξ, η, ζ, τ). Upon this quartic surface there is an involution of sets of sixteen points, each set corresponding to one of the ordinary points of the Kummer surface in the tangent solid of Σ. As for the double points of the Kummer surface, fifteen of them give rise each to two ordinary points of the new surface; for we have seen that each of the double lines of Σ gives rise to two lines of the space (ξ, η, ζ, τ). But, as will be shewn at once, the remaining double point of the Kummer surface, the point of contact of the tangent solid with Σ, gives rise to sixteen points of the new quartic surface, each of which is a double point. In fact, the new surface is also a Kummer surface. This will appear naturally below (p. 215), from another point of view. And, this being so, interesting consequences arise by identifying the space (ξ, η, ζ, τ) with the tangent solid of Σ. For clearness, however, it should be remarked that the two facts, that four of the singular solids in the Segre figure pass through any one of the fifteen lines, and that four of the ten fundamental quadric surfaces, suitably chosen, in the threefold space, have two generators in common, are geometrically distinct. This may be obscured by the adoption of the same number notation for the singular solids and for the quadric surfaces, convenient as this is. If we identify the threefold space (ξ, η, ζ, τ) with a particular tangent solid of the locus Σ, the fifteen sets of six lines, such as $12.34.56$, which are the joins of four points of the space (ξ, η, ζ, τ), will be lines which we have not considered in our discussion of Segre's figure in Chapter v.

To see that the point of contact of the tangent solid with Σ gives rise to double points of the surface in the space (ξ, η, ζ, τ), under the transformation discussed, suppose, more generally, that $\phi_1, \phi_2, \phi_3, \phi_4, \phi$ are five polynomials, of the same order, in ξ, η, ζ, τ, which, as regards their ratios, are invariants of a group of substitutions for these (homogeneous) variables ξ, η, ζ, τ; put, respectively, X, Y, Z, T, U for these polynomials. There will exist a rational homogeneous equation $F(X, Y, Z, T, U) = 0$, which is an identity in ξ, η, ζ, τ. This equation represents a locus in the fourfold space (X, Y, Z, T, U), of which the tangent solid at any point may be represented by

$$X(\partial F/\partial X)_0 + \ldots + U(\partial F/\partial U)_0 = 0.$$

The intersection of this solid with the locus corresponds to a surface, in the space (ξ, η, ζ, τ), whose equation is obtained from this by putting $\phi_1, \phi_2, \ldots, \phi$, respectively, for $X, Y, \ldots, U$. The conditions for this surface to have a double point at one of the points, say $(\xi_0, \eta_0, \zeta_0, \tau_0)$, which correspond to the point of contact of the tangent solid with the locus $F = 0$, are of the form

$$(\partial F/\partial X)_0 (\partial \phi_1/\partial \xi)_0 + \ldots + (\partial F/\partial U)_0 (\partial \phi/\partial \xi)_0 = 0,$$

wherein for ξ are to be put, in turn, ξ, η, ζ, τ. These four conditions arise, however, by differentiating the identity in ξ, η, ζ, τ, expressed by $F = 0$, and then putting ξ_0, η_0, etc., for ξ, η, etc.

Kummer's quartic surface as the locus of singular points of a quadratic complex of lines. A quadratic complex of lines, in threefold space, is an aggregate (∞^3), represented by a single quadratic relation connecting the line coordinates. The lines of such a complex which pass through an arbitrary point are the generators of a quadric cone; those which lie in an arbitrary plane are the tangents of a conic (Vol. iii, p. 99). But the quadric cone of lines through a point may break up, for suitable positions of the point, into two planes, the lines through the point then forming two flat pencils in these planes. The points, called *singular points of the complex*, for which this is so, are the points of a Kummer quartic surface. It was from this point of view that the surface was discussed by Kummer (*Berlin. Monatsber.*, 1864, pp. 246, 495; 1865, p. 288; *Berlin. Abhandl.*, 1866, p. 1). Similarly, there is a quartic envelope of planes, wherein the rays of the quadratic complex constitute two flat pencils; but it will appear that this consists of the tangent planes of the former surface. For special forms of the quadratic complex, the locus of singular points may degenerate; for instance, in the case of the tetrahedral complex, the locus breaks up into the four fundamental planes, and the envelope into the four fundamental points. Properties arising in the special cases may, however, suggest more general theorems. For instance, we know (Vol. I, p. 30) that any line meets four planes in a range which is related to that of the four planes joining the line to the four points of intersection of threes of the planes; it will be found that any line meets the general Kummer surface in a range of four points related to that of the four tangent planes which pass through the line.

We suppose the quadratic complex to be represented by a quadratic relation connecting the coordinates of a point in fivefold space; that is, by the intersection of an (∞^4) quadric of this space with the fundamental quadric, Ω, which represents the necessary relation connecting the line coordinates. Further, we suppose the quadratic complex to be so general that the two quadrics can be represented by squares of the same six linear functions of the coordinates. For the fundamental quadric, Ω, and for that, Ω', giving the quadratic complex, we may then suppose the equations to be, respectively,

$$x_1^2 + x_2^2 + \ldots + x_6^2 = 0, \quad k_1 x_1^2 + k_2 x_2^2 + \ldots + k_6 x_6^2 = 0.$$

The lines of the quadratic complex which pass through an arbitrary point of the threefold space, are represented, in the fivefold space, by the conic in which a certain plane, lying on the quadric Ω, meets the quadric Ω'. Similarly, the lines of the quadratic complex which lie in an arbitrary plane, are represented by points, of a conic on Ω', lying in a plane entirely on Ω, this plane being of the opposite system to the former. With a very slight change from the notation used immediately above (p. 203), we may, as before, (p. 46, above) represent a plane lying entirely on Ω by three equations

$$ix_4 = l_1 x_1 + m_1 x_2 + n_1 x_3, \quad ix_5 = l_2 x_1 + m_2 x_2 + n_2 x_3, \quad ix_6 = l_3 x_1 + m_3 x_2 + n_3 x_3,$$

where $l_1^2 + m_1^2 + n_1^2 = 1$, etc. The difference of notation is immaterial in equations in which l_1, m_1, etc. enter homogeneously. Then, the conic in which this plane meets the quadric Ω' corresponds to the equation

$$k_1 x_1^2 + k_2 x_2^2 + k_3 x_3^2 - k_4(l_1 x_1 + m_1 x_2 + n_1 x_3)^2 - k_5(l_2 x_1 + m_2 x_2 + n_2 x_3)^2$$
$$- k_6(l_3 x_1 + m_3 x_2 + n_3 x_3)^2 = 0.$$

The left side of this breaks up into two linear factors provided x_1, x_2, x_3 can be found to satisfy the three equations

$$-k_1 x_1 + k_4 l_1 \ p + k_5 l_2 \ q + k_6 l_3 \ r = 0,$$
$$-k_2 x_2 + k_4 m_1 p + k_5 m_2 q + k_6 m_3 r = 0,$$
$$-k_3 x_3 + k_4 n_1 \ p + k_5 n_2 \ q + k_6 n_3 r = 0,$$

where p, q, r denote, respectively, the three linear forms

$$l_1 x_1 + m_1 x_2 + n_1 x_3, \text{ etc.}$$

In virtue of the three such identities as $x_1 = l_1 p + l_2 q + l_3 r$, which follow from the definition of p, q, r, these three equations are the same as

$$(14)\, l_1 \ p + (15)\, l_2 \ \ q + (16)\, l_3 \ \ r = 0,$$
$$(24)\, m_1 p + (25)\, m_2 q + (26)\, m_3 r = 0,$$
$$(34)\, n_1 \ p + (35)\, n_2 \ \ q + (36)\, n_3 \ r = 0,$$

where (rs) is put for $k_r - k_s$. The condition for the consistence of the three equations is, therefore,

$$\begin{vmatrix} (14)\, l_1, & (24)\, m_1, & (34)\, n_1 \\ (15)\, l_2, & (25)\, m_2, & (35)\, n_2 \\ (16)\, l_3, & (26)\, m_3, & (36)\, n_3 \end{vmatrix} = 0.$$

In evaluating this determinant, we denote the product $(r4)(s5)(t6)$, where r, s, t are the numbers 1, 2, 3, in some order, simply by rst. The expansion is, then, first,

$$123. l_1 m_2 n_3 - 132. l_1 m_3 n_2 + 312. l_2 m_3 n_1 - 213. l_2 m_1 n_3$$
$$+ 231. l_3 m_1 n_2 - 321. l_3 m_2 n_1;$$

herein we substitute, from the properties of the orthogonal determinant $(l_1 m_2 n_3)$,

$$l_2 m_3 n_1 = m_3 (l_1 n_2 + \epsilon m_3), \ l_2 m_1 n_3 = n_3 (l_1 m_2 - \epsilon n_3),$$
$$l_3 m_1 n_2 = n_2 (l_1 m_3 + \epsilon n_2), \ l_3 m_2 n_1 = m_2 (l_1 n_3 - \epsilon m_2),$$

where ϵ is ± 1 according as the determinant $(l_1 m_2 n_3)$ is ± 1. Then the expansion is

$$(123 - 213 - 321)\, l_1 m_2 n_3 + (-132 + 312 + 231)\, l_1 m_3 n_2$$
$$+ \epsilon (312. m_3^2 + 213. n_3^2 + 231. n_2^2 + 321. m_2^2),$$

wherein the coefficients of $l_1 m_2 n_3$ and $l_1 m_3 n_2$ are easily verified to be equal, but of opposite sign. Thus, with the omission of a common factor ϵ, after replacing $m_2 n_3 - m_3 n_2$ by ϵl_1, the whole is

$$321. m_2^2 + 213. n_3^2 + 231. n_2^2 + 312. m_3^2 + (123 - 213 - 321)\, l_1^2.$$

Now consider the coefficients 321, 213, etc., occurring here. Each is a product of three differences of the quantities k_1, k_2, ..., k_6, and all these enter once into each coefficient. Thus, the ratios of these coefficients are unaltered if we replace k_1, k_2, ..., k_6 each by the same arbitrary linear function of itself, say k_r by

$$(p k_r + q)/(m k_r + n).$$

In particular, we may replace k_4, k_5, k_6, respectively, by 1, 0, ∞; let the values of k_1, k_2, k_3 then arising be named, respectively,

a, b, c. After this, put a', b', c' for $1 - a, 1 - b, 1 - c$, respectively, and $\xi, \eta, \zeta, \xi', \eta', \zeta'$ for $bc', ca', ab', b'c, c'a, a'b$; so that

$$\xi + \eta + \zeta = \xi' + \eta' + \zeta', \quad \xi\eta\zeta = \xi'\eta'\zeta',$$

and the reverse equations are

$$a(\xi - \xi') = \eta' - \zeta, \quad b(\eta - \eta') = \zeta' - \xi, \quad c(\zeta - \zeta') = \xi' - \eta.$$

Then the expression found becomes

$$\xi m_2{}^2 + \zeta n_3{}^2 + \xi' n_2{}^2 + \eta' m_3{}^2 + (\zeta' - \zeta - \xi) l_1{}^2.$$

The vanishing of this expression is the condition that the plane, $(l_1, m_1, \ldots)$, lying upon Ω, whether it be of the first or second kind of planes of Ω, should meet Ω' in two lines.

We may interpret this result in two ways, in four dimensions or in three; either by regarding the ratios of the squares of l_1, m_2, n_3, m_3, n_2 as coordinates in fourfold space, as above; or, by regarding the plane of Ω as representing a point, (ξ, η, ζ, τ), or a plane, (u, v, w, p), of threefold space.

In the former way, writing, as before, X, Y, Z, X', Y', W, T, respectively, for $m_2{}^2, n_2{}^2, -l_1{}^2, n_3{}^2, m_3{}^2, -l_2{}^2, -l_3{}^2$, the equation becomes

$$-aW + bX + cY + bcT - caY' - abX' = 0.$$

This, however (p. 159, above), is the equation of the tangent solid of the locus expressed by $(XY')^{\frac{1}{2}} + (X'Y)^{\frac{1}{2}} + (WT)^{\frac{1}{2}} = 0$, at a point determined by the values of a, b, c; and this locus is no other than the locus, Σ, otherwise expressed by

$$(XX')^{\frac{1}{2}} + (YY')^{\frac{1}{2}} + (ZZ')^{\frac{1}{2}} = 0.$$

(Cf. the scheme given above, p. 210.) So that we have verified the statement made (p. 211), that the transformation from the fourfold to the threefold space, changes the section of Σ into a Kummer surface. The values of the parameters a, b, c, in terms of the coordinates, at the point of contact of the tangent solid, which are given in general (p. 159, above) by

$$-2aX'Y' = WT - XY' - X'Y, \quad 2bTX' = XY' - WT - X'Y,$$
$$2cTY' = X'Y - XY' - WT,$$

are easily found to be $a = -\mu\nu$, $b = -\nu\lambda$, $c = -\lambda\mu$, where

$$\lambda = l_2/l_3, \quad \mu = m_2/m_3, \quad \nu = n_2/n_3.$$

Thus we see that the coefficients, in the equation for the quadratic complex, $k_4, k_5, k_6, k_1, k_2, k_3$, if regarded as parameters of points of a conic, lead to a range related to any one of the sixteen ranges previously considered. (Cf. p. 127, above, and Ex. 2, p. 206, above.)

In the latter way, replacing ξ, η, ζ, τ by x, y, z, t, and putting $l_1^2 = (t^2 + x^2 - y^2 - z^2)^2$, etc., the equation found represents a quartic surface in x, y, z, t. On expansion this is of the form

$$2(2\eta - \zeta')(t^2x^2 + y^2z^2) + 2(2\xi - \zeta')(t^2y^2 + z^2x^2) + 2(2\zeta - \zeta')(t^2z^2 + x^2y^2)$$
$$+ 8(\eta' - \xi')xyzt + \zeta'(x^4 + y^4 + z^4 + t^4) = 0.$$

The four ratios of the coefficients in this equation are functions of only three, and can be expressed in terms of the ratios of four quantities, p, q, r, s, so that the equation becomes

$$x^4 + y^4 + z^4 + t^4 + U\,(t^2x^2 + y^2z^2) + V\,(t^2y^2 + z^2x^2)$$
$$+ W\,(t^2z^2 + x^2y^2) - 2Mxyzt = 0,$$

with

$$U = (s^2 + p^2 - q^2 - r^2)/(qr - sp), \quad V = (s^2 + q^2 - r^2 - p^2)/(rp - sq),$$
$$W = (s^2 + r^2 - p^2 - q^2)/(pq - sr),$$

and

$$M = (pqrs)^{\frac{1}{2}} \, [\Pi\,(s + \epsilon_1 p + \epsilon_2 q + \epsilon_1\epsilon_2 r)]/(qr - sp)(rp - sq)(pq - sr),$$

the product indicated by the symbol Π being for the four factors in which $\epsilon_1 = \pm 1$, $\epsilon_2 = \pm 1$. To see this, remark that the ratios of p, q, r, s are determined when the ratios are determined of the four

$$P, = p - q - r + s, \quad Q, = -p + q - r + s, \quad R, = -p - q + r + s,$$
$$S, = p + q + r + s;$$

let these be determined so that

$$QR/PS = \eta/(\eta - \zeta'), \quad RP/QS = \xi/(\xi - \zeta'), \quad PQ/RS = \zeta/(\zeta - \zeta');$$

then the coefficients U, V, W are rightly expressed. But these equations, it may be verified, lead to

$$S^{-4}\,\Pi\,(S + \epsilon_1 P + \epsilon_2 Q + \epsilon_1\epsilon_2 R) = (\zeta')^4(\eta' - \xi')^2/(\eta - \zeta')^2(\xi - \zeta')^2(\zeta - \zeta')^2,$$

of which the left side is $256S^{-4}pqrs$; thus the coefficient M is also expressible as stated.

But the equation, with coefficients expressed in terms of p, q, r, s, represents a surface which can easily be seen to have a double point at the point for which x, y, z, t have the ratios of $p^{\frac{1}{2}}, q^{\frac{1}{2}}, r^{\frac{1}{2}}, s^{\frac{1}{2}}$; and, therefore, also, at the fifteen other points obtainable from this by, (i), change of sign of two of $p^{\frac{1}{2}}, q^{\frac{1}{2}}, r^{\frac{1}{2}}$, (ii), interchange of two of these accompanied by interchange of $s^{\frac{1}{2}}$ with the remaining one.

The history of this equation (first found by Rosenhain, or Göpel, 1846, 1847, in algebraical investigation of theta functions of two variables, arising from the study of the irrationality involved in the square root of a sextic polynomial) is very interesting, as shewing how gradual the synthesis of a simple conception may be. The reader may compare the writer's *Abel's Theorem* (Cambridge, 1897), pp. 467 and 338—340.

There is, we have seen (cf. p. 46, above), a precisely identical relation in u, v, w, p, representing the quartic envelope of singular *planes* of the quadratic complex. It may be verified by direct algebra that this consists of the tangent planes of the quartic locus; but a geometrical proof occurs below (p. 222). It is easy to see at once that, through any point, (x', y', z', t'), of the locus, there pass six planes of the envelope. For instance, that the plane

$$xt' - yz' + zy' - tx' = 0$$

is a plane of the envelope, follows, if the equation of the envelope is satisfied by $(u, v, w, p) = (t', -z', y', -x')$; and this is true, because the equation of the locus is satisfied by $(x, y, z, t) = (x', y', z', t')$. This plane is the focal plane of (x', y', z', t') in the linear complex whose equation is $l' + l = 0$. The same argument applies to the focal planes in the complexes $l' - l = 0$, $m' \pm m = 0$, $n' \pm n = 0$. When the envelope has been shewn to consist of the tangent planes of the locus, it will thus follow that the six focal planes of any point of the locus are tangent planes of the locus; dually, the six foci of any tangent plane of the locus are points of the locus.

Ex. 1. If, instead of $x_1, x_2, \ldots$, other coordinates, $x_1', x_2', \ldots$, be taken in the fivefold space, such, however, that Ω is expressed by the same form, $x_1'^2 + x_2'^2 + \ldots = 0$, the equations of a definite plane, on Ω, $ix_4 = l_1 x_1 + m_1 x_2 + n_1 x_3$, etc., may become

$$ix_4' = l_1' x_1' + m_1' x_2' + n_1' x_3', \text{ etc.}$$

The point, (x', y', z', t'), of threefold space, determined by such equations as $x'/t' = (n_1' + l_3')/(l_2' - m_1')$, will not be the same as (x, y, z, t), determined by $x/t = (n_1 + l_3)/(l_2 - m_1)$, etc. Prove, in particular, that, if $x_2' = x_3$, $x_3' = x_2$, the other coordinates being unaltered, then $x' = y + z$, $y' = -t - x$, $z' = t - x$, $t' = y - z$.

Ex. 2. A particular case of the above equation arises when the pairs of coefficients (k_1, k_4), (k_2, k_5), (k_3, k_6), interpreted as parameters of points of a range, are those belonging to an involution. Then, as is easy to see, we have $\xi' - \eta' = 0$, and the equation of the Kummer surface, in x, y, z, t, is without the term in $xyzt$. We can, correspondingly, suppose $s = 0$, and the sixteen double points consist of four such sets as

$$(0, r^{\frac{1}{2}}, -q^{\frac{1}{2}}, p^{\frac{1}{2}}), \quad (-r^{\frac{1}{2}}, 0, p^{\frac{1}{2}}, q^{\frac{1}{2}}), \quad (q^{\frac{1}{2}}, -p^{\frac{1}{2}}, 0, r^{\frac{1}{2}}), \quad (p^{\frac{1}{2}}, q^{\frac{1}{2}}, r^{\frac{1}{2}}, 0).$$

The surface then meets each of the planes $x = 0$, $y = 0$, $z = 0$, $t = 0$ in two conics, which intersect in four double points lying in this plane. This surface is the *Tetrahedroid*, or, essentially, it is the *Wave Surface*, of Fresnel.

Of the sixteen double tangent planes of this surface, each touching it along a conic, there are four through each of the four fundamental points, $(1, 0, 0, 0)$, $\ldots$, $(0, 0, 0, 1)$. Those through the last point, for example, contain the four common tangent lines of the two conics in which the surface is met by the fundamental plane $t = 0$. The eight points of contact, with these two conics, of these four common tangents, lie, with the four conics in the double tangent planes through $(0, 0, 0, 1)$, upon a quadric surface, in regard to which the four fundamental points are a self-polar tetrad. The four double tangent planes through $(0, 0, 0, 1)$ meet, in pairs, in six lines; each of these lines contains two of the sixteen double points of the surface, harmonic conjugates of one another in regard to the plane $t = 0$, and the point $(0, 0, 0, 1)$. The six double points of each of these planes are, then, evidently, in involution on the conic which contains them. Cf. pp. 127, 131.

In fact, as the equation we have given at once shews, the equation of the surface may be formed from that of a general quadric touching the four planes $x = 0$, $y = 0$, $z = 0$, $t = 0$, by replacing, therein, x, y, z, t, respectively, by x^2, y^2, z^2, t^2. The four quadric surfaces, such as that described, each containing twelve of the double points, are obtained, by the same substitution, from the polar planes of the four fundamental points in regard to this quadric surface.

The quadratic complex, of which this particular quartic surface is the singular surface, may be taken to be that corresponding to the equations, in the fivefold space, $x_1^2 + \ldots + x_6^2 = 0$, and

$$k_1 (x_1^2 - x_4^2) + k_2 (x_2^2 - x_5^2) + k_3 (x_3^2 - x_6^2) = 0 ;$$

this follows from what is said as to the involution, if we recall the forms of the coefficients, in the Kummer surface, in terms of $k_1, \ldots, k_6$. The equation $k_1 (x_1^2 - x_4^2) + \ldots = 0$, it is easily seen, is that for the complex of lines, in the threefold space, which meet two quadric surfaces in two harmonically conjugate pairs of points.

Kummer's surface in the geometry of space of five dimensions. It has been seen that Kummer's quartic surface is the locus of points for which the cone of lines of a quadratic complex breaks up into two planes. If we represent the lines by points of a quadric, Ω, in fivefold space, and the quadratic complex by the points common to Ω and a further quadric, Ω', the representation of Kummer's surface is by an aggregate of planes lying on Ω, whose conic intersection with Ω' breaks up into two lines; as we have already seen. We may follow the theory through from this point of view, without explicit reference to the lines in threefold space. Thereby we obtain a clearer view of some of the results. But we also, then, make an interesting comparison with the theory of the locus, Γ, the intersection of two quadrics in *fourfold* space, studied in the preceding chapter.

We denote the coordinates in the fivefold space by x, y, z, u, v, w.

We suppose the equations of Ω, Ω', respectively, to be
$$x^2 + \ldots + w^2 = 0, \quad k_1 x^2 + \ldots + k_6 w^2 = 0.$$
We may also denote by Ω'' the quadric whose equation is
$$k_1^2 x^2 + \ldots + k_6^2 w^2 = 0.$$

If a plane, α, lying entirely on Ω, meet Ω' in a conic which breaks up into two lines, it is convenient to have a notation, and a name, for the point of intersection of these two lines. It represents what, in the line theory in the threefold space, is called a *singular line* of the quadratic complex. We shall denote the point by P, and call it the *focus* of the plane α, in regard to Ω'. It is shewn that P lies on Ω'', as well as on Ω and Ω'; and, further, in consequence, that the tangent fourfold of Ω', at P, touches Ω, at a point, Q, lying on the plane α. If $(x_0, y_0, \ldots, w_0)$ be the coordinates of P, the line PQ is the locus of points whose coordinates are of the forms
$$(k_1 + \sigma) x_0, \ldots, (k_6 + \sigma) w_0,$$
for different values of σ. This line is then, in part, the generalisation of a line considered, for two quadrics in threefold space (Vol. III, p. 92) and, for two quadrics in fourfold space (p. 183, above). This line is of great importance here, and we may call it the *focal line* of the plane α. If the plane α be of the first kind, of planes lying on Ω, we shew that the plane of Ω of the second kind, drawn through the focal line, is likewise a *singular plane*, meeting Ω' in two lines. This will be seen to give the result, in the threefold space, that the envelope of the singular planes of the quadratic complex, in the threefold space, is the locus of the singular points. The intersections of such focal lines with the fundamental fourfolds $x = 0, \ldots, w = 0$, suggest the consideration of six quadratic *congruences*, associated with the Kummer surface, in a very natural way. Further, it is shewn that the plane, α, which meets Ω' in two lines, equally meets every quadric of equation $k_1' x^2 + \ldots + k_6' w^2 = 0$ in two lines, if k_r' be any (fractional) linear function of k_r, independent of r; and that the focal line is the locus of the foci of the plane in regard to these quadrics.

We shall speak of a fourfold, of the fivefold space, as a *prime*, suggesting this, in general, as a name for an $(n-1)$-fold planar space of a fundamental n-fold space, (for which the word *form* has been used).

Consider a plane, α, lying on Ω, usually meeting Ω' in a proper conic. Let O be any point of this plane; thus O lies on Ω, but is supposed, at first, not to lie on Ω'. Let T be the tangent prime of Ω at the point O, and Π be the polar prime of O in regard to Ω'. The prime T meets Ω in a point-cone, of vertex O, which contains α as one of its planes; this cone meets the solid (T, Π) in a quadric surface, which we denote by ω. The line in which the plane α meets the prime Π is one generator of the quadric surface ω; this line we may denote by l. The solid (T, Π) contains another quadric surface, in which this solid is met by the quadric Ω'; this quadric surface we denote by ω'.

The line l is the polar line of O in regard to the conic (α, Ω'); and meets Ω' in the two points of contact, with this conic, of the two tangent lines drawn from O. But, if the plane α meets Ω' in two lines, these two points of contact coincide. Conversely, if O does not lie on Ω', the intersections of l with Ω' can only coincide when α meets Ω' in two lines. The quartic curve of intersection of the two quadric surfaces, ω and ω', of the threefold space (T, Π), is touched, we know (Vol. III, p. 69), by four generators of ω, of either system of generators (a simple proof of this is by remarking that, on a quadric (∞^4), in fivefold space, the tangent primes of the quadric, at the points of a plane section, meet any other plane in the tangent lines of a conic; this has four tangents common with the conic section of the quadric with this other plane). Therefore, the necessary and sufficient condition that the conic (α, Ω') should be two lines, is that the generator l, or (α, Π), of the quadric surface ω, should be one of these four particular generators. Thus we see that, through a point, O, of Ω, not lying on Ω', there can be drawn four planes, of either of the two systems of planes of Ω, to meet Ω' in two lines. This is the representative of the fact, for a quadratic complex of lines in threefold space, that the locus of the singular points is of the fourth order, and the envelope of the singular planes is of the fourth class.

As was suggested, denote the intersection of the two lines of Ω', in such a plane, α, by P, and call it the focus of this plane in regard to Ω'. The tangent prime of Ω' at P contains these lines, and contains the plane α, which lies on Ω. This prime meets Ω in an (∞^3) quadric, lying in a fourfold space; such a quadric, it is easily proved, cannot contain a plane, unless the quadric be a point-cone with vertex on the plane. Thus the tangent prime of Ω' at P meets Ω in a point-cone, with vertex on the plane, α, of Ω. Therefore, this tangent prime touches Ω, at this vertex; we denote this vertex by Q. Conversely, if the tangent prime of Ω', at a point P, which lies both on Ω and Ω', touch Ω at a point Q, there can be drawn, through the line PQ, a single plane, lying on Ω, of either system of planes of Ω, to meet Ω' in two lines intersecting at P. For, the line QP, lying in the tangent prime of Ω at Q, meets Ω, there, in two coincident points, and meets Ω also at P; this line, therefore, lies on Ω. Thus there can be drawn, through this line, a definite plane of Ω of either system. Such a plane, being in the tangent prime of Ω at Q, that is, in the tangent prime of Ω' at P, meets Ω' in two lines which meet at P. It may be remarked too, considering the polar primes, in regard to Ω', of the ∞^2 points of the plane α, that, among these will be the tangent prime of Ω' at P. This contains the plane α, touching Ω at Q. The ∞^2 polar primes, by their intersection with α, thus give only ∞^1 lines, the lines of the plane passing through P. As suggested, we may call the line PQ the focal line of the plane α; it lies on Ω, and on the tangent prime of Ω' at P; but does not lie in Ω', in general.

With the notation explained above, the point P being $(x_0, ..., w_0)$, the condition that the tangent prime of Ω' at P, or

$$k_1 x x_0 + ... + k_6 w w_0 = 0,$$

should touch Ω, is $k_1^2 x_0^2 + ... + k_6^2 w_0^2 = 0$, the point of contact, Q, being $(k_1 x_0, ..., k_6 w_0)$. Thus P lies on a surface common to the three quadrics Ω, Ω', Ω''; and the general point of the focal line PQ has coordinates $(k_1 + \sigma) x_0, ..., (k_6 + \sigma) w_0$.

The tangent lines of a Kummer surface at a point. We have seen that in general there are four planes of Ω, of either system, passing through a point of Ω, which meet Ω' in two lines. We have also seen that such a plane, α, contains a focal line, through which there passes a plane of Ω, of the other system, which also meets Ω' in two lines. We consider, now, the four singular planes of Ω, of either system, passing through a general point of this focal line; we shew that these consist of the plane, α, spoken of, in which the line is the focal line, this counting doubly, together with two other planes. Let the point of the focal line be called S. We take the tangent prime, T, of Ω, at the point S, and the polar prime, Π, of S, in regard to Ω'. The tangent prime contains the plane α, and, therefore, contains the focus, P, of the plane; the tangent prime of Ω' at P is the tangent prime of Ω at a point, Q, of the focal line, and also contains the plane α. Thus the polar prime of S, in regard to Ω', likewise contains P. Next, we take the two quadric surfaces in which, respectively, Ω and Ω' meet the solid intersection of T and Π. These two surfaces have a common tangent plane at P. For, to express briefly by the symbols what can also be expressed in words, if the point S be of coordinates

$$(k_1 + \sigma) x_0, ..., (k_6 + \sigma) w_0,$$

where P is $(x_0, ..., w_0)$, the two primes T, Π are, respectively,

$$(k_1 x x_0) + \sigma (x x_0) = 0$$

and

$$(k_1^2 x x_0) + \sigma (k_1 x x_0) = 0,$$

and the tangent prime of Ω at P is $(x x_0) = 0$, where $(x x_0)$ denotes $x x_0 + ... + w w_0$, etc.; thus, the tangent plane of the former, ω, of the two quadric surfaces, at P, is given by

$$(x x_0) = 0, \quad (k_1 x x_0) = 0, \quad (k_1^2 x x_0) = 0.$$

Again, the tangent prime of Ω' at P is $(k_1 xx_0) = 0$; the tangent plane, at P, of the latter, ω', of the two quadric surfaces, is, therefore, given by the same three equations. This common tangent plane is that of the surface of intersection of the three quadrics Ω, Ω', Ω'', and is independent of the position of S on the focal line (the values $\sigma = 0$, $\sigma = \infty$ not being considered). It is, however, easily proved that, when two quadric surfaces, in threefold space, touch one another, the generating lines, of a particular system, of one of these surfaces, which meet the common curve in two coincident points, consist of the generator of this system, through the point of contact, counting doubly, together with two other generators. This proves what was said, for it was seen that the plane α gives a generator of the quadric ω. It was also seen that the plane, β, of Ω, of the other system from α, which passes through the focal line, meets Ω' in two lines; the argument shews that this also counts doubly, among the four planes of this system which can be drawn through any general point of the focal line.

If we recur to the quadratic complex of lines, in the threefold space, the points of the focal line, PQ, represent lines, in the threefold space, each of which has two coincident intersections with the singular (Kummer) surface, at the same point, this being the point represented by the plane α; that is, the points of the line PQ represent the tangent lines of the singular surface at a point. These lie in the tangent plane of this surface at this point, which is represented by the plane β through the focal line, in the fivefold space. As this, we have seen, is a singular plane, counting doubly, we see that, in the threefold space, *any tangent plane, of the surface which is the locus of singular points, is a plane of the aggregate of singular planes*; as was stated. The point P, of the focal line, whereat the tangent prime of Ω' is equally a tangent prime of Ω, is the point of intersection of the two lines in which the plane α meets Ω', and also of the two lines in which the plane β meets Ω'. Thus, in the threefold space, one of the tangent lines, at any point of the locus of singular points, is the intersection of the two planes into which the quadric cone of complex lines, through the point of contact, breaks up; and this tangent line also contains the two points through one of which passes every complex line in the tangent plane.

The inflexional directions at a point of the Kummer surface. It can be proved from the present point of view (and another proof occurs below) that, in the fivefold space, there are two points of the focal line, PQ, such that, of the four singular planes through the point, three coincide with the plane α. These two points represent the two inflexional tangents, in the threefold space of the quadratic complex of lines, at the point of the Kummer surface. It is interesting to prove this from the present point of view because, as will be seen (p. 231, below), we are then able to put the inflexional curves of the Kummer surface into correspondence with the lines of curvature of a Cyclide. We assume that, when, in threefold space, the curve of intersection of two quadric surfaces, which touch, has a cusp at the point of contact, then, of the generators, of one system, of one of these surfaces, which meet the other surface in two coincident points, three are coincident in the generator of that system which passes through the cusp. The reader may examine the case of the surfaces

$$tz = x^2 - y^2, \; tz = (a+1)\,x^2 + (b-1)\,y^2 + z\,(lx+my),$$

of which the common curve has a cusp at $(0, 0, 0, 1)$ if a or b be zero. We require then, by what has been said, that the curve of intersection of the two quadric surfaces (T, Π, Ω) and (T, Π, Ω'), lying in the solid (T, Π), in the fivefold space, should have a cusp at P, or $(x_0, y_0, \ldots, w_0)$, where T, Π are defined from a point, S, of the line PQ. Let U, U', U'' denote the tangent primes, respectively of the quadrics $\Omega, \Omega', \Omega''$, at the point P. Then the tangent prime, T, is $\sigma U + U'$, and the polar prime, Π, is $\sigma U' + U''$. The quadric cone, in the solid (T, Π), with vertex at P, which contains the curve of intersection $(T, \Pi, \Omega, \Omega')$, spoken of, may be given by T, Π and $\sigma\Omega + \Omega'$, this being the intersection of Π with the point-cone $(\sigma U + U', \sigma\Omega + \Omega')$. The two tangent lines of the curve, at P, are the intersections of this cone with U. These coincide if the plane (U, U', U'') touches the cone. This will be so if a point, $(x', y', \ldots)$, exists on the cone, such that the tangent prime of $\sigma\Omega + \Omega'$, at this point, contains P. Namely, with $(k_1 + \sigma)\,x_0 x' + \ldots = 0$, $k_1\,(k_1 + \sigma)\,x_0 x' + \ldots = 0$, $(k_1 + \sigma)\,x'^2 + \ldots = 0$, we must have $(k_1 + \sigma)\,x' x_0 + \ldots = 0$. The equations are satisfied by taking $(k_1 + \sigma)\,x' = x_0, \ldots, (k_6 + \sigma)\,w' = w_0$, provided σ be one of the roots of the equation $(k_1 + \sigma)^{-1}\,x_0^2 + \ldots + (k_6 + \sigma)^{-1}\,w_0^2 = 0$. In virtue of $x_0^2 + \ldots = 0$, $k_1 x_0^2 + \ldots = 0$, $k_1^2 x_0^2 + \ldots = 0$, this is a quadratic equation. If its roots be λ and μ, the two points of the focal line satisfying the conditions are those of coordinates $(k_1 + \lambda)\,x_0, \ldots, (k_6 + \lambda)\,w_0$ and $(k_1 + \mu)\,x_0, \ldots, (k_6 + \mu)\,w_0$.

The associated quadratic congruences. There are also, upon the focal line, six points of great importance, namely those where the line meets the fundamental primes given by $x = 0, \ldots,$ $w = 0$. Consider, in particular, that point for which $w = 0$, whose coordinates are $(k_1 - k_6)\,x_0, \ldots, (k_5 - k_6)\,v_0, 0$. This point we may denote by Q_6. A plane, lying on Ω, which meets Ω' in two lines, will likewise meet the quadric $(k_1 + \theta)\,x^2 + \ldots + (k_6 + \theta)\,w^2 = 0$, or, say, $\theta\Omega + \Omega'$, in two lines; and conversely. And the tangent prime of Ω, at the point of the focal line whose coordinates are

$(k_1 + \theta) x_0, \ldots, (k_6 + \theta) w_0$, will be the tangent prime of the quadric $\theta\Omega + \Omega'$ at the point P, or $(x_0, \ldots, w_0)$. In particular, for $\theta = -k_6$, the quadric $\Omega' - k_6\Omega$ is a point-cone, passing through the intersection of Ω and Ω'; this cone we may denote by K. The tangent prime of Ω at Q_6 is the tangent prime of the cone K at the point P, and, therefore, at all the points of the line joining P to the vertex of this cone. To obtain the planes of Ω meeting Ω' in two lines, which pass through Q_6, we consider the solid intersection of the tangent prime of Ω at Q_6, that is, the tangent prime of the point-cone K at P, with the polar prime of Q_6 in regard to this cone; this polar prime also contains the line joining P to the vertex of the cone K. Then we consider the quadric surfaces which are the intersections of this solid with Ω, and with the cone K. The latter of these quadric surfaces, say ω', consists, however, of two planes. For, the intersection of a quadric point-cone with a prime, Π, passing through its vertex, is a quadric point-cone in this prime, and the intersection of Π with a tangent prime of the former cone touches the latter cone (p. 121, above). Wherefore, the quadric surface ω' is the intersection of a point-cone, lying in the fourfold space Π, with a tangent solid of this. The intersection, we know, consists of two planes, both containing the line joining P to the vertex of this cone. But, the generators of one system, of a quadric ω, in the threefold space, which meet the intersection of this with a pair of planes, ω', in two coincident points, consist of the two generators of ω at the two points where this is met by the line of intersection of the two planes; and each of these counts doubly. Thus we see, for the figure in the space of five dimensions, that the planes of Ω, of the first system, meeting Ω' in two lines, which can be drawn through Q_6, consist of the plane α, containing the line Q_6P, counting doubly, together with another plane also counting doubly. It is clear, however, what this other plane is. For, if the line joining P to the vertex of the cone K meet Ω again in the point P', of coordinates $(x_0, \ldots, v_0, -w_0)$, the point P' is equally on the cone K, on the tangent prime of Ω at Q_6 (which touches the cone at all points of PP'); and P' is also on the polar prime of Q_6 in regard to this cone. The quadric surface ω thus passes through P', as do the two planes constituting the quadric ω'. Thus, the second singular plane, of the first kind, through Q_6, is the plane of Ω, of this kind, which passes through the line Q_6P'. And, there is a plane of the second kind through each of Q_6P and Q_6P', which is likewise a singular plane counting doubly.

In general, an arbitrary plane, lying on Ω, meets the surface expressed by the three equations

$$w = 0, \quad x^2 + \ldots + v^2 = 0, \quad (k_1 - k_6)^{-1} x^2 + \ldots + (k_5 - k_6)^{-1} v^2 = 0,$$

in two points. This surface, however, has the property that the tangent prime of Ω, at any point of this surface, touches the cone K, whose equation is $(k_1 - k_6) x^2 + \ldots + (k_5 - k_6) v^2 = 0$. Thus, the tangent primes of Ω, at the two points where the plane meets this surface, both of which contain this plane of Ω, are two tangent primes of the cone K; they are, therefore, the two tangent primes of K which can be drawn from the solid joining the given plane to the vertex of K. Take the case when, as above, the plane of Ω considered is a plane, α, which meets the cone K in two lines; then, as may be proved, the plane is such that the two tangent primes of the cone through the plane are coincident. We thus infer that the plane α meets the surface in two coincident points.

In the threefold space of the quadratic complex of lines we thus have the following results: Among the tangents at any general point of the singular (Kummer) surface, there are six which touch this surface again (these are the tangent lines from this point to the quartic curve in which the tangent plane cuts the surface). Each of these bitangents, beside belonging to the fundamental quadratic complex, belongs to one of six linear complexes, these complexes being such that any two are conjugate (or apolar). Of the lines of the quadratic congruence given by the quadratic complex and one of the linear complexes, there pass two through an arbitrary point; but each of the bitangents spoken of consists of two coincident lines. The singular surface thus appears as the *focal* surface, or *caustic* surface, of each of six quadratic congruences (cf. for instance, Darboux, *Théorie...des Surfaces*, 2me Partie, Livre iv). The natural form for the equation of the Kummer surface when regarded as the focal surface of a quadratic congruence is that given above, Ex. 4, p. 138, Chap. v. The theory, in the space of five dimensions, applies equally to both systems of planes of Ω; so that there are also results in the threefold space which are the dual of those we have stated.

The common singular points and planes of the six congruences. In the fivefold space, the surface given by

$$w = 0, \quad x^2 + \ldots + v^2 = 0, \quad (k_1 - k_6)^{-1} x^2 + \ldots + (k_5 - k_6)^{-1} v^2 = 0$$

is the intersection of two (∞^3) quadrics of the fourfold space $w = 0$. This surface has been studied in Chap. vi. It appeared that it contains sixteen lines. Through each of these lines, which lies on Ω, there passes a plane of Ω, of each of two systems. It follows, from what is said above, that each of these two planes lies in infinitely many tangent primes of the cone which we denoted by K; this also appears from a previous remark (p. 48, above), that the two planes of Ω through the line are the intersection of Ω with the polar solid of the line in regard to Ω; the two planes, therefore, lie on the tangent primes of Ω at all the points of the line, and these primes are all tangent primes of the cone K. Thus, there are sixteen planes of the first system of Ω, each of which is a singular plane of special character, having the property of containing, not two but, an infinite number of points of the surface given by

$$w = 0, \quad x^2 + \ldots + v^2 = 0, \quad (k_1 - k_6)^{-1} x^2 + \ldots + (k_5 - k_6)^{-1} v^2 = 0.$$

For greater clearness we may give the equations of these planes. Denote by $f(t)$ the sextic polynomial whose roots are $t = k_1, \ldots, k_6$, and by f_1 the value, for $x = k_1$, of the differential coefficient $\partial f / \partial t$, etc. We use the fact that, for any value of λ, the sum,

$$(k_1 + \lambda)^r / f_1 + \ldots + (k_6 + \lambda)^r / f_6,$$

vanishes, for $r = 0, 1, \ldots, 4$; it is so, in particular, for $\lambda = -k_6$. Denote one of the two square roots of f_r, that is, of $\partial f(k_r) / \partial k_r$, by g_r. When λ, μ vary, the point whose coordinates are

$$g_1^{-1} (k_1 + \lambda)(k_1 + \mu), \ldots, g_6^{-1} (k_6 + \lambda)(k_6 + \mu)$$

describes a plane. This plane meets the prime $w = 0$ in the line which joins the two points

$$[g_1^{-1}(k_1 - k_6)^2, \ldots, g_5^{-1}(k_5 - k_6)^2, 0], [g_1^{-1}(k_1 - k_6), \ldots, g_5^{-1}(k_5 - k_6), 0].$$

This line lies on the quadric $(k_1 - k_6)^{-1} x^2 + \ldots + (k_5 - k_6)^{-1} v^2 = 0$. The plane itself lies on the quadric Ω. The tangent primes of Ω, at the points of the line where the plane meets $w = 0$, are the tangent primes of the cone K, or $(k_1 - k_6) x^2 + \ldots + (k_5 - k_6) v^2 = 0$, at the points of the line joining the two points

$$[g_1^{-1}(k_1 - k_6), \ldots, g_5^{-1}(k_5 - k_6), 0], [g_1^{-1}, \ldots, g_5^{-1}, 0].$$

The line joining these two points lies on Ω and on the cone, so that it lies on Ω and Ω'; in fact it lies also on Ω'', the coordinates of every point of the line satisfying the equation

$$k_1^2 x^2 + \ldots + k_5^2 v^2 = 0.$$

By taking all possible combinations of the ambiguous signs in $g_1, \ldots, g_6$, the plane given by the points

$$[g_1^{-1}(k_1 + \lambda)(k_1 + \mu), \ldots, g_6^{-1}(k_6 + \lambda)(k_6 + \mu)]$$

becomes, in turn, thirty-two different planes, sixteen of one system of planes of Ω, and sixteen of the other. If we denote by (pqr) the product $(k_q - k_r)(k_r - k_p)(k_p - k_q)$, the equations of the plane can be shewn to be the three

$$(123) g_m x_m = (23m) g_1 x + (31m) g_2 y + (12m) g_3 z,$$

where $m = 4, 5, 6$ and $x_m = u, v, w$. The ratios of the coefficients of the coordinates in these equations are such as are unaffected by any, the same, fractional linear transformation of $k_1, k_2, \ldots, k_6$.

In the threefold space of the quadratic complex of lines, for any one of the six quadratic congruences which have been referred to, there are, therefore, sixteen points (and sixteen planes) through which there pass (in which there lie), not two lines but, a pencil of lines, of the congruence; and these sixteen elements are the same for all the six congruences. These points and planes are the double elements of a Kummer surface. (Cf. Hudson, *Kummer's Quartic Surface*, p. 59.)

The confocal quadrics. It appears from what has been said (p. 214, above) that a plane of Ω, which meets Ω' in two lines, equally meets in two lines any quadric with the equation

$$K_1 x^2 + \ldots + K_6 w^2 = 0, \quad \text{where} \quad K_r = (ak_r + b)/(ck_r + d).$$

When $c = 0$, this is the same as $K_r = d^{-1}b + d^{-1}ak_r$; when c is not zero, it is the same as $K_r = A + B(k_r + \theta)^{-1}$, where $\theta = c^{-1}d$, and A, B are independent of the suffix r. It will, therefore, be sufficient to consider quadrics of the form

$$(k_1 + \theta)^{-1}x^2 + \ldots + (k_6 + \theta)^{-1}w^2 = 0.$$

Of such quadrics there are four which pass through an arbitrary general point of Ω; conversely, four arbitrary values of θ determine a set of thirty-two points of Ω. There are also four planes of Ω, through a general point of Ω, which meet Ω' in a pair of lines. We may easily see that these two facts are connected.

For, consider a plane of Ω, say of the *second* system, which meets Ω' in a pair of lines. Let $(x_0, y_0, \ldots, w_0)$ be the focus of this plane, in regard to Ω'. The quadric

$$(k_1 + \sigma)^{-1}x^2 + \ldots + (k_6 + \sigma)^{-1}w^2 = 0$$

contains the point of coordinates $(k_1 + \sigma)x_0, \ldots, (k_6 + \sigma)w_0 = 0$, lying on the focal line of the plane. The tangent prime of this quadric, at this point, is the tangent prime of Ω at the point $(x_0, \ldots, w_0)$, and, thus, contains the plane considered. The quadric, therefore, touches every line of the plane through the point of coordinates $(k_1 + \sigma)x_0, \ldots, (k_6 + \sigma)w_0$. Now, let $(x', \ldots, w')$ be any point of the plane, and $(k_1 + \sigma)^{-1}x^2 + \ldots + (k_6 + \sigma)^{-1}w^2 = 0$ be a quadric through this point. Then, the line, joining $(x', \ldots, w')$ to the point of coordinates $(k_1 + \sigma)x_0, \ldots, (k_6 + \sigma)w_0$, lies entirely on this quadric. Through this line, lying in a plane of Ω of the second system, can be drawn a plane of Ω of the first system; and, this plane, containing one line of the quadric $(k_1 + \sigma)^{-1}x^2 + \text{etc.} = 0$, contains another line of this quadric. Therefore, as has been seen, and will be proved again immediately, this plane is one of the planes of Ω, of the first system, passing through $(x', \ldots, w')$, which cut Ω' in two lines. Whence we see that the four singular planes, of the first system, which can be drawn through the point $(x', ..., w')$, meet the focal line, of any singular plane, of the second system, drawn through $(x', ..., w')$, in four points of coordinates

$$(k_1 + \sigma_p)\, x_0, ..., (k_6 + \sigma_p)\, w_0,$$

where σ_p has the four values belonging to the quadrics,

$$(k_1 + \sigma)^{-1} x^2 + \text{etc.} = 0,$$

which can be drawn through $(x', ..., w')$. This result includes also the fact, incidentally referred to above (p. 213), that, in the three-fold space, the points, in which a general line meets the Kummer surface, constitute a range which is related to the pencil of four tangent planes of the surface passing through the line. (Cf. also p. 240, below.)

But the quartic equation for θ,

$$(k_1 + \theta)^{-1} x'^2 + ... + (k_6 + \theta)^{-1} w'^2 = 0,$$

giving the four quadrics of the system which pass through the point $(x', ..., w')$, reduces to a quadratic when $(x', ..., w')$ is a point, P, or $(x_0, ..., w_0)$, for which, beside $x_0^2 + \text{etc.} = 0$, we have

$$k_1 x_0^2 + \text{etc.} = 0 \text{ and } k_1^2 x_0^2 + \text{etc.} = 0.$$

And, more generally, when, with such values of $(x_0, ..., w_0)$, the point $(x', ..., w')$ is such that $x' = (k_1 + \sigma)\, x_0, ..., w' = (k_6 + \sigma)\, w_0$, the quartic equation reduces to

$$(\theta - \sigma)^2 \left[(k_1 + \theta)^{-1} x_0^2 + ... + (k_6 + \theta)^{-1} w_0^2 \right] = 0,$$

as we see by writing $(k_1 + \sigma)^2 (k_1 + \theta)^{-1}$ in the form

$$k_1 - \theta + 2\sigma + (\theta - \sigma)^2 (k_1 + \theta)^{-1}, \text{ etc.}$$

The four roots of the quartic equation thus consist of the root σ, repeated, together with the roots of the quadratic

$$(k_1 + \theta)^{-1} x_0^2 + ... + (k_6 + \theta)^{-1} w_0^2 = 0.$$

If the roots of this quadratic, which are independent of σ, be denoted by λ and μ, it follows that the two quadrics,

$$(k_1 + \lambda)^{-1} x^2 + \text{etc.} = 0 \text{ and } (k_1 + \mu)^{-1} x^2 + \text{etc.} = 0,$$

contain the line PQ, joining $(x_0, ..., w_0)$ and $(k_1 x_0, ..., k_6 w_0)$, completely; and, further, that the other two, of the four quadrics, $(k_1 + \theta)^{-1} x^2 + ... = 0$, which pass through any point, S, of coordinates $(k_1 + \sigma)\, x_0, ..., (k_6 + \sigma)\, w_0$, lying on this line PQ, coincide with one another, both being the quadric $(k_1 + \sigma)^{-1} x^2 + \text{etc.} = 0$. The tangent prime of this quadric, at this point S, is the tangent prime of Ω at P, as remarked above; and this shews, from what has been said, that the plane, α, of the quadric Ω, of the first (or second) system, which passes through the line PS, meets the quadric $(k_1 + \sigma)^{-1} x^2 +$ etc. $= 0$ in two lines, these intersecting at S. Thus S is the focus of the plane in regard to this quadric, just as P is its focus in regard to Ω'. There are, however, two particular points of the line PS for which three of the quadrics,

$$(k_1 + \theta)^{-1} x^2 + \text{etc.} = 0,$$

passing through the point, coincide with one another; namely, those for which $\sigma = \lambda$ or $\sigma = \mu$. For each of these two points, of the four planes of Ω through the point which meet Ω', and, therefore, any of the quadrics $(k_1 + \theta)^{-1} x^2 +$ etc. $= 0$, in two lines, there are three which coincide with one another. (Cf. Vol. III, p. 92; and p. 185, Chap. VI, above.)

Singular planes through a more particular point. In the preceding we consider planes of Ω, meeting a quadric,

$$(k_1 + \theta)^{-1} x^2 + \text{etc.} = 0,$$

in two lines, which are drawn from a point of Ω. Consider now, briefly, planes of Ω meeting Ω' in two lines, which are drawn from a point, O, which lies on Ω' as well as on Ω. This we excluded above. In this case, any plane of Ω, drawn through O, meets Ω' in a conic passing through O; if this conic break up into two lines, one of these must pass through O. Such a line lies on Ω, and on Ω', and on the tangent primes of Ω and Ω' at O. Conversely, the solid which is the intersection of the tangent primes of Ω and of Ω' at O, is met, by Ω and Ω', in two conical sheets of lines, with vertex at O. These have, in general, four lines in common, through O. Through any one of these four lines there passes a definite plane of Ω, of the first system, (as also one of the second system); this plane meets Ω' in another line, beside this one, intersecting this, say, in the point P; this is the focus, in regard to Ω', of the plane of these two lines. We thus have, as before, four singular planes, of one system of planes of Ω, passing through O.

But, it may happen that the plane, α, of the first system, through one of the four generators common to the two conical sheets, contains another of these four generators. Then the plane of these two generators, which are lines of Ω' meeting in O, lies in the tangent prime of Ω' at O. This prime, then, containing a plane of Ω, is a tangent prime of Ω, at some point, say P', of the plane. Then O is the focus of the plane, in regard to Ω', and P coincides with O. The plane then counts doubly among the four singular planes which can be drawn from O, as passing through two of the four generators. Through the line OP', lying on Ω, can be drawn a plane, β, of the second system of planes of Ω; this plane is on the tangent prime of Ω at P', which is the tangent prime of Ω'

at *O*. Thus the plane β meets Ω' in two lines which intersect at
O; and these, lying on Ω and Ω', are the other two generators, of
the four originally proved to pass through *O*, beside the two which
lie in the plane α. Through each of these lines, in the plane β,

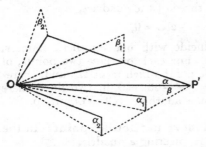

can be drawn a single plane, of
the first system of planes of Ω;
say these are α_1 and α_2. In the
same way, through each of the
two lines lying in the plane α,
can be drawn a singular plane,
of the second system of planes
of Ω; say these are β_1 and β_2.
Thus we see that if, of the four
lines originally drawn through
O, two lie in a plane of the first

system, counting doubly among the four singular planes of that
system which pass through *O*, then the other two of these lines lie
in a plane of the second system passing through *O*.

In this figure, the line *OP'* meets Ω' in two coincident points
at *O*. But it may happen that *OP'* lies entirely on Ω'. Then it is
one of the two lines of Ω' through *O* lying in the plane α; in
which case, the plane of the second system of Ω, say β_1, drawn
through this line, coincides with the plane β. But, also, *OP'* is
then one of the two lines of Ω' through *O* lying in the plane β,
and the plane of the first system, say α_1, drawn through this line,
coincides with the plane α. In this case, the four singular planes
of the first system, passing through *O*, consist of the plane α,
counting three times, together with the plane α_2; and, similarly,
the four singular planes of the second system, passing through *O*,
consist of the plane β, counting three times, together with the
plane β_2. In general, the solid, which is the intersection of the
tangent primes of Ω and Ω' at *O*, was met by Ω in a conical
sheet. In this particular case, this conical sheet consists of the two
planes α, β; and the line of intersection, *OP'*, of these planes, is
on the conical sheet of lines of Ω', through *O*, which lie in this
solid.

**The Inflexional (or Asymptotic) curves of the Kummer
Surface.** We have considered, in the fivefold space, the surface of
points $(x_0, \ldots, w_0)$, for which $x_0{}^2 + \ldots + w_0{}^2 = 0, k_1 x_0{}^2 + \ldots + k_6 w_0{}^2 = 0$,
$k_1{}^2 x_0{}^2 + \ldots + k_6{}^2 w_0{}^2 = 0$. The points of this surface may be represented, in terms of two parameters, λ and μ, by the equations
$x_0 = (k_1 + \lambda)^{\frac{1}{2}}(k_1 + \mu)^{\frac{1}{2}}/[f'(k_1)]^{\frac{1}{2}}$, etc., where $f(t) = (t - k_1) \ldots (t - k_6)$.
The parameters belonging to $(x_0, \ldots, w_0)$ are the two roots of the
equation $(k_1 + \theta)^{-1} x_0{}^2 + \ldots + (k_6 + \theta)^{-1} w_0{}^2 = 0$. Through the point
$(x_0, \ldots, w_0)$ there passes a line of which any point has coordinates

$(k_1 + \sigma) x_0, \ldots, (k_6 + \sigma) w_0$, this line lying entirely on the two quadrics $(k_1 + \lambda)^{-1} x^2 + \text{etc.} = 0$, $(k_1 + \mu)^{-1} x^2 + \text{etc.} = 0$. The two points of this line, for which $\sigma = \lambda$, $\sigma = \mu$, have been shewn to represent inflexional tangents of the Kummer surface (p. 223). For clearness, we call this line the focal line through $(x_0, \ldots, w_0)$. Consider now the curve, of points $(x_0, \ldots, w_0)$, of the surface given by $x_0^2 + \ldots = 0$, $k_1 x_0^2 + \ldots = 0$, $k_1^2 x_0^2 + \ldots = 0$, for which μ is constant; draw the focal lines through the points of this curve; and, upon the line drawn through the point (λ) of this curve, take the points of coordinates $(k_1 + \lambda) x_0, \ldots, (k_6 + \lambda) w_0$. Similarly, upon the focal line drawn through the point $(\lambda + d\lambda)$ of this curve, take the point of which the first coordinate is $(k_1 + \lambda + d\lambda)(x_0 + dx_0)$; with $dx_0 = \frac{1}{2} x_0 (k_1 + \lambda)^{-1} d\lambda$, this is the same as

$$(k_1 + \lambda) x_0 + 3 x_0 d\lambda/2 + \tfrac{1}{2} x_0 (k_1 + \lambda)^{-1} (d\lambda)^2.$$

Thus, neglecting the term in $d\lambda^2$, this point lies on the former focal line. From this we can infer that the focal lines, of the points of the curve in question, along which μ is constant, form a developable surface (cf. Vol. III, p. 132), whose cuspidal edge is the locus, as λ varies, of the point whose coordinates are such as $(k_1 + \lambda) x_0$, or $(k_1 + \lambda)^{\frac{3}{2}} (k_1 + \mu)^{\frac{1}{2}}/[f'(k_1)]^{\frac{1}{2}}$. The points of this cuspidal edge represent, in the original threefold space of the quadratic complex of lines, a sequence of inflexional tangents of the Kummer surface, so taken that, to any one of them there follows, in a limiting sense, another, lying in the same tangent plane of the surface. Thus, the cuspidal edge, in the fivefold space, represents the tangent lines of an inflexional curve of the Kummer surface.

Ex. Shew that this curve, in the threefold space, is of order 16, and of genus 17, when μ is general. There are, however, six particular *principal* inflexional curves, for which μ has one of the values $-k_1, \ldots, -k_6$; these are of order 8, and of genus 5.

It is of interest now to shew that the inflexional curves of the Kummer surface are in correspondence with the lines of curvature of a Cyclide. Consider the two planes of the quadric, Ω, in fivefold space, of the two systems, say α and β, which pass through the focal line joining the points $(x_0, \ldots, w_0)$, $(k_1 x_0, \ldots, k_6 w_0)$. The solid determined by these planes lies in the tangent prime of Ω at every point of the focal line. This solid meets any fixed prime in a plane, determined by the two lines in which the prime is met by the two planes. Or, this plane may be defined as the intersection of the three primes consisting of: the fixed prime; the tangent prime of Ω at the point, say Q_6, where the focal line meets this prime; and, the tangent prime of Ω at some other specified point of the focal line, say the point P, or $(x_0, \ldots, w_0)$, for which

$x_0{}^2 + \ldots = 0$, $k_1 x_0{}^2 + \ldots = 0$, $k_1{}^2 x_0{}^2 + \ldots = 0$. It will be sufficient here to take, for the fixed prime, the prime $w = 0$. As P varies, the point Q_6 describes a surface, in $w = 0$; as Q_6 is on Ω, this surface is on Ω; as Q_6 is of coordinates $(k_1 - k_6)\, x_0, \ldots, (k_5 - k_6)\, v_0, 0$, this surface lies also on $(k_1 - k_6)^{-1} x^2 + \ldots + (k_5 - k_6)^{-1} v^2 = 0$. Thus, the tangent plane of this surface at Q_6, which is given by $w = 0$, $(k_1 - k_6)\, xx_0 + \ldots + (k_5 - k_6)\, vv_0 = 0$, and $xx_0 + \ldots + vv_0 = 0$, is the plane spoken of, in which the prime $w = 0$ is met by the solid (α, β). We put $X_0 = (k_1 - k_6)\, x_0, \ldots, V_0 = (k_5 - k_6)\, v_0$, and denote the quadrics $(k_1 - k_6)^{-1} x^2 + \ldots = 0$, $x^2 + \ldots = 0$ by Φ and Φ', respectively. The general solid passing through the tangent plane of the surface (w, Φ, Φ'), at the point $(X_0, \ldots, V_0)$, namely, the tangent solid of the quadric $[w, (k_6 + \theta)\, \Phi + \Phi']$, whose equations are $w = 0$, with $(k_1 + \theta)\, xx_0 + \ldots + (k_5 + \theta)\, vv_0 = 0$, is the intersection of $w = 0$ with the tangent prime of Ω at the point of the focal line whose coordinates are $(k_1 + \theta)\, x_0, \ldots, (k_6 + \theta)\, w_0$. This solid meets the surface (w, Φ, Φ') in a quartic curve having a double point at $(X_0, \ldots, V_0)$, and there are two values of θ for which this double point is a cusp (p. 188, Chap. VI, above). These values of θ, as we see at once from what was said in Chap. VI, are given by $(k_1 - k_6)\, x_0{}^2 (k_1 + \theta)^{-1} + \ldots + (k_5 - k_6)\, v_0{}^2 (k_5 + \theta)^{-1} = 0$, that is, by $x_0{}^2 (k_1 + \theta)^{-1} + \ldots + v_0{}^2 (k_5 + \theta)^{-1} + w_0{}^2 (k_6 + \theta)^{-1} = 0$. We have seen that there are two points of the focal line with the property that, of the four planes of Ω drawn through each, to meet Ω' in two lines, three are coincident; the tangent prime of Ω at such a point, by its intersection with Ω, Ω' and the polar prime of the point in regard to Ω', gives a quartic curve with cusp at the point P (p. 223, above). It now appears that this same tangent prime, by its intersection with $w = 0$, Ω, and the quadric

$$(k_1 - k_6)^{-1} x^2 + \ldots + (k_5 - k_6)^{-1} v^2 = 0$$

which, together, represent, in fivefold space, a congruence of bitangents of the Kummer surface—gives also a quartic curve with a cusp at the point, Q_6, where the focal line meets $w = 0$. Either of the two particular points of the focal line, in the former case, represents an inflexional tangent of the Kummer surface; in the latter case, as we have seen (p. 186, above), the intersection of Ω with the corresponding tangent prime gives, in $w = 0$, a surface which projects, from any point of Ω in $w = 0$, into a sphere, of an arbitrary threefold space, determining a tangent line of a line of curvature of a Cyclide. We recall, now, the method, previously explained (p. 55, above), by which we pass from a line, of threefold space, to a sphere; this consists in taking the threefold intersection, with an arbitrary prime, of the tangent prime, of the quadric Ω in fivefold space, at the point of this which represents the line, and then projecting the surface in which this threefold meets Ω. We see then that, by this method, the inflexional line of the Kummer surface is transformed into the sphere having stationary contact with the Cyclide along a direction of curvature. Moreover the cuspidal edge spoken of above, whose points are given by $x = (k_1 + \lambda)^{\frac{3}{2}} (k_1 + \mu)^{\frac{1}{2}} / [f'(k_1)]^{\frac{1}{2}}$, etc., with λ variable but μ constant, which represents the inflexional lines of the Kummer surface along an inflexional curve, corresponds to the line of curvature on the Cyclide, represented, also with λ variable but μ constant, by the curve in $w = 0$ whose points are given by

$$x = (k_1 - k_6)(k_1 + \lambda)^{\frac{1}{2}} (k_1 + \mu)^{\frac{1}{2}} / [f'(k_1)]^{\frac{1}{2}}, \text{ etc.}$$

Further, in Lie's transformation, by which the correspondence is established, there are two lines corresponding to a sphere. The second point, in our figure, thus associated with the point

$$x = (k_1 + \lambda)^{\frac{3}{2}} (k_1 + \mu)^{\frac{1}{2}} / [f'(k_1)]^{\frac{1}{2}}, \text{ etc.,}$$

is the point with the same coordinates $x, \dots, v$, but with a change of sign of w. Thus (p. 224, above), there are two inflexional tangents of the same inflexional curve, of a Kummer surface, corresponding to the same sphere of stationary contact of the line of curvature of the Cyclide; and the points of contact of these are the two points where a bitangent of the Kummer surface touches the surface. The aggregate of these bitangents is a ruled surface in the threefold space, which is of order 8, represented, in the fivefold space, by the curve

$$w = 0, \ x^2 + \dots = 0, \ (k_1 - k_6)^{-1} x^2 + \dots = 0, \ (k_1 + \mu)^{-1} x^2 + \dots = 0.$$

The complete intersection of this ruled surface with the Kummer surface is the inflexional curve, counted twice, which is of order 16 (cf. Klein, *Ges. Math. Abh.*, 1921, pp. 91, 126, 147, etc. Also Hudson, *Kummer's Quartic Surface*, 1905, p. 61, etc.).

The theory is capable of generalisation. By Lie's method of transformation, the tangent lines of a surface (S), in space of three dimensions, correspond to the spheres touching another surface (S'). The tangents at a point, M, of (S), correspond to the spheres touching (S') at a point M'. When the point M describes an inflexional curve of (S), the corresponding point M' describes a line of curvature of (S'). (Lie, *Math. Annal.* v, 1871, pp. 145—256.)

Ex. In the correspondence between a Kummer surface and the Weddle surface which has been obtained above (p. 158), prove that the inflexional directions at any point of either surface correspond to conjugate directions on the other (cf. the writer's *Multiply-Periodic Functions*, 1907, p. 127). In particular, shew that, corresponding to one of the six principal inflexional curves of the

Kummer surface, there is a curve upon the Weddle surface which is of order 7, being the intersection of two cubic curves touching along a generator (*loc. cit.*, p. 322. Also Humbert, *Journ. École Polytechn.*, LXIV, 1894, p. 123).

The rationality of the Quadratic Complex, and of the Quadratic Congruence. We pass now to various theorems, mainly of line geometry, which seem called for by a comparison of some of the results which have been given in this Volume.

The representation, in space of five dimensions, of a quadratic complex of lines of threefold space, is by the (∞^3) locus which is common to two quadrics of this fivefold space. This locus is, in fact, obtainable by considering, in a threefold space, the cubic surfaces which can be drawn through a certain quintic curve. For, consider, in space of five dimensions, any two quadrics of general character. Let U be any point common to both. The lines, through U, lying on both quadrics, lie on the solid intersection of the tangent primes at U, and, therefore, on two conical sheets. We can thus draw four lines through U lying on both quadrics. Let V be any point of one of these lines. The line lies on the tangent primes of both quadrics at V, as well as at U. Denote the tangent primes of the first quadric at U and V, respectively, by $x = 0$, $y = 0$; and the tangent primes of the second quadric, at these two points respectively, by $z = 0$, $t = 0$; also, let U, V, beside the zero coordinates x, y, z, t, have the respective coordinates $u = 1$, $v = 0$ and $u = 0$, $v = 1$. Then the equations of the two quadrics will be, respectively,

$$\phi + ux + vy = 0, \quad \psi + uz + vt = 0,$$

where ϕ, ψ are homogeneous quadratic polynomials in x, y, z, t. These equations lead to $u(xt - yz) = \psi y - \phi t$, $v(xt - yz) = \phi z - \psi x$. Hence, in terms of parameters ξ, η, ζ, τ, if Φ, Ψ denote the same polynomials in these that ϕ, ψ are in x, y, z, t, the coordinates, x, y, z, t, u, v, of any point of the ∞^3 locus common to the two quadrics, are in the ratios of

$$\xi(\xi\tau - \eta\zeta), \ \eta(\xi\tau - \eta\zeta), \ \zeta(\xi\tau - \eta\zeta), \ \tau(\xi\tau - \eta\zeta), \ \Psi\eta - \Phi\tau, \ \Phi\zeta - \Psi\xi.$$

Regarding ξ, η, ζ, τ as coordinates in a threefold space, each of these six functions, equated to zero, represents a cubic surface. Moreover, the quintic curve which is the intersection, other than the line $\eta = 0$, $\tau = 0$, of the quadric surface $\xi\tau - \eta\zeta = 0$ with the cubic surface $\Psi\eta - \Phi\tau = 0$, is the same as the quintic curve which is the intersection, other than the line $\zeta = 0$, $\xi = 0$, of this quadric, $\xi\tau - \eta\zeta = 0$, with the cubic surface $\Phi\zeta - \Psi\xi = 0$. The two cubic surfaces have in common, beside this quintic curve, the quartic curve of intersection of the two quadric surfaces $\Phi = 0$, $\Psi = 0$.

The quintic curve, common to a quadric surface and a cubic surface having a line in common, meets an arbitrary plane drawn through this line, in the two points common to the residual intersections of this plane with the two surfaces—which are, a line and a conic. Thus the quintic curve meets the common line of the two surfaces in three points. It, therefore, meets every generator of the quadric surface, of one system, in three points, and the other generators each in two points, and its genus, p, is given by $p = (3-1)(2-1), = 2$ (see, for example, *Proc. Lond. Math. Soc.*, xi, 1912, p. 286). Conversely, for a surface, of order n, to contain the quintic curve of genus p, the number of linear conditions, for the coefficients in the equation of this surface, when n is large enough, is $5n - p + 1$. For $n = 3$ and $p = 2$, this is 14. This gives $20 - 14$, or six, as the number of linearly independent cubic surfaces containing the curve. The ratios of the polynomials expressing the equations of these surfaces can be used as coordinates in a space of five dimensions.

Interpreted in the original space of five dimensions, the process is the simple one of projecting the original ∞^3 locus, by means of planes passing through a line of the locus, given above by

$$x = y = z = t = 0,$$

upon an arbitrary threefold space, say Π. Such a plane is determined by one point of the locus, and meets Π in a point. Conversely, such a plane, determined by any point of Π, meets each of the given quadrics in another line, beside the common line of these through which the plane is drawn; the intersection of these residual lines is a point of the locus, corresponding to the point taken in Π. An arbitrary prime of the fivefold space, say

$$Ax + By + Cz + Dt + Pu + Qv = 0,$$

meets the locus in a quartic surface; this is projected from the line $x = y = z = t = 0$ by the line-cone given by

$$Ax(xt - yz) + \ldots + Dt(xt - yz) + P(\psi y - \phi t) + Q(\phi z - \psi x) = 0,$$

which is of the third order.

This rationalisation of the locus is given by Klein, *Ges. Math. Abhandl.*, i, p. 89, as due to Noether (1869). By combining the two equations of the locus, we may evidently regard one of them as having the form

$$z't' + ux' + vy' = 0,$$

usual for the relation connecting the coordinates of a line in threefold space. The intersection of two (∞^{n-1}) quadrics, in n-fold space, may be represented by the points of an $(n-2)$-fold, in a similar way, by planes through a line common to the two quadrics (the quadrics not being general when $n = 3$). The aggregate of (planar) p-folds existing on the (∞^{n-2}) locus of intersection of two (∞^{n-1}) quadrics, in n-fold space, can be shewn to be of dimension $(p+1)(n-2p-2)$. In particular, when $n = 2p+2$, the number of such p-folds is 2^n, in the most general case; for example, the number of lines on the quartic surface so obtained in fourfold space is sixteen.

We pass now to the case of a Quadratic *Congruence* of lines, defined as the aggregate of the lines common to a quadratic complex and a linear complex. We suppose this to be represented, in the space of five dimensions, by the surface which is the intersection of the preceding ∞^3 locus with a prime of this space. This surface, being the intersection of two quadrics of a fourfold (planar) space, is the same as that studied in Chap. VI. It was shewn to be representable rationally by two parameters (p. 165, above). Any one of the sixteen lines, which were shewn to lie upon it, may be taken to be the line UV of the preceding discussion. Then the prime, whose intersection, with the ∞^3 locus above, determines the congruence, as it contains all the points representing the congruence, and therefore contains this line UV, will have an equation of the form $Ax + By + Cz + Dt = 0$. Now, the intersection of a plane, in a threefold space, with a cubic surface passing through a quintic curve, is a cubic curve passing through five fixed points. Thus we see that the line coordinates of a line of a quadratic congruence, in threefold space, are proportional to homogeneous cubic polynomials, in three coordinates, which, equated to zero, represent plane cubic curves passing through five given points. This has been verified in several cases previously examined (pp. 137, 141, 165). Conversely, take five arbitrary points in a plane, and consider what are the relations connecting the polynomials which, equated to zero, represent cubic curves through these five points. If we take six such polynomials, as there are only five linearly independent cubic curves through the five points, these six polynomials are connected by a linear relation. Further, the five ratios of these polynomials, being dependent upon the two ratios of the coordinates in the plane, must be connected by two other algebraic relations. That these may be regarded as arising from two *quadratic* equations connecting the six polynomials, can be made clear by considering six particular polynomials, which are connected by two quadratic relations; for any other six polynomials are expressible linearly in terms of these.

Two ways of obtaining the necessary relations have presented themselves in foregoing work. If, referred to three of the five points, the conic containing these five points be $S=0$, and the line joining the other two points be $P=0$, where $S=\eta\zeta+\zeta\xi+\xi\eta$, and $P=a\xi+b\eta+c\zeta$, then six cubic polynomials which vanish at the five points are x, y, z, x', y', z', given by

$$x=\xi S, \quad y=\eta S, \quad z=\zeta S, \quad x'=(q-r)\eta\zeta P, \quad y'=(r-p)\zeta\xi P, \quad z'=(p-q)\xi\eta P,$$

in which p, q, r are quite arbitrary constants. These are connected by the relations

$$xx' + yy' + zz' = 0, \quad pxx' + qyy' + rzz' = 0,$$
$$-ax - by - cz + (q-r)^{-1}x' + (r-p)^{-1}y' + (p-q)^{-1}z' = 0.$$

Or again, if, referred to three of the points, the other two points be $(1, 1, 1)$

and (a, b, c), then, also with three arbitrary constants, p, q, r, six cubic polynomials vanishing at the five points are given by

$$x = (\eta c - \zeta b)(\zeta - \xi)(\xi - \eta), \quad y = (\zeta a - \xi c)(\xi - \eta)(\eta - \zeta), \quad z = (\xi b - \eta a)(\eta - \zeta)(\zeta - \xi),$$
$$x' = (q - r)(\zeta a - \xi c)(\xi b - \eta a)(\eta - \zeta), \quad y' = (r - p)(\xi b - \eta a)(\eta c - \zeta b)(\zeta - \xi),$$
$$z' = (p - q)(\eta c - \zeta b)(\zeta a - \xi c)(\xi - \eta).$$

These are connected by the three equations

$$(b - c)[ax + (q - r)^{-1}x'] + (c - a)[by + (r - p)^{-1}y'] + (a - b)[cz + (p - q)^{-1}z'] = 0,$$
$$xx' + yy' + zz' = 0, \quad pxx' + qyy' + rzz' = 0.$$

Caporali's theorem, given below (p. 240), also deals with the representation of a quadratic congruence.

The generation of a quadric in fivefold space. The tetrahedral complex in threefold space. Let $\alpha, \beta, \gamma, \delta$ be four planes, in fivefold space, meeting in pairs in six points, (β, γ), (γ, α), (α, β), (α, δ), (β, δ), (γ, δ). These points we denote, respectively, by X, Y, Z, X', Y', Z'; and we denote coordinates relative to these points by x, y, z, x', y', z'. It is easy to see that the general quadric containing these four planes has an equation of the form $Axx' + Byy' + Czz' = 0$; and that the quadric is made definite by the condition of containing a given fifth plane, which meets each of $\alpha, \beta, \gamma, \delta$ in a point. Let the points of the plane $X'Y'Z'$ be related to the lines of the plane XYZ, so that the point

$$lX' + mY' + nZ',$$

namely the point of coordinates $(0, 0, 0, l, m, n)$, corresponds to the line $Alx + Bmy + Cnz = 0$, $x' = 0$, $y' = 0$, $z' = 0$, for all values of l, m, n. Then the quadric is the locus of the plane joining any point of the plane $X'Y'Z'$ to the corresponding line of the plane XYZ.

For, the equation of the quadric, containing the points X, Y, Z, X', Y', Z', will not contain the squares of the coordinates. As the quadric contains the plane α, or YZX', given by $y' = 0$, $z' = 0$, $x = 0$, the equation will not contain terms in yz, $x'y$, $x'z$; for such terms are not reduced to zero by $y' = z' = x = 0$. Similarly, the equation will not contain terms in zx, $y'z$, $y'x$, or terms in xy, $z'x$, $z'y$. Lastly, as the quadric contains the plane δ, or $X'Y'Z'$, given by $x = 0$, $y = 0$, $z = 0$, the equation will not contain terms in $y'z'$, $z'x'$, $x'y'$. The equation is, therefore, of the form

$$Axx' + Byy' + Czz' = 0,$$

any quadric of this form, conversely, obviously containing the planes $\alpha, \beta, \gamma, \delta$.

Again, a plane meeting each of $\alpha, \beta, \gamma, \delta$ in a point, is, it is easily seen, a plane given by three points with symbols of the forms $a'X' + bZ - cY$, $b'Y' + cX - aZ$, $c'Z' + aY - bX$. The general point of this plane has coordinates of the forms

$$qc - rb, \quad ra - pc, \quad pb - qa, \quad pa', \quad qb', \quad rc' \; ;$$

and this point lies on the quadric $Axx' + Byy' + Czz' = 0$, for all values of p, q, r, if, and only if, $Aa' = Bb' = Cc'$. Thus, a, b, c, d being arbitrary, the quadric contains the ∞^3 planes of which one is defined by three points, in α, β, γ, respectively, with symbols

$$A^{-1}dX' + bZ - cY, \quad B^{-1}dY' + cX - aZ, \quad C^{-1}dZ' + aY - bX.$$

When one of these planes is given, the ratios of A, B, C are thereby determined.

Finally, the general point of the plane joining the point $(0, 0, 0, l, m, n)$ to the line in $x' = y' = z' = 0$ given by $Alx + Bmy + Cnz = 0$, has coordinates, for variable values of p, q, r, given by

$$p\,(0, 0, 0, l, m, n) + q\,(-Cn, 0, Al, 0, 0, 0) + r\,(-Bm, Al, 0, 0, 0, 0);$$

for instance, $x = -qCn - rBm$, $x' = pl$. Thus

$$Axx' + Byy' + Czz' = 0,$$

and this plane lies on the quadric whatever l, m, n may be.

Now consider the four planes $Y'Z'X$, $Z'X'Y$, $X'Y'Z$, XYZ, denoting them, respectively, by α', β', γ', δ'. Each of these meets three of the four planes α, β, γ, δ in a line; for instance, δ' contains the points (β, γ), (γ, α), (α, β). Thus these planes lie on the quadric, being planes of the other system from α, β, γ, δ. With these they may be said to constitute a *double four* of planes of the quadric. When two planes meet in a line, a general prime meets the solid, which is defined by these two planes, in a plane; and the prime meets the planes in two intersecting lines lying in this plane. Thus an arbitrary prime meets the planes, of a double four of planes, in eight lines of this fourfold space, forming a system constituted by four lines and the four transversals of threes of these four; or say, in a *double four* of lines in this prime. Let O be any point of the quadric. The solid $O\alpha$, for example, obtained by joining O to the points of the plane α, meets the quadric in another plane beside α, of the opposite system from this, and meeting this in a line; this other plane passes through O, and the line is the intersection of the plane α with the tangent prime of the quadric at O. By considering the solids $O\alpha$, $O\beta$, $O\gamma$, $O\delta$, we thus have four planes through O, of the opposite system of planes of the quadric from α, β, γ, δ, meeting these, respectively, in lines. These are planes of the point-cone in which the quadric is met by the tangent prime at O, of one system; and they meet any plane of this cone, of the other system, in four lines through O. Similarly, four planes of the quadric, meeting the planes α', β', γ', δ' each in a line, pass through O, and meet any plane of the

Representation of a tetrahedral complex 239

quadric through O, of the same system of planes as α', β', γ', δ', in four lines. These two pencils of four lines, in two planes, of different systems, through O, are *related* pencils.

For, we have remarked that the tangent prime of the quadric, at O, meets the double four of planes in a double four of lines in this prime. Also, it was proved (p. 144, above) that, if a double four of lines, in a fourfold space, be joined to a point, O, of that space, so taken that the eight joining planes belong to a quadric point-cone of vertex O, then the two sets of four joining planes, regarded as belonging to the two systems of planes of the cone, are related. Thus the result stated is obvious. It is the representative, in the fivefold space, of the fact that, in a threefold space, the planes, joining an arbitrary line to four points, are related to the range, on the line, of the four points in which the line meets the planes containing the threes of the four points (Vol. I, p. 30).

Denote the quadric by Ω; suppose that its points represent the lines of a threefold space. Consider a tetrahedral complex of lines of this threefold space. The lines of this complex will be represented, in the fivefold space, by the points common to the quadric Ω and to another quadric, say Ω'. When the equation connecting the coordinates of a line, in the threefold space, is written

$$ll' + mm' + nn' = 0,$$

point coordinates being relative to the four points by which the tetrahedral complex is defined, then the equation for the lines of the tetrahedral complex is of the form $all' + bmm' + cnn' = 0$. Thus it appears, from comparison of the equations, that a tetrahedral complex is represented by two quadrics, Ω and Ω', which have in common four planes of Ω, of the same system; this is evident, otherwise, from the fact that, in the threefold space, every line through any one of the four fundamental points is a line of the tetrahedral complex. In the fivefold space, the four planes of one system common to Ω and Ω', imply four other common planes, of the other system, making, with the first four, a double four of planes; correspondingly, in the threefold space, every line in one of the four planes which are fundamental for the tetrahedral complex, belongs to this complex.

We can, however, interpret, in the fivefold space, the property of a line of the tetrahedral complex, that the planes joining this line to the four fundamental points form a pencil which is related to a definite range. Taking the two quadrics, Ω, Ω', in fivefold space, with the four common planes α, β, γ, δ, meeting in pairs in six points, consider any other plane of Ω, of the same system as α, β, γ, δ. This plane, say ϵ, meets Ω' in a conic. The four points, say A, B, C, D, in which the planes α, β, γ, δ meet the plane ϵ, lie on this conic. Let O be any point of this conic; it lies on both Ω and Ω'. Planes of Ω can be drawn through O, of the system other than that of α, β, γ, δ, to meet these planes each in a line; these planes meet ϵ in lines. Four lines so obtained are OA, OB, OC, OD; and the pencil formed by these is related to the range, *on the conic*, of the four points A, B, C, D. If another plane of Ω, say ϵ', also of the same system as α, β, γ, δ, be drawn through O, meeting α, β, γ, δ, respectively in A', B', C', D', we obtain another pencil OA', OB', OC', OD', likewise lying in the four planes of Ω drawn through O to meet α, β, γ, δ in lines. As the planes of Ω, of the two systems, drawn from O, are planes of the quadric point-cone in which Ω is met by the tangent prime at O, the two pencils $O(A, ..., D)$ and $O(A', ..., D')$ are related. Thus on the two conics, on Ω', in which this quadric is met by the two planes, ϵ and ϵ', drawn through O, the two ranges $(A, ..., D)$ and $(A', ..., D')$, are related. A plane of Ω such as ϵ' might, however, have been drawn equally through any point, other than O, of the conic in the plane ϵ. Thus we have the result: Let two quadrics, Ω and Ω', in fivefold space, have four planes of the same system in common, and any plane of Ω, of this system, be drawn, meeting Ω' in a conic; consider the four points of this conic, lying on the four planes common to Ω and Ω', regarded as belonging to the range of the points of this conic. Let a second plane of Ω of the same system be drawn, and consider the four points of the conic, in which this plane meets Ω', which are similarly obtained from the four common planes. Then, the range of this last four points is related to the range of the first four. Moreover, by what is said above, if we take the other four planes, of the opposite system, common to Ω and Ω', say α', β', γ', δ', and a variable plane of Ω, of this system, meeting Ω' in a conic, then the planes α', β', γ', δ' meet this variable plane in four points, of the range of points of this conic on Ω', likewise related to the range of points A, B, C, D, above.

The singular points and planes of a Quadratic Congruence. Caporali's theorem. It has been remarked that a quadratic congruence, consisting of the lines common to a quadratic complex and a linear complex of lines, in threefold space, is represented in fivefold space by the surface which is the intersection of two quadrics and a fourfold space. Any one of the sixteen lines, known to lie on this surface (p. 166 above), represents a plane pencil of lines of the threefold space of the quadratic congruence. In general, only two lines of the congruence pass through an arbitrary point of this threefold space; but, it appears from the present point of view, there are sixteen points through each of which pass an infinity of lines, forming a flat pencil in a plane through this point. Each of the sixteen lines of the quartic surface, in the four-fold space, meets five others; thus, in the threefold space, each of the singular pencils of lines of the quadratic congruence has a ray in common with five others of these pencils; namely, the singular plane of any one of the pencils contains five other singular points, of which the corresponding planes, passing respectively through these points, all pass through the centre of the first pencil. Thus, there are six singular points in any singular plane, and six singular planes through any singular point. It was seen, moreover (p. 171, above), that the sixteen lines of the quartic surface, in the fourfold space, give rise to twenty double-fours of lines, each consisting of four non-intersecting lines, with four others, transversal to the threes of these. The first four lines of such a double-four represent four pencils of lines, in the threefold space; each pencil has a centre, and lies in a plane through this centre. The second four lines of the double-four give rise, similarly, to four pencils, in the threefold space, of which each has rays through three of the four preceding centres, these rays lying in the respective planes through these centres. The double-four of lines, on the quartic surface of the fourfold space, thus gives rise, in the threefold space, to a pair of mutually inscribed Moebius tetrads (Vol. I, p. 61), the rays through one of the points of one tetrad being in a plane containing three points of the other tetrad. From the singular points and planes of the congruence, corresponding to the double-fours of lines for the quartic surface in fourfold space, twenty such pairs of Moebius tetrads can be formed. No two of these pairs have a common tetrad, as we may see from the notation previously given for the double-fours (p. 171, above).

Let us represent the two quadrics, and the prime, in the fivefold space, which define the quadratic congruence, by $\Omega = 0$, $\Omega' = 0$, $\Pi = 0$, respectively, $\Omega = 0$ being the representative of the lines of the threefold space. We may replace Ω', for the definition of the congruence, by a quadric, Ω'', passing through the common points of Ω' and Π; and this may, in fact, be so chosen that the poles of Π in regard to Ω and Ω'' are the same point. For, let $(x_0, ..., w_0)$ be the pole of Π in regard to Ω, let Π_0 be the result of substituting $x_0, ..., w_0$ for $x, ..., w$, in Π, and let D_0 denote the operator $x_0 \partial/\partial x + ... + w_0 \partial/\partial w$. Then we may take $\Omega'' = \Omega'\Pi_0 - \Pi D_0 \Omega'$. Now consider the common self-polar hexad of Ω and Ω''. The six principal primes of the hexad represent six linear congruences of lines in the threefold space, of which every two are conjugate or apolar; and of these Π is one. Thus the twenty pairs of Moebius tetrads obtained above are clearly seen to be part of the figure previously discussed (p. 133, above), containing eighty tetrads, with sets of four tetrads of which every two are mutually inscribed.

The quartic surface, given by $\Omega = 0$, $\Omega' = 0$, $\Pi = 0$, has also (p. 171, above) ten sets of conics. Thus we may infer the existence, in the threefold space, of ten sets of quadric surfaces, of which the generators are lines of the quadratic congruence. And this may be generalised by considering curves of higher order on the quartic surface.

From the present point of view we may also prove easily that the lines of a quadratic congruence, in threefold space, belong to forty *tetrahedral* complexes (Caporali, *Memorie...Lincei* (3°), II, 1878, p. 756). Let $\Omega = 0$, $\Omega' = 0$ be two quadrics, and $\Pi = 0$ be a prime, in fivefold space; we shew that, in forty ways, a linear function of the six coordinates, say A, can be chosen, so that the quadric $\Omega' + A\Pi = 0$ contains four planes of $\Omega = 0$, of the same system; this is sufficient, after what has been said above. To prove this, take one of the double-fours of lines of the surface $\Omega = 0$, $\Omega' = 0$, $\Pi = 0$; through each line, of one of the fours of this, draw the definite plane of Ω, belonging to the first system of planes of Ω. These four planes meet, in pairs, in six points. Let the coefficients, in the linear function A, be chosen so that the quadric $\Omega'' = 0$, where $\Omega'' = \Omega' + A\Pi$, contains these six points of intersection. Then this quadric, $\Omega'' = 0$, wholly contains these four planes. For, any one of these planes contains, beside the three points of Ω'' in which this plane is met by the other three planes, also the line, belonging to the double-four, through which the plane is drawn; and this line, lying on $\Omega' = 0$, $\Pi = 0$, lies on $\Omega'' = 0$. As the intersection of a plane with a quadric is a conic, proper or degenerate, it follows, if we assume that the three points in which this plane is met by the other three planes do not lie in line, that this plane lies on $\Omega'' = 0$; and the same argument applies to the other three planes. This quadric also contains the planes of Ω of the *second* kind, drawn through the four complementary lines of the double-four. Next, beginning again with the four lines of the double-four through which the four planes of the first system of Ω were drawn, we may draw through these lines the four planes of the second system of Ω. Thence we similarly derive a quadric, $\Omega_2 = 0$, containing four, and, therefore, a double-four of planes of Ω. As there are twenty double-fours of lines on the quadric surface $\Omega = 0$, $\Omega' = 0$, $\Pi = 0$, we can derive, in this way, 2.20, or forty, quadrics, each having common with Ω a double-four of planes. Conversely, let Σ be any quadric containing four planes of the same system of Ω. There are then four planes of the other system of Ω, each meeting three of the first four in lines, which also lie on Σ. Any prime, Π, as has been remarked, meets two planes of Ω which meet in a line, in two lines which intersect. The prime Π thus meets the common double-four of planes of Ω and Σ in a double-four of lines. Thus it appears that the only quadrics containing a double-four of planes of Ω which can be found are those constructed as were Ω'' and Ω_2; so that forty is the total number of such quadrics.

A particular case of the theorem would be that in which the quadratic congruence of lines consists of the lines of a quadratic complex which meet a given line. Then the prime Π touches the quadric Ω. It may be seen that the quartic surface given by Ω, Ω', and Π, is not essentially modified.

Ex. 1. The points of a line, in space of three dimensions, are represented, in the fivefold space, by the planes, of the first kind, of the quadric Ω, which pass through a point of this. These planes meet any fixed plane, of the first kind, of Ω, in points lying on a line; these points form a range related to that of the points of the original line in the threefold space.

Consider, in particular, the range, in the threefold space, of points given by $\eta = \zeta = 0$, $\xi = a\tau$, as a varies. These points are represented by planes through the point $(1, 0, 0; 1, 0, 0)$ of the quadric $x^2 + y^2 + z^2 = u^2 + v^2 + w^2$, in the fivefold space, the planes being given by

$$u = x, \quad v = y \cos \theta - z \sin \theta, \quad w = y \sin \theta + z \cos \theta,$$

where $\tan \tfrac{1}{2}\theta = a$. Putting

$$p = m_3 + n_2, \quad q' = n_1 - l_3, \quad r = l_2 + m_1, \quad k = l_1 - m_2 + n_3 - 1,$$

these planes meet the plane of Ω which is given by

$$u = l_1 x + m_1 y + n_1 z, \quad v = l_2 x + m_2 y + n_2 z, \quad w = l_3 x + m_3 y + n_3 z,$$

in the range of points whose coordinates, x, y, z, u, v, w, are the six expressed by

$$a(q', -p, -k, q', p, k) - (r, -k, p, r, -k, p),$$

so that, for example, $x = aq' - r$, $w = ak - p$.

Ex. 2. Given two ·lines in threefold space, which do not intersect, any point, P, determines another point, P, on the transversal from P to the lines, harmonically conjugate to P with respect to these. Thus any line, not meeting the first two, on which P moves, determines a fourth line as the locus of P'. Shew that the representation of four lines so related, in the fivefold space, is by a harmonic range of four points of a conic, lying on the fundamental quadric Ω.

Ex. 3. Let the lines of a quadratic congruence, in threefold space, be given by $ll' + mm' + nn' = 0$, $all' + bmm' + cnn' = 0$, $A'l' + A'l + Bm' + B'm + Cn' + C'n = 0$. Let θ, θ' be the two roots of the quadratic equation

$$(a + \phi)^{-1} AA' + (b + \phi)^{-1} BB' + (c + \phi)^{-1} CC' = 0;$$

put $\alpha = (a + \theta)^{-1} AA'$, $\beta = (b + \theta)^{-1} BB'$, etc., and $\alpha' = (a + \theta')^{-1} AA'$, etc., so that $\alpha + \beta + \gamma = 0$, $\alpha' + \beta' + \gamma' = 0$. Then, with $\lambda = A'l$, $\lambda' = Al'$, etc., we have

$$\lambda\lambda'/\alpha\alpha' = \mu\mu'/\beta\beta' = \nu\nu'/\gamma\gamma', \quad \lambda + \lambda' + \mu + \mu' + \nu + \nu' = 0.$$

Further, with $L = A^{-1}l(a + \theta)$, $M = B^{-1}m(b + \theta)$, etc., we have

$$L\lambda' + M\mu' + N\nu' = 0, \quad L\lambda'\alpha/\alpha' + M\mu'\beta/\beta' + N\nu'\gamma/\gamma' = 0,$$
$$\alpha L + \lambda' + \beta M + \mu' + \gamma N + \nu' = 0,$$

whereby the lines of a quadratic congruence are reduced to the lines of a tetrahedral complex which meet a line (cf. also, Ex. 10, p. 142, above).

Ex. 4. Find the geometrical transformations in fivefold space corresponding, respectively, to the following transformations in threefold space: (1) Harmonic inversion with a fixed point and a fixed plane; (2) Harmonic inversion with two fixed lines; (3) Polarity in regard to a quadric surface; (4) Polarity in regard to a focal system (or linear complex).

CORRECTIONS TO VOLUME III

p. 9, l. 8. *After* ρ *add* and θ.
p. 22, l. 8. *For* (x, y, z, t) *read* (x, y, z, t).
p. 46, l. 1. *For* AXX′ *read* OXX′.
p. 47, l. 30. *For* pole, D, of *read* pole, D′, of.
p. 48, l. 8. *For* DFF′ *read* DFE.
p. 55, l. 7. *For* Ulp *read* Ulp+....
p. 69, l. 11 f. b. *For* quadrics, the *read* quadrics, four of the. Of either system of generators of either quadric there are four touching the curve.
p. 82, l. 8. *For* point *read* points.
p. 87, l. 14. *For* a+b+c+$\Sigma\lambda_r$ *read* $t^2[a+b+c+\Sigma\lambda_r]$.
p. 94, l. 16. *For* P=(A+σ)/(1+σ) *read* P=(A+σ)/(1+σ), etc.
p. 102, l. 4 f. b. *For* 193 *read* 163.
p. 112, l. 24. *For* normals of *read* normals at.
p. 120, l. 9 f. b. *For* point *read* points.
p. 124, l. 3. *For* (0, 1, 0, 0) *read* (1, 0, 0, 0).

p. 147, ll. 12 and 11 f. b. *For* A′ and B′ *read* the quadric.
p. 147, l. 8 f. b. *For* these *read* the quadric.
p. 148, ll. 19, 20. *For* the two points where the chord meets the curve *read* any quadric constructed as above for the seven points.
p. 148, l. 26. *For* point-pair constituted by *read* two points of the quadric on the join of. The matter is correctly stated in Ex. 5, p. 155.
p. 177, l. 25. *For* c_{12} *read* c_{34}.
p. 210, l. 23. *For* ten *read* five.
p. 221, l. 15. *For* Bull. Sc. Acad. *read* Bull. Acad.
p. 224, under Vol. II, *for* p. 96 *read* p. 97; also *add* p. 199, l. 4. *For* (−ξ, −η′) *read* (−ξ′, −η′).
p. 225, under Cayley and Cremona, *add* 142.
p. 226, under Moebius tetrads, *add* 143.

INDEX

Anchor Ring, or Tore, and Dupin's Cyclide, 193, 199

Angle between two circles, in projection from a quadric, 7; unaltered by inversion, 14

Apolar, or conjugate, linear complexes, 42; apolar relation for point of contact of Hart section, 76

Apollonius's problem for circles, 66

Associated lines, five, in fourfold space, 114; all met by planes, 120; planes meeting these through any point, 122; associated system of five planes, 118

Bauer, on a quartic surface, 20

Bennett, theorem on circles through threes of four points, 17; Hart section, equations for, 86

Bertini, on Veronese's surface, 54

Bitangents of Kummer's surface, from lines of Segre's cubic locus, 156

Bobillier, theorem for pedal circles, 22

Burkhardt, the group of the lines of a cubic surface, 105

Burnside : a figure in fourfold space, 105; the group of the lines of a cubic surface, 105; invariants of a linear group, 210

Caporali, tetrahedral complexes containing a quadratic congruence, 242

Cardioid, 99

Cartesian, 99

Castelnuovo and Kronecker, surface with double infinity of reducible plane sections, 55

Caustic, or focal, surface of a quadratic congruence, 225

Cayley, representation of a line by a point of fivefold space, 40; on Veronese's surface, 54; algebraic solution of Malfatti's problem, 68

Centre of a circle, in projection from a quadric, 7

Chords common to two curves, 51

Circle, obtained by projection of plane section of a quadric, 1; circle and two inverse points changed into same by inversion, 14; circle of similitude, obtained by projection, 15; circles of similitude of the pairs of four circles, 15; circles in a plane, chain of theorems for, 29; circle represented by a line of fourfold space, 39; Hart and Feuerbach circles, Chapter II, p. 65; circles meeting given circles at equal, or at given angles, 67; three circles with three concurrent common tangents, 67

Clifford, a theorem for circles, 1, 31, 64, 105

Complex, tetrahedral, determined by planes in fourfold space, 32; linear, determined by planes in fourfold space, 34; linear, represented in fivefold space, 42; six linear complexes mutually conjugate, 42; lines common to three, or to four linear complexes, 43, 44; linear, point and polar plane represented in fivefold space, 48; linear, six conjugate, constructed from six points, 139; linear, six conjugate, from lines of Segre's cubic locus, 155; linear, six conjugate, poles of a plane lie on a conic, 204; quadratic, represented in fivefold space, 213; quadratic, condition for singular point, 214; quadratic, rationality of, 234; tetrahedral, in fivefold space, 239. See also *Focal System*

Confocal Cyclides, cutting at right angles, 181; equation for, 183

Congruence of lines, its order and class, 49; for chords of a curve, 50; number of lines common to two, 51; quadratic, in tangent solid of Segre's quartic locus, 137; six congruences associated with a Kummer surface, 219, 223; quadratic, caustic or focal system of, 225; six quadratic, equations for common singular points, 226; quadratic, singular points and planes, 240; quadratic, lies in

CAMBRIDGE : PRINTED BY W. LEWIS AT THE UNIVERSITY PRESS